MW01700080

Steam Engineering: A Textbook - Primary Source Edition

William Richard King

Nabu Public Domain Reprints:

You are holding a reproduction of an original work published before 1923 that is in the public domain in the United States of America, and possibly other countries. You may freely copy and distribute this work as no entity (individual or corporate) has a copyright on the body of the work. This book may contain prior copyright references, and library stamps (as most of these works were scanned from library copies). These have been scanned and retained as part of the historical artifact.

This book may have occasional imperfections such as missing or blurred pages, poor pictures, errant marks, etc. that were either part of the original artifact, or were introduced by the scanning process. We believe this work is culturally important, and despite the imperfections, have elected to bring it back into print as part of our continuing commitment to the preservation of printed works worldwide. We appreciate your understanding of the imperfections in the preservation process, and hope you enjoy this valuable book.

WORKS OF W. R. KING

PUBLISHED BY

JOHN WILEY & SONS

The Elements of the Mechanics of Materials and of Power Transmission.
8vo, v + 266 pages, 140 figures. Cloth, $2.50 net.

Steam Engineering.
8vo, vii + 450 pages, 177 figures. Cloth, $4.00 net.

STEAM ENGINEERING

A TEXT BOOK

WILLIAM R. KING, U. S. N., Retired
PRINCIPAL, BALTIMORE POLYTECHNIC INSTITUTE

FIRST EDITION
FIRST THOUSAND

NEW YORK
JOHN WILEY & SONS
43 and 45 East 19th St.
London: CHAPMAN & HALL, Limited
1913

Copyright, 1913,
by
WILLIAM R. KING

Stanhope Press
F. H. GILSON COMPANY
BOSTON, U.S.A.

173551
APR 22 1913
TH
.K58

6 9 4 2 1 4 5

PREFACE

THE production of this volume was prompted by the apparent need of a text-book that would present concisely to students of universities and of advanced secondary technical institutes the principles and constructive details of the make-up of a steam power plant, together with such calculations as are necessary for a practical knowledge of steam engineering. The manifest mistake is too commonly made of introducing the subject to engineering students through the medium of texts that treat it thermodynamically, accompanied with such abstruse calculations and consequent maze of higher mathematics that the result is bewildering rather than instructive.

It has been the aim in this text to present in a clear and systematic manner, involving no greater degree of knowledge of physics and of higher mathematics than the class of students for which it is intended should possess, the principles and action of the reciprocating steam engine, of the steam turbine, of the steam boiler, and of their accessories. Its mastery may be followed profitably by the study of the excellent advanced texts now extant.

No claim of originality is made for the matter contained in the book, but merit is claimed for its systematic arrangement and for the simplicity of treatment of the subjects involved.

I acknowledge having drawn from sources of engineering information other than those credited in the text, and express obligation to my assistants in engineering, Charles Ernest Conway and Samuel P. Platt — to the former for valuable aid and suggestion and to the latter for the care in which he made the tracings for many of the cuts.

WILLIAM R. KING.

BALTIMORE, January 1, 1913.

CONTENTS

STEAM ENGINEERING

CHAPTER I

INTRODUCTORY

1. **Kinetic Theory of Gases.** — A gas may be defined as a fluid of such nature that if a certain volume of it be admitted into a vessel it will distribute itself throughout the vessel, whatever the volume of the vessel may be. According to the kinetic theory, the molecules of a gas are in a state of rapid motion in straight lines, and as a result of their collision with each other and with the walls of the containing vessel pressure is exerted.

2. **Boyle's Law.** — The law of Boyle, sometimes attributed to Mariotte, may be stated as follows:

The pressure of a mass of gas at constant temperature varies inversely as its volume.

In other words:

The product of the pressure and volume of a mass of gas is a constant quantity as long as the temperature is constant.

The algebraic expression for Boyle's law is

$$PV = C,$$

in which P is the pressure in any units, usually in pounds per square foot, V the volume in cubic feet, and C a constant quantity to be determined experimentally for a particular gas.

3. **Charles's Law.** — The law of Charles, sometimes attributed to Gay Lussac, asserts that all gases have the same coefficient of expansion, and this coefficient is the same whatever the pressure supported by the gas. Hence, for each degree of rise or fall

of temperature of a gas its volume will be increased or diminished by a fixed fraction of its original volume, the pressure remaining unchanged. This fraction has been computed to be $\frac{1}{273} =$ 0.003663 on the Centigrade scale, and $\frac{1}{493} = 0.0020284$ on the Fahrenheit scale, the temperature of the original volume being that of melting ice.

4. **Absolute Zero of Temperature.** — If a column of air whose height is unity and temperature that of melting ice be confined in a tube and separated from the atmosphere by a piston which is free to move in the tube, and it then be exposed to the steam of boiling water, the pressure remaining constant, the temperature of the inclosed air will rise to 100° C, or to 212° F. The volume of the air will then be, by Charles's law,

$$1 + (100 \times 0.003663) = 1.3663,$$

or

$$1 + (212 - 32)\,0.0020284 = 1.36511.$$

This unit volume of air having expanded the fraction 0.3663 of itself by raising its temperature 100° C, or the fraction 0.36511 by raising its temperature 180° F, it follows that the expansion will reach the volume of unity, or the original volume will be doubled, when the temperature is raised through x degrees, thus:

$$0.3663 : 1 :: 100 : x, \quad \text{whence } x = 273^\circ \text{ C},$$

or

$$0.36511 : 1 :: 180 : x, \quad \text{whence } x = 493^\circ \text{ F}.$$

It follows by the law of Charles that if the unit volume had been cooled through 273° C, or through 493° F, there would be no existing volume. This point is called the ***absolute zero*** of temperature, and temperatures reckoned from this point are called absolute temperatures.

The absolute zero is then 273° below the zero of the Centigrade scale, and 493° − 32° = 461° below the zero of the Fahrenheit scale. To convert temperatures measured on the

C or F scales to absolute temperatures, we have only to add 273°, or 461°, as the case may be. Absolute temperatures will be denoted by the capital letter T.

It must be observed that if the volume of a gas be kept constant and its temperature raised, the pressure will be increased or diminished in the same ratio that the volume is increased or diminished when the pressure is constant. For, let P and V be the pressure and volume of a portion of gas at 32° F, and 1.36511 V the volume when heated to 212° F under the constant pressure P. If the temperature of the gas be now kept constant at 212° while it is being compressed back to its original volume V, it will then have a certain pressure P', and we shall have, by Boyle's law,

$$P'V = P \times 1.36511\, V, \quad \text{whence } P' = 1.36511\, P.$$

In other words, if a volume V of a gas at 32° and pressure P be heated to 212°, the volume remaining at V, its pressure will be 1.36511 P.

We may now regard this absolute zero from another viewpoint. Let us consider a mass of a perfect gas of volume v_0, pressure p_0, and a temperature 0° C, or 32° F. If the volume be kept constant while its temperature is raised or lowered $t°$ the resulting pressure p will be, by Charles's law, $p = p_0 \pm p_0 \alpha t = p_0 (1 \pm \alpha t)$, where α is the coefficient of expansion of the gas. If the gas now be cooled until its temperature is reduced to $-\frac{1^\circ}{\alpha}$, we shall then have

$$p = p_0\left(1 - \frac{\alpha}{\alpha}\right) = 0.$$

That is, at a temperature $-\frac{1^\circ}{\alpha}$ the gas would exert no pressure on the walls of the containing vessel. If the kinetic theory of gases, or the theory that heat is a form of energy, or of motion, be accepted, then when the point is reached where molecular motion ceases, all heat and pressure must be non-existent, and as it is impossible to imagine a body colder than one devoid of heat, that is, one at a temperature $-\frac{1^\circ}{\alpha}$, this temperature is called the *absolute zero.*

5. Combination of the Laws of Boyle and Charles. — The interrelation of the pressure, volume, and absolute temperature of gases is shown by the combination of the laws of Boyle and Charles.

Let v_0 be the volume and p_0 the pressure of a portion of gas at 0° temperature. Let v denote its volume when the temperature is raised to t', the pressure remaining at p_0, and let v' be the volume under the pressure p', the temperature remaining t'. Then, by Boyle's law, we shall have $p'v' = p_0v = C$. By Charles's law we have $v = v_0(1 + \alpha t')$. Hence

$$p'v' = p_0v_0(1 + \alpha t') = p_0v_0\alpha\left(\frac{1}{\alpha} + t'\right).$$

But $\frac{1}{\alpha} = T_0 =$ the absolute zero of temperature; hence $\alpha = \frac{1}{T_0}$, and $\frac{1}{\alpha} + t' = T'$; therefore $p'v' = \left(\frac{p_0v_0}{T_0}\right)T'$, or $\frac{p'v'}{T'} = \frac{p_0v_0}{T_0}$. In other words, the product of the volume and pressure of a gas is proportional to the absolute temperature.

By Boyle's law we have, $PV =$ a constant in foot pounds, where $P =$ pressure in pounds per square foot, and $V =$ volume in cubic feet. The constants for different gases have been calculated with accuracy by Regnault.

The constant for air may be found as follows:

It is known that water is 773 times as heavy as air when at the temperature of 32° F and pressure of 14.7 pounds per square inch; and since a cubic foot of water weighs 62.4 pounds, it will take $\frac{773}{62.4} = 12.387$ cubic feet of air to weigh one pound. This pound of air may be assumed to be contained in a cylinder one square foot in area and 12.387 feet in height. The pressure of the air being 14.7 pounds per square inch, we may assume at the bottom of the cylinder a piston weighing $144 \times 14.7 = P$ pounds. The movement of this piston through the height of the cylinder will displace 12.387 cubic feet $= V$, and perform $144 \times 14.7 \times 12.387$ foot pounds of work. Hence, $PV = 144 \times 14.7 \times 12.387 = 26{,}220$ foot pounds.

We have shown that the pressure multiplied by the volume of a portion of gas is proportional to the absolute temperature; so that if T_0 be the absolute zero on the Fahrenheit scale (493° below 32°) corresponding to P_0V_0, then for any other pressure and volume, as P_1V_1, we shall have

$$\frac{P_1V_1}{T_1} = \frac{P_0V_0}{T_0} = \frac{26{,}220}{493} = 53.2 \text{ nearly,}$$

whence $P_1V_1 = 53.2\,T_1$.

This equation for a perfect gas is usually written $PV = RT$, where R is a constant depending on the density of the gas and the units in which P and V are measured. For air we have found it to be 53.2 when P is in pounds per square foot and V in cubic feet. For superheated steam it is 85.5.

6. Kinetic Theory of Heat. — The *caloric* or *material* theory of heat was advanced by Black in 1798, but in 1802 it was shown that heat could be produced to an unlimited extent by friction, thus proving its identity with motion and establishing the *kinetic* theory of heat. Under this theory heat is regarded as a form of motion consisting of the agitation of the molecules of matter, either by rotation or vibration, and the hottest bodies are those whose molecules are agitated at the greatest velocities.

Another statement of the kinetic theory of heat is that heat is a form of energy, or that it is energy in the invisible form of molecular motion.

Since the motion of the molecules of a body is invisible, the exact nature of molecular motion is unknown, but the application of the laws of motion and energy to phenomena produced by heat has given us the knowledge upon which is founded the science of *thermodynamics* or the *mechanical theory of heat.*

7. **Energy.** — Energy is defined as the capacity to do work. A force acting through a distance exerts energy, and the resistance overcome is the work done. Energy is of two kinds — *kinetic* and *potential.* The energy of a body due to its motion is kinetic and is measured by the work done by the body in being brought to rest. The work done in raising a body of weight w through a height h is wh, and if v be the velocity acquired by the body in falling a distance h, then $v^2 = 2gh$, whence $wh = \frac{wv^2}{2g} = \frac{mv^2}{2}$, which is the expression for the kinetic, or actual, energy of a body of weight w, moving at a velocity v. It is, in fact, the work stored up in a body of weight w moving with a velocity v.

If a body be raised a distance above the surface of the earth, a certain amount of work is performed in overcoming the attraction of gravity, and this work, stored in the body, gives to it the power of doing work. Energy thus possessed by a body, due to its position and not its velocity, is called potential energy. A body may possess potential energy due to the action of forces other than gravity, as, for example, when a watch is wound work is done on the spring by which its energy is increased, giving to it potential energy which causes a reappearance of the work in driving the watch while the spring unwinds.

8. **Temperature.** — Temperature refers to the condition or a body as regards its sensible heat and this condition is measured by the thermometer. Two bodies are said to be of the same temperature when there is no tendency of heat to pass from one to the other.

9. **Thermometer.** — A thermometer is an instrument for measuring temperatures. The lower fixed point on thermo-

metric scales is the temperature of melting ice under atmospheric pressure. The upper fixed point is the temperature of boiling water under the normal pressure of the atmosphere. The scale universally used for scientific work is that known as Centigrade, but in England and in the United States the Fahrenheit scale is in general use. In Russia the Reaumur scale is used.

The temperature of melting ice is marked 0° on the Centigrade and Reaumur scales, but on the Fahrenheit scale it is marked at 32°. The temperature of boiling water is marked 80° on the Reaumur, 100° on the Centigrade, and 212° on the Fahrenheit scales. Denoting the three scales by C, F, and R, we shall have 0° C = 32° F = 0° R. The zero of the Fahrenheit scale is then 32° below the temperature of melting ice, so if 32° be deducted from the Fahrenheit reading we shall get the number of degrees on each scale by which any given temperature differs from that of melting ice. Let the given temperature be that of boiling water; then

$$C : F - 32 : R :: 100 : 180 : 80 :: 5 : 9 : 4,$$

whence

$$\frac{C}{5} = \frac{F - 32}{9} = \frac{R}{4}.$$

By means of these equations temperatures on one scale may readily be converted into either of the others.

10. Calorimetry. — In order to measure the quantity of heat which a body absorbs or yields in passing through a given range of temperature, or in changing its state, we adopt as a unit that quantity of heat which, acting on a given weight of water, changes its temperature a definite amount. The apparatus in which the measurement is made is called a calorimeter. The specific heats, the latent heats of vaporization, and the heats of combustion of substances are determined by the calorimeter.

11. British Thermal Unit. — The unit for the measurement of heat, or the *British thermal unit* (*B.t.u.*), is defined by Marks and Davis as $\frac{1}{180}$ of the quantity of heat required to raise the temperature of one pound of water from 32° F to 212° F.

12. Specific Heat. — The specific heat of a substance is the quantity of heat required to raise the temperature of the unit weight of the substance through 1° of temperature.

Water is taken as the standard in the determination of specific heats, its average specific heat being unity, so that the specific heat of a substance may be defined as the ratio of the quantity of heat required to raise the temperature of a given weight of the substance through 1° to the quantity required to raise the temperature of an equal weight of water through 1°. In this country the quantity of heat is measured in thermal units and the temperature reckoned in degrees Fahrenheit.

13. Thermodynamics. — The science which treats of the mechanical or dynamical theory of heat, or of the relation between heat and work, is called thermodynamics.

14. First Law of Thermodynamics. — *Heat and mechanical energy are mutually convertible, and heat requires for its production or produces by its disappearance mechanical energy equivalent to 778 foot pounds for each thermal unit.*

15. Mechanical Equivalent of Heat. — The mean result of many experiments has shown that the production of one thermal unit of heat is equivalent to the performance of 778 foot pounds of mechanical work, and this quantity, 778 foot pounds, is known as the mechanical equivalent of heat.

16. The Conservation of Energy. — The first law of thermodynamics expresses the principle of the conservation of energy — that the total energy of the universe is constant, that energy can neither be created nor destroyed by any process of nature, and that every gain in form of energy corresponds exactly to a loss in some other form. While no heat is absolutely lost in the transformation of energy of one form into that of another, a considerable amount is lost as available energy, being transformed into useless heat and dissipated or diffused among surrounding bodies and thus becoming unavailable. It is only when a body is hotter than surrounding objects that its heat energy can be utilized in doing work. If all bodies were of the same temperature no work of any kind would be possible.

A general statement of the conservation of energy is

$$\text{Kinetic Energy} + \text{Potential Energy} = \text{Constant},$$

but the statement in a mechanical sense means that the total energy deposited in a machine is equal to the useful work given out plus the work lost in resistance.

17. Second Law of Thermodynamics. — *Heat cannot pass from a cold body to a hot one by a self-acting process unaided by external agency.*

It follows from this law that no engine can convert into work the whole of the heat supplied to it; for as soon as the heat supplied has fallen in temperature to that of the surrounding atmosphere the heat remaining is no longer available for doing useful work. It follows also from this law that it is impossible to convert any part of the heat of a body into mechanical work, except by allowing it to pass to another body at a lower temperature.

18. Definitions. — *Force* is that which moves or tends to move a body, or which changes or tends to change the motion of a body. The unit of force is a pound, and forces are expressed as being equal to so many pounds.

Work is the overcoming of resistance through space, and its measure is the product of the resistance and the space through which it is overcome, the space being reckoned in feet. Since the unit of force is a pound and the unit of distance a foot, the product of a foot by a pound is the *unit of work*, called a *foot pound*. Other units of work are sometimes used as a matter of convenience. For example, an ***inch ton*** represents the work performed in moving 2240 pounds one inch, and is expressed as 2240 inch pounds, or by $\frac{2240}{12}$ foot pounds. A *foot ton* represents an expenditure of work equal to 2240 foot pounds. It is thus seen that the different units of work are readily converted, one to another.

Power is the rate of doing work, and is measured by the number of foot pounds of work done in a unit of time.

Horse-power (*H.P.*) is the unit employed to represent the rate of work of an engine or machine, and its measure is the quantity of work equivalent to the moving of 33,000 pounds through the distance of one foot in one minute. The performance of 33,000 foot pounds of work in one minute is then a horse-power, and an engine performing this amount of work in a second would be working at the rate of 60 horse-power.

Brake Horse-power (*B.H.P.*) is the power which the engine transmits for useful work, that is, it is the total or *indicated horse-power* (*I.H.P.*) minus the power absorbed in driving the engine itself.

PROBLEMS

1. The ratio of the volumes of air under the same pressure at temperatures of 0° C and of 100° C, or at 32° F and at 212° F, is 1 to 1.3663. Find the absolute zero of temperature on the Centigrade and Fahrenheit scales.

2. A volume of air at temperature of 32° F, and pressure of 14.7 pounds per square inch, is heated to a temperature of 100° F, the volume remaining the same. What will be its pressure? *Ans.* 16.73 pounds.

3. A pound of steam at temperature of 233° F and absolute pressure of 22 pounds per square inch has a volume of 18.37 cubic feet. What will be its volume at temperature of 341° F and pressure of 120 pounds?

Ans. 3.875 cubic feet.

CHAPTER II

THE APPLICATION OF HEAT TO WATER

19. General Effects. — The effect of the application of heat to water is of the greatest importance in the study of the steam engine.

Radiant heat is given off from hot bodies in straight lines, and the rays of heat are subject to the same laws as are the rays of light. As far as the generation of steam in a boiler is concerned the useful radiation is confined to the furnace, the crowns and sides of which become heated by intercepting the rays of heat from the burning fuel, the heat passing through them to the water in the boiler. A considerable amount of heat is given off by radiation from burning coal and it is very important to intercept this and to insure, as far as possible, that the whole of the heat diffused in this way be transmitted, directly or indirectly, to the water in the boiler and not wasted on the external air.

Radiation is an important item to be considered with reference to the economical employment of steam, for it always causes a certain loss of heat, and unless proper precautions are taken this loss may become very considerable. The surfaces of the boilers, steam pipes, cylinders, etc., when the engines are at work, are very much hotter than the surrounding bodies, and in order that the loss of heat by radiation may not be excessive all these surfaces should be clothed, or lagged, with some non-conducting material.

The transfer of heat by convection is the manner in which liquids and gases are heated. When heat is applied to the bottom of a vessel containing a liquid or gaseous fluid, the particles in contact with the bottom are first heated and become less dense in consequence. They therefore rise, allowing cooler particles to take their places, which, in turn, becoming heated, rise and circulate through the mass in a similar manner.

It is essential that circulation and mixture of all the particles of a fluid should take place to cause the temperature to be uniform throughout the mass. In order that heat may be efficiently transmitted through boiler plates and flues each of the fluids in contact with them, viz., the water on one side and the heated gases on the other, should have free circulation so that the particles in contact with the plates should not be considerably different in temperature from those at some distance from the plates. Boilers are sometimes fitted with circulating plates to set up currents in the water, and with bafflers in the flues to break up the currents of hot gases and form eddies, in order to promote circulation and mixture in the respective fluids.

To render most efficient the transfer of heat from one fluid to another through a plate the general movement of the two fluids should be in opposite directions so that the hottest parts of the two fluids are opposed to each other and the difference of temperature between them a minimum. Consequently, in order that the best results may be obtained in a boiler, the feed pipes and circulating plates should be so arranged that the general movement given to the water should be as far as possible in the opposite direction to that of the hot gases on their way to the funnel. For the same reason the action of surface condensers is most efficient when the cold water for condensing the steam enters the condenser at the end at which the condensed water leaves it, so that the entering steam will be opposed to the water which has been heated to some extent by its passage through the condenser.

Upon the application of heat to the water in the boiler the temperature is at first raised. The particles of water in contact with the heating surface, which consists of the tubes and the plates of the furnace and combustion chamber, become heated and rise and circulate through the mass of the water, their places being taken by cooler particles, till at length the whole of the water is raised in temperature to the boiling point by the process of convection.

20. Sensible Heat. — The heat added to the water up to the temperature at which boiling occurs is called sensible heat, its effect being simply to change the temperature and not the state of the water, and its amount is readily calculated; for the quantity of heat of a body, or the amount of heat energy which a body gains or loses in passing through a given range of temperature, is measured in thermal units, and is given by the continued product of its weight, range of temperature, and specific heat. The specific heat of water being taken as unity, it follows that in calculations involving thermal units where the weight of water considered is one pound, the two factors of unity may be omitted and the temperature be taken as the expression for thermal units.

21. Latent Heat. — After the temperature of boiling is reached, and in order to convert the boiling water into steam, a large quantity of heat has to be expended which does not produce any increase in the temperature. This is known as latent heat. It was supposed at an early period that heat was a substance, and that the heat thus expended became hidden or latent during the change from the liquid to the gaseous state, and that it again became sensible on the reverse process being performed. The science of thermodynamics, however, has shown that heat is not a substance but a form of mechanical energy, and that this heat is expended in performing the internal work of overcoming the molecular cohesion of the particles of water which resist the change of state, and also performing the external work of overcoming the resistance of external bodies to the change of volume which ensues. Work is therefore done on changing the water into steam and is stored in the steam as mechanical energy which can be given back either in work, or in heat, when the steam returns to its original state of water.

It is now seen that this latent heat of evaporation performs the work which is necessary in the interior of the water and exterior to it, the external work being a measure of the useful effect directly obtained.

In the case of the boiler and the steam engine the external work of evaporation is represented by the work performed by the piston while the cylinder is in communication with the boiler. It will be seen later on that this external work is the equivalent of but a small part of the latent heat, the greater part being expended in internal work, and is represented by the energy acquired by the water during the evaporation.

The term *latent heat* has been retained for convenience, but it is to be understood as an expression that means simply *the quantity of heat that must be added to or subtracted from a body in a given state to change it into another state without altering its temperature.*

22. The Boiling Point. — The boiling point, or the temperature of ebullition, of any liquid may be defined as that stage in the addition of heat to the liquid at which the pressure on it is just overcome by the pressure of vapor due to the temperature.

The temperature of the boiling point depends on the pressure under which the liquid is evaporated. The greater the pressure the higher the temperature at which the liquid boils. The temperature depends upon the *total*, or *absolute* pressure, i.e., the pressure including that of the atmosphere, so that in ascertaining the temperature corresponding to any given pressure as shown on the ordinary pressure gauge the atmospheric pressure must be added.

The boiling point of a liquid is affected by its density. Ordinary sea water contains about one thirty-second of its weight in salt, and this raises the temperature of the boiling point by 1.2° F, so that the boiling point of sea water under the atmospheric pressure is 213.2° F.

23. **The Generation of Steam.** — The generation of steam under constant pressure and at constant volume will best be understood by the consideration of the two following examples respectively. These examples are, substantially, as given in Northcott's "The Steam Engine," and in Sennett's "The Marine Steam Engine."

Example I. — **The Generation of Steam under Constant Pressure.** — Imagine one pound of water at a temperature of 32° F contained in an upright tube of indefinite height and 1 square foot, or 144 square inches, cross-sectional area. The pound of water has a volume of $\frac{1728}{62.4} = 27.7$ cubic inches nearly, and will occupy $\frac{27.7}{144} = 0.1923$ inch of the height of the tube, or the water will expose an area of surface equal to one square foot and will be 0.1923 inch deep. The tube is to be open to the atmosphere, so that each square inch of the surface of the water will be pressed upon by the average atmospheric pressure of 14.7 pounds, making the total pressure $14.7 \times 144 = 2116.8$ pounds. On heat being applied to the tube the temperature of the water rises until 212° F is reached when vaporization commences, and there is no further rise in temperature until the whole of the water is converted into steam, the temperature of which will also stand at 212° F.

Immediately previous to the commencement of vaporization a quantity of heat will have been imparted to the water equal to $212° - 32° = 180$ thermal units, or to $180 \times 778 = 140{,}040$ foot pounds. When the pound of water is all converted into steam a further quantity of heat will have been imparted equal to 965.7 thermal units, or to $965.7 \times 778 = 751{,}315$ foot pounds;

and yet, notwithstanding this large absorption of heat, the temperature remains at 212° F. Since this heat has changed the state of the water from that of a liquid to that of a gas without raising its temperature, it is, from our definition, latent. We know, however, that the only manner in which heat energy can disappear as heat is by its being transformed into some other form of energy or in the performance of work; and in order to understand what has become of this large quantity of heat we must inquire as to what work has been done.

In the first place 180 units were absorbed in raising the temperature of the water from 32° to 212°, and this we know as the *sensible heat.* This change of temperature is practically the only change effected up to that point of the process, as the expansion of the water in rising from 32° to 212° is extremely small and may be disregarded. The effect of the heat abstracted from the source of heat and communicated to the water up to this point has been only to raise the temperature of the water.

Steam of a pressure of 14.7 pounds per square inch has been found to occupy a volume 1644 times that of the water from which it was generated, that is, a cubic inch of water is converted into 1644 cubic inches of steam at atmospheric pressure of 14.7 pounds per square inch. The steam from one pound of water will thus have a volume of $\frac{1644 \times 27.7}{1728} = 26.36$ cubic feet. In expanding to this volume under atmospheric pressure it will necessarily have performed external work to the extent of $26.36 \times 2116.8 = 55{,}798$ foot pounds. This, perhaps, may be more clearly seen if we disregard the pressure of the atmosphere, and assume that the steam expands in the tube under a piston weighing $14.7 \times 144 = 2116.8$ pounds, which it will evidently have to lift in the tube through a height of 26.36 feet.

The heat required to convert the water at 212° F into steam of the same temperature has been found by experiment to be

965.7 thermal units, or to $965.7 \times 778 = 751,315$ foot pounds, and of this, 55,798 foot pounds are accounted for in the external work as shown above, leaving $751,315 - 55,798 = 695,517$ foot pounds as the mechanical equivalent of the internal work, or of the heat absorbed in effecting the expansion against the internal or molecular forces.

Of the total quantity of heat energy imparted to the water $\frac{140,040}{778} = 180$ thermal units have been expended in increasing the temperature of the water by 180°; $\frac{695,517}{778} = 893.98$ thermal units have been expended in the internal work of expanding the steam against molecular forces; and $\frac{55,798}{778} = 71.72$ thermal units have been expended in the external work of overcoming the resistance of the atmosphere which opposed the expansion of volume. The sum of these three quantities of heat, that is, the sum of the units due to the *sensible heat*, the *internal work*, and the *external work*, is known as the *total heat* of steam, or the total heat of vaporization, and is, in this case, $180 + 893.98 + 71.72 = 1145.7$ thermal units, which is the total heat of one pound of steam at temperature of 212° F; that is, it is the units of heat required to evaporate one pound of water from 32° F into steam at 212° F. A more accurate determination of this quantity makes it 1146.6 B.t.u.

The results of very careful experiments have shown that the total heat of steam increases but slightly as the temperature increases, the increment of increase being 0.305 of a thermal unit for each degree of increase in temperature. This fact enables us to deduce a formula for ascertaining the total heat of steam of any temperature. Thus, knowing the total heat of steam at 212° F to be 1146.6 units, we will have for the total heat, H_t, at any other temperature t, reckoning from 32°, the formula

$$H_t = 1146.6 + 0.305\,(t - 212).$$

It will be convenient to change the form of this expression as follows:

$$H_t = 1146.6 + 0.305\,(t - 212) = 1146.6 + 0.305\,[t - (32 + 180)]$$
$$= 1091.7 + 0.305\,(t - 32).$$

The sum of the two quantities of heat that disappears, representing the internal and external work, is equal to the total heat minus the sensible heat, above 32°, and is known as the latent heat of vaporization, or the latent heat of steam under the given pressure, and equals in this case 893.98 + 71.72 = 965.7 thermal units.

It has been experimentally determined that the increment of decrease in the latent heat of steam is 0.7 of a thermal unit for each degree of increase of temperature. We can, therefore, deduce a formula for ascertaining the latent heat of steam at any temperature.

Thus, knowing the latent heat of steam at 212° to be 965.7 units, we will have for the latent heat, H_l, at any temperature t, reckoning from 32°,

$$H_l = 965.7 - 0.7\,(t - 212).$$

We may change the form of this formula in order to make it very similar to that for the total heat and, consequently more easily remembered, thus

$$H_l = 965.7 - 0.7\,(t - 212) = 965.7 - 0.7\,[t - (32 + 180)]$$
$$= 1091.7 - 0.7\,(t - 32).$$

The total and latent heats are among the properties of steam which have been tabulated for convenience in calculations. The results obtained by the formulas given above do not agree exactly with those tabulated, but in the absence of the table they may be used without very serious error. The properties of saturated steam, so very essential in many calculations in steam engineering, were originally determined by Regnault, and as tabulated by Marks and Davis are the most accurate available and are used in this book. In the same volume Marks and Davis give a table of properties of superheated steam.

An important modification of the conditions under which steam is generated in the above example will now be considered.

Example II. — **The Generation of Steam at Constant Volume.** — In Example I the piston was free to move and the steam was allowed to expand gradually as the water evaporated. We will now suppose this not to be the case, and that the piston is fixed in the cylinder at a certain distance above the water, as shown in Fig. 1, so that whatever steam is formed from the water must occupy the same constant volume. Suppose further that only the water occupies the space *AB* and that there is no pressure therein. If heat now be applied to the water, instead of the formation of steam taking place only when a certain temperature is reached, as in the first example, steam immediately commences to form, and both the temperature and pressure gradually increase as more of the water is evaporated, although the temperature and pressure are connected by the same law as before.

FIG. 1.

The pressure and temperature of the steam when all the water has just been evaporated can be ascertained from our knowledge of the weight of water and the known volume of the steam; for the weight of the steam is the same as that of the water from which it was formed, so that its volume per pound can be calculated and its pressure and temperature then ascertained from the table of the properties of steam.

It will be seen that this can occur in practice when steam is being raised in a boiler, with the stop valve and other outlets closed, except that instead of there not being any pressure on the water prior to the application of heat, there is the pressure of the atmosphere. In this case the temperature of the water will gradually rise, but no steam will be formed, and the pressure will not increase until the temperature reaches that corresponding to the atmospheric pressure, viz., 212°, when steam commences to be given off, and both temperature and pressure rise. It is thus seen that the absolute pressure of boiler steam is in excess of the gauge pressure by an amount equal to the atmospheric pressure. In our practical illustration the evaporation does not continue until all the water is evaporated, as the amount of water is so large compared with the volume of the boiler — from 65 per cent to 75 per cent of the volume of the boiler — that the desired pressure is obtained when only a small portion of the water has been evaporated.

24. Formula Connecting Pressure and Temperature. — The relations between the pressure and temperature of steam are very complicated, and for practical purposes these relations are best ascertained from the table of the properties of saturated steam. Many formulas have been devised to represent the relations, but most of them are of only theoretical interest, and those of logarithmic form the most accurate.

Rankine gives the following equation connecting the temperature and pressure of saturated steam, which gives fairly accurate results:

$$\log p = A - \frac{B}{T} - \frac{C}{T^2},$$

in which T is the absolute temperature. For a value of p in pounds per square inch the values of A, B, C are: $A = 6.1007$, $\log B = 3.43642$, and $\log C = 5.59873$.

25. Formula Connecting Pressure and Volume of Steam. — As will be seen later on, the law governing the connection of the pressure and volume of gases is expressed by the equation $pv^n = C$, a constant quantity. For saturated steam Rankine gives for n the value $\frac{17}{16}$, and for C the value 482, so that for saturated steam we shall have $pv^{\frac{17}{16}} = 482$, in which p is given in pounds per square inch, and v in cubic feet per pound. This formula gives results that are very approximately correct, but values for p and v should be taken from the table of the properties of saturated steam when available.

Example I. — Find the volume of a pound of steam at 280 pounds absolute pressure.

Solution. — $pv^{\frac{17}{16}} = 482$, whence $v^{\frac{17}{16}} = \frac{482}{280}$.

$$\tfrac{17}{16}\log v = \log 482 - \log 280.$$

$$\begin{array}{rlrl}
482 \log & 2.68305 & & \\
280 \log & \underline{2.44716} & & \\
\log & 0.23589 & \text{llog} & 9.37271 - 10 \\
 & & 16 \log & 1.20412 \\
 & & 17 \text{ colog} & \underline{8.76955} - 10 \\
v = 1.6674 \log & 0.22202 & \text{llog} & 9.34638 - 10
\end{array}$$

Example II. — Find the volume of a pound of steam at 500 pounds absolute pressure.

Solution. — $pv^{\frac{17}{16}} = 482$, whence $\frac{17}{16} \log v = \log 482 - \log 500$.

$$\begin{array}{rlrl}
482 \log & 2.68305 & & \\
500 \log & \underline{2.69897} & & \\
\log & 9.98408 - 10 & & \\
\log (-) & 0.01592 & \text{llog } (-) & 8.20194 - 10 \\
 & & 16 \log & 1.20412 \\
 & & 17 \text{ colog} & \underline{8.76955} - 10 \\
 - & 0.01498 & \text{llog } (-) & 8.17561 - 10 \\
v = 0.9661 \log & 9.98502 - 10. & &
\end{array}$$

PROBLEMS

1. The specific heat of mercury is 0.033. How many pounds of it at a temperature of 240° will be necessary to raise the temperature of 12 pounds of water from 50° to 58°? *Ans.* 16 pounds.

2. Work at the rate of a horse-power is expended for an hour in creating friction, the heat from which is all communicated to 10 cubic feet of water contained in a non-conducting tank. The original temperature of the water was 60°. What will be its temperature at the end of the hour? *Ans.* 64.1°.

3. Find the volume of one pound of steam at the absolute pressure of 515 pounds. *Ans.* 0.9396 cubic foot.

CHAPTER III

FUELS AND COMBUSTION

26. Fuels. — The fuels commonly used for generating steam are coal, coke, wood, and oil.

27. Coal. — Coal is an insoluble combustible mineral widely distributed over the earth and of incalculable value in the application of the sciences. Owing to its extreme abundance coal is the fuel generally used in the furnaces of boilers, and in order that precautions may be taken to insure economy in its use the processes through which it passes during combustion should be understood.

Coal consists of carbon, hydrogen, oxygen, nitrogen, sulphur, ash, and moisture in varying proportions by weight. The heat-producing elements are carbon, hydrogen, and sulphur. The carbon appears in both the free state and in combination in proportions by weight of from 50 per cent to 90 per cent; the hydrogen in proportions of from 4 per cent to 5 per cent in combination with both carbon and oxygen, combining with the latter in the proportions to form water; and the sulphur in small quantities in combination with iron. The presence of sulphur is, on the whole, detrimental, its small heat-producing quality being more than balanced by its deleterious influence in promoting spontaneous combustion and in rendering coal unfit for use in the forge when its proportion in the coal exceeds 0.75 per cent.

The nitrogen of coal, either free or in combination, is inert in influence, but as it must be heated to the temperature of the products of combustion it tends to lower the furnace temperature.

The presence of oxygen in coal really detracts from its heat value, for it is known to combine with one-eighth of its weight of hydrogen to form water — existing as moisture in the fuel — and only such hydrogen in excess of one-eighth of the oxygen can be counted upon in estimating the heat value of coal.

The ash of coal, in quantities varying from 3 per cent to 10 per cent, is the residue left after combustion, and is due to the presence in the coal of inert mineral compounds.

28. **Classification of Coals.** — There is some diversity in the classification of coals, but in general terms they are determined by the proportions of volatile hydrocarbons in their compositions.

Coal containing from 5 per cent to 12 per cent of volatile matter is classed as *anthracite*, from 12 per cent to 25 per cent as *semi-bituminous*, from 25 per cent to 50 per cent as *bituminous*, above 50 per cent as *lignite*.

29. **Anthracite Coal.** — The hard coal known as anthracite makes an intense fire and when dry burns without smoke and with little flame. It contains from 80 per cent to 90 per cent of carbon, and is much more expensive and far less abundant than the soft bituminous coal.

30. **Semi-bituminous Coal.** — The coal known as semi-bituminous contains from 70 per cent to 80 per cent of carbon, but is richer in hydrogen than the anthracite variety. It is usually low in sulphur, ash, and moisture, and is regarded as the best steaming coal.

31. **Bituminous Coal.** — Bituminous coal contains from 50 per cent to 70 per cent of carbon. Its high percentage of volatile matter is rich in oxygen which injuriously affects its heating value. Bituminous coal contains no bitumen, its name being due to the fact that its behavior when heated is similar to that of bitumen, that is, it generally fuses, or cakes, when giving off its volatile matter.

32. Lignite. — The brown variety of coal known as lignite is the lowest in heat value. It contains a high percentage of moisture and weathers rapidly on exposure.

33. Heat Value of Coal. — The heat value of coal depends of course on the percentage of combustible matter it contains and the value in thermal units of a pound of the combustible. The heat value per pound of anthracite coal lies between 14,500 and 15,000 B.t.u., of semi-bituminous between 15,500 and 15,700 B.t.u., and of bituminous from 13,000 to 15,000 B.t.u.

34. Coke. — Coke is the solid product resulting from the destructive distillation of bituminous coal. It has more value for its use in the metallurgic process of reducing iron than for the purpose of generating steam. Coke is carbonized bituminous coal, the volatile constituents having been driven off by the application of heat, and it bears to coal the same relation that charcoal bears to wood.

35. Wood. — Wood is used for generating steam only in regions where it is abundant, or where coal is scarce or difficult to obtain. Wood on the average contains about 50 per cent of carbon, from 2 per cent to 5 per cent of ash, and the remainder hydrogen and oxygen in the proportions to form water. The average heat value of a pound of wood is less than one-half that of a pound of coal.

36. Oil. — Crude petroleum, the source of oil fuel, contains many hydrocarbons as it comes from the well and varies in physical and chemical properties according to the locality from which it comes. Its production in the United States for the year 1910 exceeded 200,000,000 barrels, more than two-thirds of the world's supply. The fuel oil is the residual product of a fractional distillation by which the crude petroleum is relieved of its water and the lighter hydrocarbons, such as naphtha and gasoline, the presence of which makes the crude oil highly inflammable and dangerous. The residue, which is used for fuel, varies from 6 per cent to 50 per cent.

The temperature to which fuel oil must be heated to cause it to give off vapor, which produces an explosive mixture when in contact with air, is known as the *flash point*, and the temperature at which the vapor given off will burn continuously is known as the *firing point*. Experience has shown that a flash point between 150° F and 160° F is amply high for safe storage, though for vessels of war the limit of 200° is fixed. The firing point is usually 50° F above the flash point. As an aid to economical combustion, fuel oil should be heated immediately before being burned, as heat deprives the oil of its viscidity and tends to deprive it of any water it may contain.

A sample of good fuel oil shows by analysis an average composition of: Carbon, 87.6 per cent; hydrogen, 10.8 per cent; oxygen, 1.25 per cent; and a corresponding heat value of about 19,400 B.t.u.

For the complete combustion of fuel oil it is absolutely necessary that the oil be reduced to a fine spray and be brought into contact with the proper amount of air. The usual arrangement for firing is to use a steam jet to force the oil into the furnace through burners in the shape of spray, and at the same time induce an air supply to mix intimately with the oil. Compressed air, instead of steam, is sometimes used to atomize the oil at the burner, but this method requires about as much steam to operate the air compressor as would be necessary to furnish the atomizing jet, and involves the additional complication of an air compressor.

Another system of mixing the oil and air, known as the mechanical spray, consists in forcing the oil in a conical form, and with a whirling motion, through the burner, the oil being subjected to a pressure of from 100 to 200 pounds. The air enters through a cone enveloping the burner and mixes mechanically with the oil in the furnace.

Among the advantages claimed for fuel oil are:

1. The higher heat value per unit of weight than that of coal.

2. A decrease in the loss of heat through the smoke pipe, due to cleaner tubes and the smaller amount of air that passes through the furnace to effect combustion.

3. The ease with which the fire can be regulated, extinguished in an emergency, and relighted.

4. No cleaning of fires, and the elimination of smoke.

5. Great saving in labor, one man with oil doing the work of about four men operating with coal.

37. **Combustion.** — Combustion in a furnace is a rapid oxidation caused by the chemical union of the oxygen of the air with the hydrogen, carbon, and sulphur of the fuel, heat being produced in the process. It has been seen that coal consists largely of carbon, part of which is free and part in combination with hydrogen, forming hydrocarbons. These volatile hydrocarbons are solid at ordinary temperatures, but when heated to a degree below that necessary for combustion they take the form of gaseous combinations of carbon and hydrogen which readily decompose under the action of heat, the elements becoming more distinctly separated as the temperature rises. When under the influence of a temperature sufficiently high the hydrogen and carbon become entirely separated, and as each of these elements has a stronger affinity for oxygen than for each other they will, if the supply of air be just sufficient, combine with oxygen separately and burning ensues. Under such conditions combustion would be perfect, with no formation of soot or smoke, the results being carbonic acid (CO_2), superheated steam in process of falling to water, sulphurous acid gas (SO_2), and free nitrogen. The earthy impurities, varying in amount from 3 per cent to 10 per cent, form the residue of ash and clinker.

If the supply of air through the ash pit be insufficient for complete combustion, or if it be not perfectly mixed with the gases, the result of the union will be carbonic oxide (CO), which produces less than one-third the heat due from the production of CO_2; and unless provision is made for the admission of air above the fuel to supply oxygen to convert this CO into CO_2, much of the carbon escapes unconsumed, resulting in a loss of fuel, the production of smoke, and the deposit of soot in the tubes and uptakes. The prevention of smoke is possible only when the supply of air to the gases is sufficient when they are at the temperature of ignition, and as the gases pass off very quickly from the furnace it has been found to be necessary in practice to supply 1.5 times the quantity of air theoretically necessary for combustion if the draft be artificial and 2 times the theoretical amount if the draft be natural.

38. Air Necessary for Combustion. — The quantity of air theoretically necessary for the combustion of a pound of carbon may be estimated. In the formation of CO_2 the *C* and *O* combine by weight in the proportion of 12 to 32, or as 1 to $2\frac{2}{3}$. It follows that every pound of *C* requires $2\frac{2}{3}$ pounds of *O* for its combination to form CO_2. This *O* must come from the air containing oxygen and nitrogen in the proportion very approximately of 23 parts of *O* and 77 parts of *N* in every 100 parts by weight, and since a cubic foot of air weighs very nearly 0.08 pound, it contains $0.23 \times 0.08 = 0.0184$ pound of *O*. Then, to produce $2\frac{2}{3}$ pounds of *O*, the amount required for the combustion of one pound of carbon, we must have $\frac{2\frac{2}{3}}{0.0184} = 145$ cubic feet of air.

The formula given by Rankine for the weight of air theoretically necessary for the complete combustion of a pound of fuel is

$$A = 12\left[C + 3\left(H - \frac{O}{8}\right)\right],$$

where *A* is the weight of air in pounds, and *C*, *H*, and *O*, the percentages of carbon, hydrogen, and oxygen contained in a pound of the fuel. This formula has a rational basis, as may be seen from the following:

For the complete combustion (CO_2) of one pound of carbon there will be required $\frac{32}{12} = 2\frac{2}{3}$ pounds of oxygen, and since each pound of air contains but 0.23 pound of oxygen, there will be required $\frac{2\frac{2}{3}}{0.23} = 11.6$ pounds of air for the combustion of the pound of carbon. For the complete combustion of one pound of hydrogen there are required 8 pounds of oxygen, which must also be abstracted from the air, requiring $\frac{8}{0.23} = 35$ pounds of air to furnish it, which is three times as much as was required for the combustion of the pound of carbon. In estimating the thermal value of a fuel only such H as is unbalanced by O is taken as available for the generation of heat, and therefore $H - \frac{O}{8}$ is the expression for this free hydrogen; that is, the H in the fuel is diminished by one-eighth of the O, because in the formation of the fuel by natural processes that amount has already combined with O to form water — existing as such in the fuel — and is therefore no longer available for generating heat. In the formula, 12 is substituted for the approximation 11.6, and all within the brackets is expressed in terms of carbon.

39. **Total Heat of Combustion of Fuels.** — The two methods of determining the theoretical heat value of fuels are: (a) By burning a sample of the fuel in a coal calorimeter; (b) by chemical analysis. The calorimeter method is the more accurate.

Mahler's bomb calorimeter consists essentially of a strong steel vessel, or bomb, immersed in a known weight of water. A small but known weight of the sample is placed in the bomb and oxygen gas admitted under a pressure of from 300 to 350 pounds. An electric spark ignites the coal explosively, the combustion being complete and instantaneous. The heat generated is conveyed to the water surrounding the bomb and its quantity is determined from the rise in temperature of the water, correction being made for the heat capacity of the bomb.

The method based on chemical analysis is not very reliable because of the assumption that the heat values of the elements as existing in the fuel are the same as when existing in the separate and uncombined condition, and that no heat is absorbed in separating the carbon and hydrogen of the hydro-carbons.

The amounts of heat liberated in the combustion of a pound of each of the heat-producing elements of fuel have been determined experimentally, and are as follows:

A pound of hydrogen on entering into combination with 8 pounds of oxygen liberates about 62,000 B.t.u.

A pound of carbon in combination with $2\frac{2}{3}$ pounds of oxygen, falling to CO_2, liberates about 14,500 B.t.u.

A pound of carbon in combination with $1\frac{1}{3}$ pounds of oxygen, falling to CO, liberates about 4500 B.t.u.

A pound of sulphur in combination with a pound of oxygen, falling to SO_2, liberates about 4050 B.t.u.

Knowing the chemical composition of a fuel, its heat value per pound, as determined from its combustible elements, is given with a fair degree of approximation by the formula

$$h = 14{,}500\left[C + 4.28\left(H - \frac{O}{8}\right)\right] + 4050\,S,$$

in which h is the total heat in B.t.u. of a pound of the fuel, and C, H, O, and S are the percentages of carbon, hydrogen, oxygen, and sulphur contained in the fuel.

Regarding $\left(H - \frac{O}{8}\right)$ as free hydrogen, its value in terms of carbon is $\frac{62{,}000}{14{,}500} = 4.28$, so that all within the brackets is carbon, the coefficient of which is 14,500 B.t.u. The value of sulphur as a heat-producing element is small, and often neglected.

Example in the Combustion of Coal. — Coal containing 90 per cent of carbon, 4.7 per cent of hydrogen, 2.8 per cent of oxygen, and 1 per cent of sulphur is burned in a furnace with a supply of air at 60° temperature of 20 pounds per pound of coal. The specific heats of N, O, CO_2, SO_2, and superheated steam are 0.249, 0.220, 0.222, 0.170, and 0.462 respectively. It is required to find by calculation the temperature of the furnace.

Solution. — The combining weights of *O* and *N* in air are as 23 to 77, therefore in 20 pounds of air there are $20 \times 0.23 = 4.6$ pounds of *O*, leaving $20 - 4.6 = 15.4$ pounds of *N* free, and making a total of $4.6 + 0.028 = 4.628$ pounds of *O*.

The combining weights of *C* and *O* to form CO_2 are 12 to 32, or 1 to $2\frac{2}{3}$, therefore $2\frac{2}{3} \times 0.9 = 2.4$ pounds of *O* are required to unite with the 0.9 pound of *C* in the coal to produce $2.4 + 0.9 = 3.3$ pounds of CO_2.

The combining weights of *H* and *O* to form water are 2 to 16, or 1 to 8, therefore $0.047 \times 8 = 0.376$ pound of *O* unites with the 0.047 pound of *H* in the coal, producing $0.376 + 0.047 = 0.423$ pound of superheated steam.

The combining weights of *S* and *O* to form SO_2 are 32 to 32, or 1 to 1, therefore 0.01 pound of *O* unites with the 0.01 pound of *S* in the coal to form 0.02 pound of SO_2.

Thus, $2.4 + 0.376 + 0.01 = 2.786$ pounds of *O* are used, leaving $4.628 - 2.786 = 1.842$ pounds of *O* free.

The gases to be heated are:	The gases entering the furnace were:
15.400 pounds of N	20.000 pounds of air
3.300 " " CO_2	0.900 " " C
0.423 " " superheated steam	0.047 " " H
0.020 " " SO_2	0.028 " " O
1.842 " " O	0.010 " " S
20.985 pounds	20.985 pounds

Since the amount of heat each gas gained in having its temperature raised is the product of its weight, specific heat, and range of temperature, it follows that 1° of elevation of the temperature of all the gases will occasion an expenditure of $15.4 \times 0.249 + 3.3 \times 0.222 + 0.423 \times 0.462 + 0.02 \times 0.017 + 1.842 \times 0.22 = 5.168$ B.t.u.

The total heat of combustion of a pound of the coal is

$$14{,}500\left[0.9 + 4.28\left(0.047 - \frac{0.028}{8}\right)\right] + 4050 \times 0.01 = 15{,}790 \text{ B.t.u.}$$

The elevation of the temperature of the gases above their initial temperature will therefore be $\frac{15{,}790}{5.168} = 3055°$, and the temperature of the furnace will be $3055 + 60 = 3115°$.

40. **Evaporative Power of Fuel.** — From the steam table of Marks and Davis the total heat of a pound of steam at atmospheric pressure is 1150.4 B.t.u., reckoning from a feed-water temperature of 32°, so that if all the heat of the coal of the example of Art. 39 were utilized we would have an evaporation per pound of coal of $\frac{15{,}790}{1150.4} = 13.72$ pounds of water from a temperature of 32° and at a temperature of 212°.

All the heat of the coal is not utilized, however, in the generation of steam. The temperature of the escaping gases through the smoke pipe is commonly 600°, so that in the example there would be $(600 - 60)\ 5.168 = 2790$ thermal units lost through the chimney, the equivalent of 17.5 per cent of the total heat of combustion. Experience has shown that on the average these additional losses may be expected: — From unburned fuel falling through the grate, 4 per cent; from imperfect combustion, 1 per cent; and from radiation, 7 per cent, making a total in losses of 30 per cent. The evaporation to be expected in the example would then be $\frac{15{,}790 \times 0.7}{1150.4} = 9.6$ pounds of water per pound of coal.

It has been shown in Art. 23 that the expression $1091.7 + 0.305\ (t - 32)$ gives approximately the thermal units required to evaporate a pound of water from a feed-water temperature of 32° into steam of temperature t. If the feed water were taken at any other temperature denoted by t_f, there would be $t_f - 32$ thermal units in every pound of feed water that was not supplied by the fuel, so that when the temperature of the feed water is t_f there must be subtracted $t_f - 32$ units from the amount given by the formula in order to obtain the heat units to be supplied by the fuel to evaporate a pound of water of temperature t_f into steam of temperature t.

If we denote by H_w the units of heat necessary to evaporate a pound of water *from* a temperature t_f into steam *at* a temperature t, we will have the expression, general in its application,

$$H_w = 1091.7 + 0.305\,(t - 32) - (t_f - 32).$$

If t and t_f each equal 212°, we shall have

$$H_w = 1091.7 + 0.305\,(212 - 32) - (212 - 32) = 966.6 \text{ B.t.u.}$$

as the amount of heat necessary to convert a pound of water of a temperature of 212° into steam at the same temperature. In this instance there has been no expenditure of heat in raising the temperature of the feed water to the boiling point, so that the quantity 966.6 B.t.u. is the total heat minus the sensible heat, or the latent heat, of steam at temperature 212° and atmospheric pressure of 14.7 pounds. The latent heat of steam at 212° is tabulated by Marks and Davis as 970.4 B.t.u., but for brevity it will be taken as 970 in the examples that follow in this book.

41. The Unit and the Factor of Evaporation. — In their revision in 1899 of the code for conducting steam boiler tests, the American Society of Mechanical Engineers adopted "*one pound of water at 212° F evaporated into dry steam at the same temperature*" as the *unit of evaporation*. This unit, as was shown in the preceding article, is equivalent to 970 B.t.u. Accordingly it is the common practice to reduce the results of boiler tests to the standard of weight of water evaporated per pound of coal *from* and *at* the temperature of 212°. The heat available for the generation of steam per pound of fuel divided by H_w will give the number of pounds of water evaporated per pound of fuel under the conditions of pressure and temperature of the test, and to express this in terms of the unit of evaporation we have simply to multiply by the ratio $\frac{H_w}{970}$, which ratio is known as the *factor of evaporation*.

42. Boiler Horse-power. — The term *boiler horse-power* must not be confused with the horse-power unit expressed in foot pounds which is used in measuring the work of an engine. The work of a boiler in converting water into steam is expressed in heat units which, though having an equivalence in foot pounds, is not the result of dynamic effort as in the case of a steam engine. The arbitrary standard of boiler horse-power adopted by a Commission of the Centennial Exhibition in 1876 is defined to be "*an evaporation of* 30 *pounds of water per hour from a feed-water temperature of* 100° *F into steam at* 70 *pounds pressure per gauge*," or at the absolute pressure of 84.7 pounds. From the steam table we find the total heat of steam at 84.7 pounds pressure to be 1183.3 B.t.u., so that $H_w = 1183.3 - (100 - 32) = 1115.3$ B.t.u.;

hence $$\text{Factor of evaporation} = \frac{H_w}{970} = \frac{1115.3}{970} = 1.1498.$$

Therefore, a boiler horse-power is equivalent to the evaporation of $30 \times 1.1498 = 34.5$ pounds of water *from* and *at* 212°.

The practical method of determining the boiler horse-power is to ascertain by test the number of pounds of water evaporated per hour under the working conditions of the boiler, and then, by means of the factor of evaporation, find the equivalent number of pounds from and at 212°. The latter quantity divided by 34.5 will give the horse-power of the boiler.

Or, since it requires $970.4 \times 34.5 = 33,480$ B.t.u. to evaporate 34.5 pounds of water from and at 212°, the boiler horse-power may be obtained very conveniently by dividing the number of thermal units absorbed by the water in the boiler per hour by 33,480.

Example. — Semi-bituminous coal containing 85 per cent of C, 5 per cent of H, 3.2 per cent of O, and 0.8 per cent of S is burned in a furnace, and it is estimated that the losses from radiation, incomplete combustion, unburned coal falling through the grate bars, and escaping gases through the smoke pipe will aggregate 28 per cent. Steam is generated at an absolute pressure of 110 pounds per square inch from feed water of temperature of 160°. The engine is required to develop 125 I.H.P. at an expenditure of 30 pounds of steam per I.H.P. per hour. Find: (a) The number of pounds of water evaporated per pound of coal; (b) the evaporation per pound of coal from and at 212°; (c) the necessary boiler horse-power.

Solution. —

$$h = 14,500\left[0.85 + 4.28\left(0.05 - \frac{0.032}{8}\right)\right] + 4050 \times 0.008$$

$$= 15,211.36 \text{ B.t.u.}$$

Heat available for generating steam $= 15,211.36 \times 0.72 =$ 10,952.18 B.t.u.

From the table of the properties of saturated steam it is found that the total heat of steam at 110 pounds pressure is 1188 B.t.u., so that $H_w = 1188 - (160 - 32) = 1060$ B.t.u.

$$\text{Steam evaporated per pound of coal} = \frac{10,952.18}{1060} = 10.33 \text{ pounds.}$$

$$\text{Evaporation per pound from and at } 212^\circ = \frac{10,952.18}{970} = 11.29$$

pounds.

Steam required per hour for the engine is $125 \times 30 = 3750$

pounds, so that the water in the boiler would have to absorb $3750 \times 1060 = 3,975,000$ B.t.u. per hour. Hence

$$\text{Boiler horse-power} = \frac{3,975,000}{33,480} = 119.$$

Or,
$$\text{Factor of evaporation} = \frac{1060}{970} = 1.0928.$$

$$\text{Boiler horse-power} = \frac{3750 \times 1.0928}{34.5} = 119.$$

PROBLEMS

1. Find the number of pounds of air necessary for the complete combustion of one pound of hydrogen. *Ans.* 35.

2. Explain the derivation of the formula for finding the total heat of combustion of a pound of fuel.

3. Coal containing 91 per cent of C, 3.5 per cent of H, 4 per cent of O, and 0.75 per cent of S is burned in a furnace with a supply of air of 28 pounds per pound of coal. Temperature of the entering air, 80°; temperature of the discharged gases, 600°; temperature of feed water, 140°. The specific heats of N, CO_2, O, SO_2, and of superheated steam are 0.249, 0.222, 0.220, 0.170, and 0.48 respectively. It is expected that the losses through the smoke pipe, from incomplete combustion, from unburned fuel, and from radiation will aggregate 32 per cent of the heat value of the coal used. Find: (a) The temperature of the furnace; (b) the heat units lost through the smoke pipe; (c) the heat available for generating steam per pound of coal; (d) the number of pounds of coal required to evaporate 50 pounds of water into steam at 324° temperature.
Ans. (a) 2557°; (b) 3166.8 B.t.u.; (c) 10,259 B.t.u.; (d) 5.25 pounds.

4. An analysis of coal shows its composition to be 91.5 per cent of C, 3.5 per cent of H, and 2.6 per cent of O. How much air per pound will be theoretically necessary for its combustion? *Ans.* 11.714 pounds.

5. Supposing 72 per cent of the heat value of the coal of problem 4 to be available, how many pounds of water will a pound of the coal evaporate into steam of 373° temperature, the feed-water temperature being 132°? What is the evaporative power of the coal from and at 212°?
Ans. 10 pounds and 11.31 pounds.

6. Assuming the whole of 14,500 B.t.u. of a pound of carbon were available for conversion into useful work, what weight of it would be required per I.H.P. per hour? *Ans.* 0.176 pound.

7. A manufacturing establishment requires 225 engine horse-power to operate its plant. The engine is to use steam of 331° temperature, generated from feed water of 142°, and it is deemed advisable to allow for an expenditure of 34 pounds of steam per I.H.P. per hour. What boiler horse-power will be required? *Ans.* 245.

CHAPTER IV

STEAM BOILERS, ATTACHMENTS AND ACCESSORIES

43. **The Function of the Boiler.** — The heat energy produced by the combustion of fuel is transferred to the engine through the medium of water in the condition of steam. The metallic vessel containing the water is the boiler, and it is the particular function of the boiler to convey the heat energy to the water and then to store that energy for use in the available form of steam which results from the evaporation of the water.

A boiler consists essentially of: — A furnace in which the fuel is burned; a receptacle for the water which is to be evaporated; a space in which to store the steam resulting from the evaporation of the water; heating surface through which the heat from the burning fuel is transmitted to the water; and a smoke pipe to carry away the furnace gases and to produce the draft for the furnace. In addition to its essential parts, a boiler must have these appendages: — A feed pump to supply the water; a gauge to indicate the pressure of the steam; a water column to indicate the height of the water in the boiler; a safety valve to permit the escape of the steam into the atmosphere when the pressure becomes excessive; and a valve and suitable piping to convey the steam to the engine.

44. **Classification of Boilers.** — There are two general classes of boilers, viz., those in which the hot gases from the furnace pass through tubes which are surrounded with water — known as fire-tube boilers; and those in which the hot gases circulate between tubes containing water — known as water-tube or tubulous boilers.

Marine boilers differ in form and setting from those of stationary boilers used on land. The high steam pressures demanded by modern engineering conditions have completely displaced the rectangular type of boiler commonly in use for marine purposes as late as the year 1875. The spherical form being impossible, the cylinder was naturally suggested as the most practical form to give strength to boilers, and this led to the development of the type known as the Scotch cylindrical return-fire-tubular boiler which is now largely used in marine practice. Special types of the Babcock & Wilcox and of other water-tubular boilers are being used advantageously on board ships, and it seems likely that some form of the sectional water-tube boiler will eventually displace the boiler of the cylindrical or tank form.

A sectional water-tube boiler, of which the Babcock & Wilcox boiler is a type, may be defined as one in which the contained water is divided into numerous small masses connected with each other by passages sufficiently large to permit free circulation, but not large enough to permit so sudden a release of pressure, in case of rupture in one of the sections, as to cause a disastrous explosion. Generally speaking, sectional boilers may be said to consist of one or more cylindrical drums on top which contain steam and water and are connected by a series of tubes with a mud drum at the bottom. The heating surface is made up almost entirely of the tubes which are of moderately small diameter and therefore very strong, though made of thin metal. These boilers may be made light and of great power, and, by reason of the small volume of water they contain, steam may be raised very quickly, but for the same reason variations in pressure quickly occur in cases of unsteady feed or irregular firing.

45. Relative Advantages of Fire-tube and Water-tube Boilers. — The water-tube boiler has the marked advantage of being much lighter than the fire-tube type for equal powers. Expressing the weight in terms of heating surface, the weight of the fire-tube boiler varies roughly from 25 pounds to 30 pounds per square foot of heating surface, while that of the water-tube boiler varies from 12 pounds to 18 pounds. When filled to the water level, the weight of contained water in the fire-tube boiler varies from 12 pounds to 14 pounds per square foot of heating surface, and for the water-tube boiler from 1.5 pounds to 2.5 pounds. The water-tube boiler, from the nature of its construction, is unquestionably the stronger of the two types, and requires only from one-quarter to one-half hour to raise steam as against from three to four hours with the fire-tube boiler.

The principal disadvantage charged to the water-tube boiler is that it requires to be fed with fresh water, and its rate of feed must be uniform because of the small water space it contains. To show the necessity for a steady feed for the water-tube boiler, we will assume a boiler having 50 square feet of grate surface and burning 25 pounds of coal per hour per square foot of grate. The consumption of coal would then be $25 \times 50 = 1250$ pounds per hour. A consumption of 3.5 pounds of coal per I.H.P. per hour would mean the development of $\frac{1250}{3.5} = 357$ I.H.P. Assuming the consumption of steam per I.H.P. per hour to be 30 pounds, an evaporation of $\frac{357 \times 30}{2240} = 4.8$ tons of water per hour would be necessary. Assuming the ratio of grate surface to heating surface to be 1 to 36, there would be $50 \times 36 = 1800$ square feet of heating surface; and taking the weight of contained water as 1.5 pounds per square foot of heating surface, the weight of the water in the boiler would be 2700 pounds, or about 1.2 tons. Therefore, the boiler would have to be filled $\frac{4.8}{1.2} = 4$ times in an hour, so that a cessation of the feed for 15 minutes would completely empty the boiler.

46. The Scotch Boiler. — A type of fire-tube boiler largely used in marine practice is shown in Figs. 2 and 3. It is variously known as the Scotch marine boiler, a tank boiler, or a

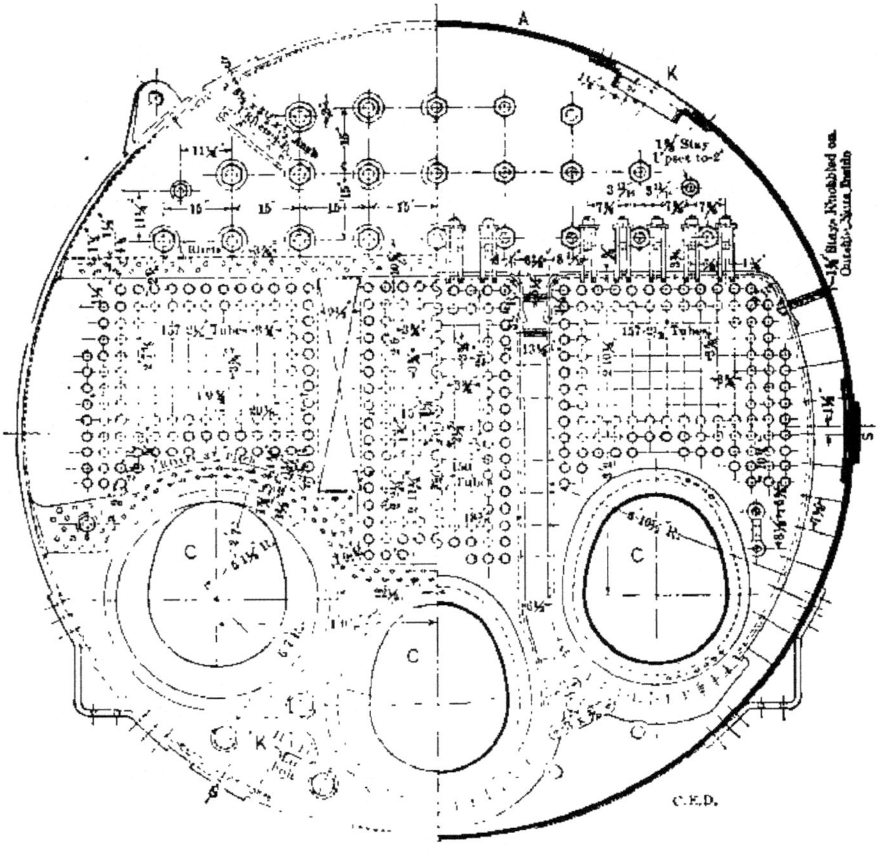

FIG. 2.

shell boiler, and is made with two, three, and four furnaces. It is cylindrical in shape with flat ends or heads, and since the furnace gases, after passing into a combustion chamber at the rear of the furnace, return through the tubes to the front, and thence through the uptake to the smoke pipe; it is of the type known as return-tubular boiler.

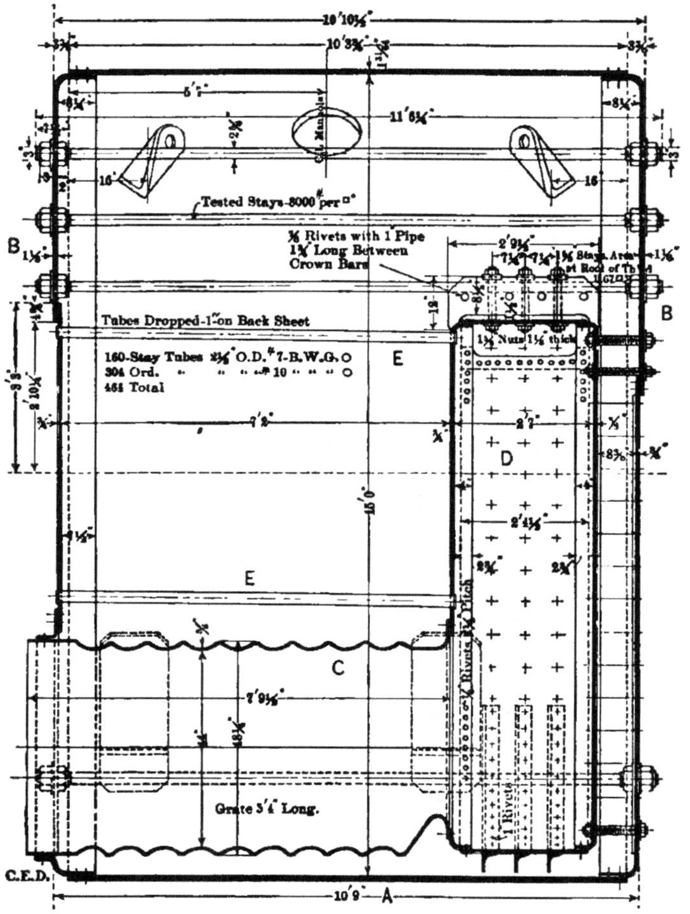

FIG. 3.

Figure 2 shows a three-furnace Scotch boiler in cross section, and Fig. 3 in longitudinal section. The cylindrical shell is shown at *A*, *A*, and the flat heads at *B*, *B*. Inside the shell are the three corrugated furnaces *C*, *C*, *C*, the corrugations adding much to the strength of the furnace. A plane cylindrical furnace flue would collapse under the high steam pressure of recent practice, unless made of prohibitive thickness. The fire grate is placed at about midway in the height of the furnaces, the fire and hot gases passing over the bridge wall (not shown) at the rear end of the furnace into the combustion chamber *D*, thence through the tubes *E* into the uptake (not shown) at the front of the boiler, and thence to the smoke pipe. The water level is maintained at a height of from 6 to 8 inches above the top of the combustion chamber, the remaining space at the top of the boiler being occupied by the steam. The combustion chambers, or back connections, one for each furnace, are stayed at their tops by girder braces, and to each other and to the shell and back head by screw stay bolts, while the heads are stayed to each other by longitudinal braces, as shown. There is a total of 364 tubes of 2½ inches outside diameter, 160 of which are stay tubes, the purpose of which is to brace together the front and back tube sheets. These stay tubes, shown in Fig. 2 in thicker section than the ordinary tubes, are enlarged at their ends and screw into the front tube sheet and are expanded into the back tube sheet, the back ends being beaded and protected by cast-iron ferrules from flame in the combustion chamber. The manholes *K* permit access to the boiler for cleaning and repairs. Sometimes there is but one combustion chamber for two, or even three, furnaces instead of one for each furnace.

A double-ended Scotch boiler consists of a shell of double length, having furnaces at each end with a common combustion chamber or a separate combustion chamber for each end. Such an arrangement is equivalent to a pair of single boilers with their back heads removed and their shells united.

47. Vertical Boilers. — Boilers of the vertical fire-tube type are cylindrical in shape with the furnace, or fire box, at the bottom. The top of the furnace, or crown sheet, forms the lower tube sheet, the upper tube sheet forming the upper head of the shell. The products of combustion pass directly through the tubes from the furnace to the smoke box at the base of the chimney. The water in the shell surrounds the tubes, and there is a water-leg around the fire box to protect the metal from the flame. The vertical boiler occupies small floor space, and is a convenient type to move from place to place for temporary work.

48. The Locomotive Boiler. — One type of the locomotive boiler, as built by the American Locomotive Works, is shown in longitudinal section and in several cross sections in Fig. 4. The different views are as follows: (*A*) is the longitudinal section, the tubes being removed; (*B*) is a half cross section showing the front tube sheet and the opening into the smoke box from the dry pipe; (*C*) is a half cross section through combustion chamber showing the rear tube sheet and the manner of bracing the fire box to the shell; (*D*) is a half cross section through the fire box showing the side water-leg, the recessed combustion chamber in outline, and the bracing; (*E*) is a half cross section just inside the furnace door; and (*F*) is the rear end of the boiler showing furnace-door opening.

The boiler consists essentially of a fire box of approximate rectangular section; a cylindrical shell which envelops the fire box and contains the tubes, the water and the steam; and a smoke box at the front into which the fire tubes discharge the products of combustion on their way to the smoke stack. That part of the front tube sheet above the tubes is stayed by diagonal braces running to the inside of the top of the shell, and similar braces run from the rear head of the shell. There is a recessed combustion chamber between the fire box proper and the rear tube sheet, and connecting the tube sheets are 234 tubes of $2\frac{1}{4}$ inches diameter and 40 superheater flues of $5\frac{1}{2}$ inches diameter. The tubes and flues are expanded into the sheets and

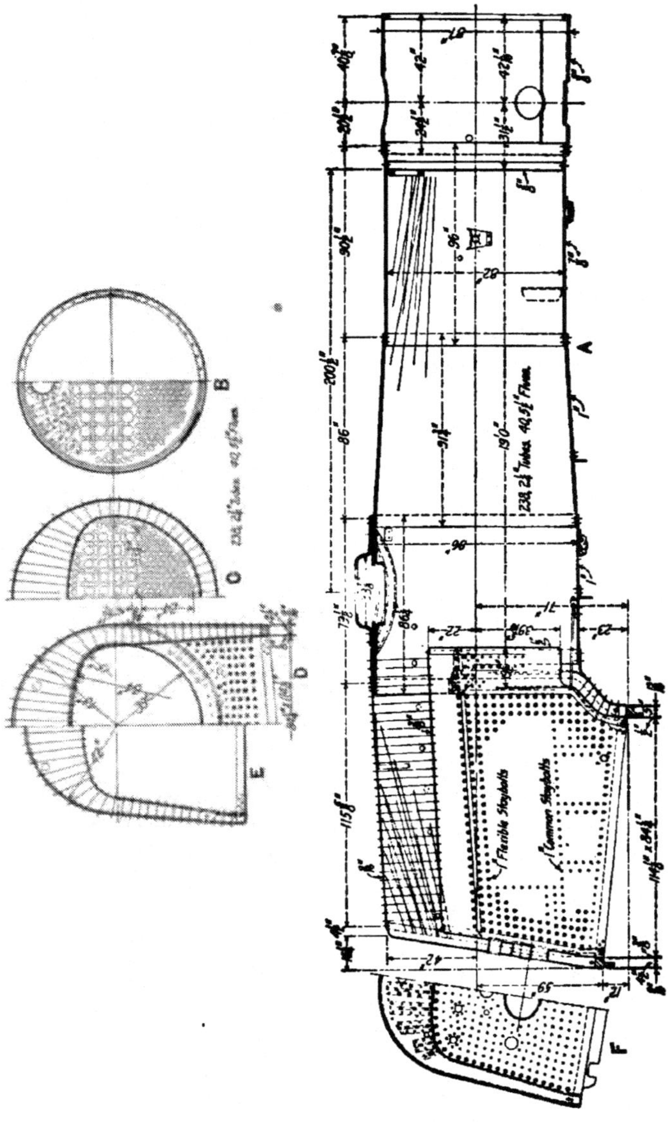
Flexible Staybolts
Common Staybolts
A
B
C
D
E
F

beaded over at their ends. The crown sheet of the fire box is braced to the shell by rows of radial stays, and the sides of the fire box are braced to the outside shell by ordinary stay bolts, but at the points of greatest expansion are placed flexible stay bolts to allow of some movement between the stayed sheets. The inner end of the flexible stay bolt is screwed directly into the fire-box sheet, Fig. 5, its outer end having a spherical head which bears in a *spud* which is screwed into the outer sheet, the open end of the spud being covered by a screwed cap, leaving a small expansion space between the head of the bolt and the crown of the cap. This device allows of some play between the sheets during expansion and contraction.

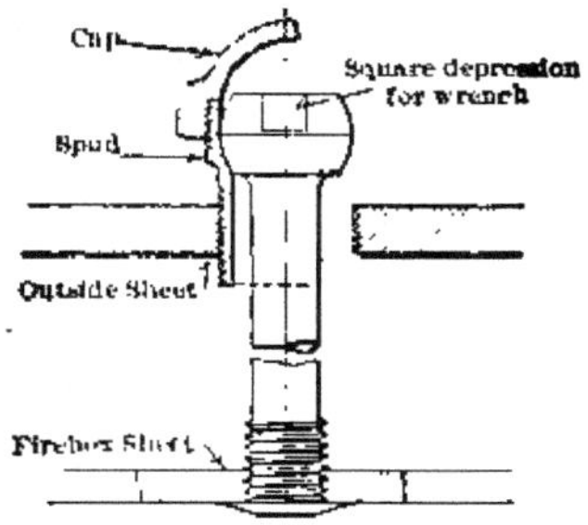

FIG. 5.

The boiler works under a strong induced draft occasioned by the cylinders exhausting into the smoke stack, and in consequence the rate of combustion is very rapid, not uncommonly exceeding one hundred pounds of coal per square foot of grate per hour. To meet this rapid combustion the tubes present a large heating surface for the evaporation of the water into steam. The steam is conveyed from the dome at the top of the shell by the dry pipe into a header placed in the smoke box. The steam passes from the header through a series of tubes of $1\frac{1}{2}$-inch diameter each one of which enters a superheater flue and passes back to within about twenty inches of the rear tube sheet and then, by means of a U-fitting, returns to the fire-box end of the flue, delivering the steam in a superheated state to the pipes leading to the cylinders. This boiler ordinarily produces steam containing about 175° of superheat.

Another type of locomotive boiler, made by the American Locomotive Works, is shown in Fig. 6. It differs from the

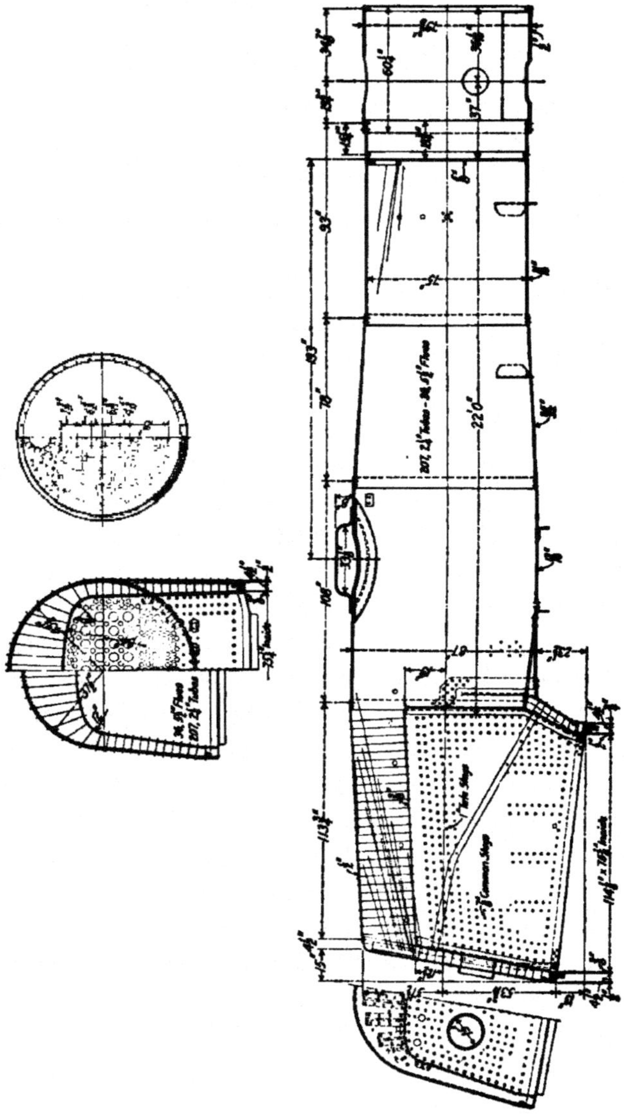

FIG. 6.

boiler just described in these particulars only: The fire box has twin fire doors, but no recessed combustion chamber, and from the throat sheet there are four arched water tubes passing through the fire box to the inside fire-door sheet.

One of these water tubes is shown in the fire box of the longitudinal section. On top of these tubes is laid a brick arch, or baffle, extending from the throat sheet to about one-half the length of the tubes in one direction, and to within a short distance of each side sheet of the fire box in the other direction. The object of this arch is to protect the rear tube sheet from the direct action of the furnace flame by deflecting it so that it passes through the opening between the end of the arch and the fire-door sheet before passing through the tubes. This course of the hottest gases increases the efficiency of the crown of the fire box as heating surface, and the rapid heating of the water in the tubes greatly facilitates the circulation of the water in the boiler.

49. **The Babcock & Wilcox Boiler.** — Perhaps the most widely known of the water-tube boilers is that of the Babcock & Wilcox Co. It is very extensively used in this country and abroad, and embraces in its make-up these advantageous features of boiler design:

Thin heating surface; joints removed from fire; efficient water circulation; quick steaming; dryness of steam; freedom for expansion; safety from explosions; and ease of transportation.

The boiler is made in two types, *stationary* and *marine*, both embracing the same general principles in construction and operation, but differing materially in form. The stationary type will here be noticed with some degree of particularity, as it is with land boilers that we are mostly concerned.

The end connections, or headers, Fig. 7, are made of forged steel, and of such sinuous form as to *stagger* the tubes, that is, make of the one vertical row of tubes virtually two parallel rows, an arrangement that affords a more efficient contact of the hot gases with the surface of the tubes.

FIG. 7

The sections are usually composed of hot finished open-hearth steel tubes, placed in an inclined position and connected with each other, and with a horizontal steam and water drum, by vertical passages at each end, while a mud drum is connected to the rear and lowest point in the boiler.

The holes are accurately sized, being counterbored, and the tubes fixed therein by an expander. The sections thus formed are connected with the drum and with the mud drum also by short tubes expanded into bored holes, doing away with all bolts, and leaving a clear passage way between the several parts. The openings for cleaning opposite the end of each tube are closed by handhole plates, the joints of which are made in the most thorough manner, by milling the surfaces to accurate metallic contact, and are held in place by wrought steel clamps and bolts. They are tested and made light under a hydrostatic pressure of 300 pounds per square inch, the joints being metal to metal, and without rubber packing, or other perishable substances.

The steam and water drums are made of flange iron or steel, of extra thickness, and double riveted. They can be made for any desired pressure, and are always tested at 50 per cent above the pressure for which they are constructed. The mud drums are made of wrought steel and are provided with blow-off connections and ample means for cleaning.

In erecting this boiler, it is suspended entirely independent of the brickwork, from wrought-iron girders resting on iron columns. This prevents any straining of the boiler from unequal expansion between it and its enclosing walls, and permits the brickwork to be repaired or removed, if necessary, without in any way disturbing the boiler. All the fixtures are extra heavy and of neat designs.

A distinctive feature of the boiler is its ease of transportation. Being made in sections, which are readily put together with an expanding tool, these boilers may be transported easily and cheaply.

FIG. 8.

A side view of the boiler with vertical headers is shown in Fig. 8. In operation the fire is usually located under the front and higher end of the tubes and the products of combustion are compelled by the bridge wall and front baffles to rise through the forward portion of the tubes into a chamber under the steam and water drum; thence, the second baffle compels their descent between the tubes, and finally they pass upward between the tubes at their rear ends and off to the chimney. The water inside the tubes, as it is heated, tends to rise toward the higher end, and as it is converted into steam — the mingled column of steam and water being of less specific gravity than the solid water at the back end of the boiler — rises through the vertical passages into the drum above the tubes, where the steam separates from the water and the latter flows back to the rear and down again through the tubes in a continuous circulation. As the passages are all large and free, this circulation is very rapid, sweeping away the steam as fast as formed, and supplying its place with water; absorbing the heat of the fire to the best advantage; causing a thorough commingling of the water throughout the boiler and a consequent equal temperature, and preventing, to a great degree, the formation of deposits or incrustations upon the heating surfaces, sweeping them away and depositing them in the mud drum, whence they are blown out.

The steam is taken out at the top of the steam drum near the back end of the boiler after it has thoroughly separated from the water, and to insure dry steam a perforated dry-pipe is connected to the nozzle inside the drum.

The boiler possesses to a marked degree the inherent advantage of all water-tube boilers, that is, the small diameter of the tubes admits of their being made thin, and as they are in contact with the hottest gases they afford such ready transmission of the heat that even the fiercest fire cannot injure them, provided that water is in intimate contact with the metal of the tubes. This is in marked contrast with the conditions of the fire-tube boiler where the thick plates of the furnaces in immediate contact with the fire not only hinder the transmission of heat to the water, but admit of overheating and even burning of the side next to the fire with the consequent dangerous tendency to rupture.

An advantage claimed for the Babcock & Wilcox boiler is that its tube-heating surface is of the most efficient character from the fact that the hottest gases come in contact with it nearly at right angles, impinging on the surface instead of gliding over it in lines parallel with it.

As all the water in the boiler tends to circulate in one direction there are no interfering currents, and the steam is carried quickly to the surface, which tends to equalize the temperature of all parts of the boiler. The water too is divided into many small streams flowing in their envelopes through the hottest part of the furnace, an arrangement which admits of steam being raised very rapidly, not more than 30 minutes being necessary ordinarily to raise steam to a working pressure from cold water.

The design of the boiler is such as to accomplish these two essential features for the production of dry steam: (1) Time is given for the steam and water to separate thoroughly at the disengaging orifice. (2) Sufficient space is provided between the steam outlet and the disengaging point to insure the steam passing from the boiler in a dry state without entraining or again picking up any particles of water in its passage.

Referring to the partial vertical section of Fig. 9, it is seen that the rapidly moving mixture of steam and water, as it enters from the headers through the disengaging orifice into the drum, comes in contact with a curved deflecting plate which extends almost across the drum, leaving spaces at its ends and the sides of the drum. The mixture coming in contact with the plate, the water is deflected downward and mixes with the main body of the water, the steam passing around the ends of the plate into the steam space in which is located the dry pipe. The figure shows the position of the manhole in the drum head and the method of securing the manhole plate, also the entrance of the feed pipe beyond the deflection plate, and the manner of attaching the water column.

The superheating attachment to the Babcock & Wilcox boiler is shown in Fig. 10. The superheating arrangement consists of

Fig. 9.

a series of tubes bent into a U-shape and connected at the ends with manifolds, the upper one of which receives the natural steam from the drum; thence the steam passes through the tubes, which are exposed to the products of combustion, and delivered in a superheated state into the lower manifold, and

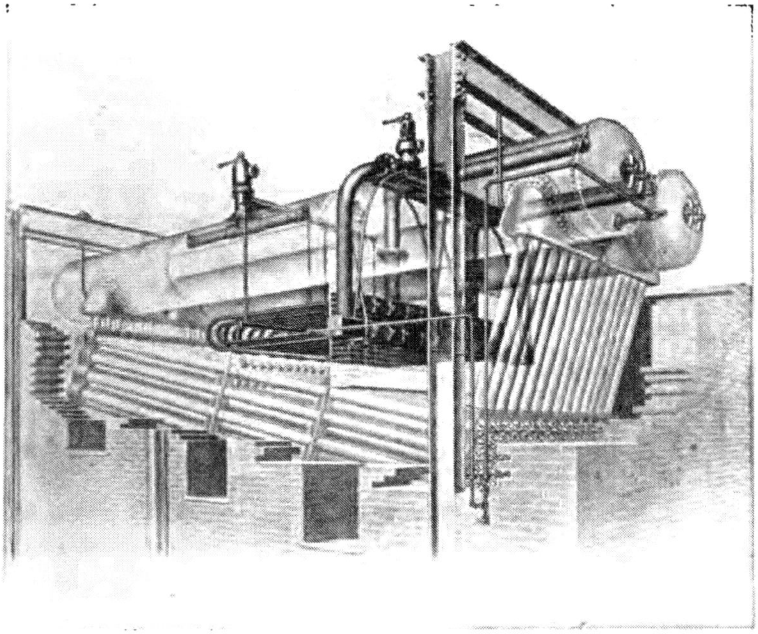

FIG. 10.

thence through the large bent pipes to the valve placed above the boiler in the steam main.

To prevent overheating while steam is being raised, a flooding device is usually supplied by which the superheater is flooded with boiler water. The arrangement, shown in the small pipes of Fig. 10, consists merely of a connection with the water space of the drum, and a three-way cock, by which, at will, the water enters the lower manifold and fills the superheater and connections up to the water level. Any steam formed in the superheater tubes is returned into the drum through the collecting pipe, which, when the superheater is at work, conveys the saturated or natural steam into the upper manifold. Prior to opening the superheater stop valve and using superheated steam, the water is drained from the manifolds by the flooding pipe.

Safety valves are placed on the outlet of the superheater. These valves are set to blow slightly below the boiler safety valves, the object being that if the demand for steam suddenly ceases the fact that the superheater valves blow first insures a flow of steam through the superheater and prevents burning.

The superheater is so placed as not to expose it to the immediate action of the fire, as the furnace gases must first pass through the front part of the boiler where there is considerable heating surface to lower their temperature before reaching the superheater. There are no flanged joints to the superheater as all the tube joints are expanded. Freedom for expansion is provided for by the tubes being free at one end, and by the manifolds not being rigidly connected with each other. The proportion of superheating surface to boiler heating surface may be made such as to provide practically any desired degree of superheat.

50. The Stirling Boiler. — The Stirling water-tube boiler, manufactured by the Babcock & Wilcox Co., shown in Figs. 11 and 12, consists of three upper or steam drums, each connected by a *bank* of tubes to a single lower or mud drum. Suitably disposed fire-tile baffles between the banks so direct the gases that their course from the furnace is upward and in contact with the surfaces of the front bank of tubes, then downward among the tubes of the middle bank, then upward again among the tubes of the rear bank, and thence to the smoke pipe.

Short tubes connect the steam spaces of all the upper drums, and also the water spaces of the front and middle drums. The boiler is supported on a structural steel framework, around which is built a brick setting whose only office is to provide furnace space and serve as a housing to confine the heat. The entire front is of metal.

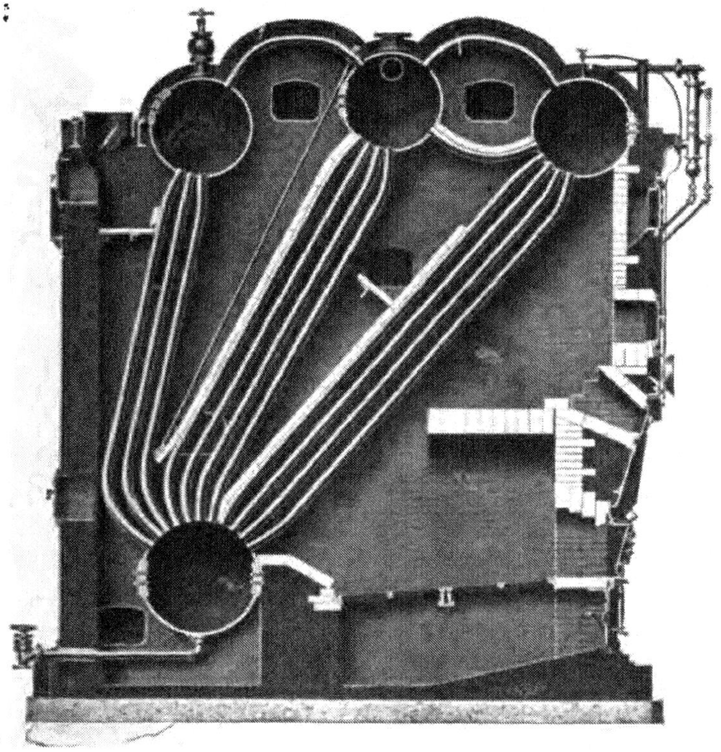

FIG. 11.

These parts, together with the usual valves and fittings, constitute the completed boiler. The drums vary in diameter from 36 inches to 54 inches and are made of the best open-hearth flange steel. The plates extend the entire distance between the heads, so that there are no circular seams. The longitudinal seams — which are double or triple riveted according to the

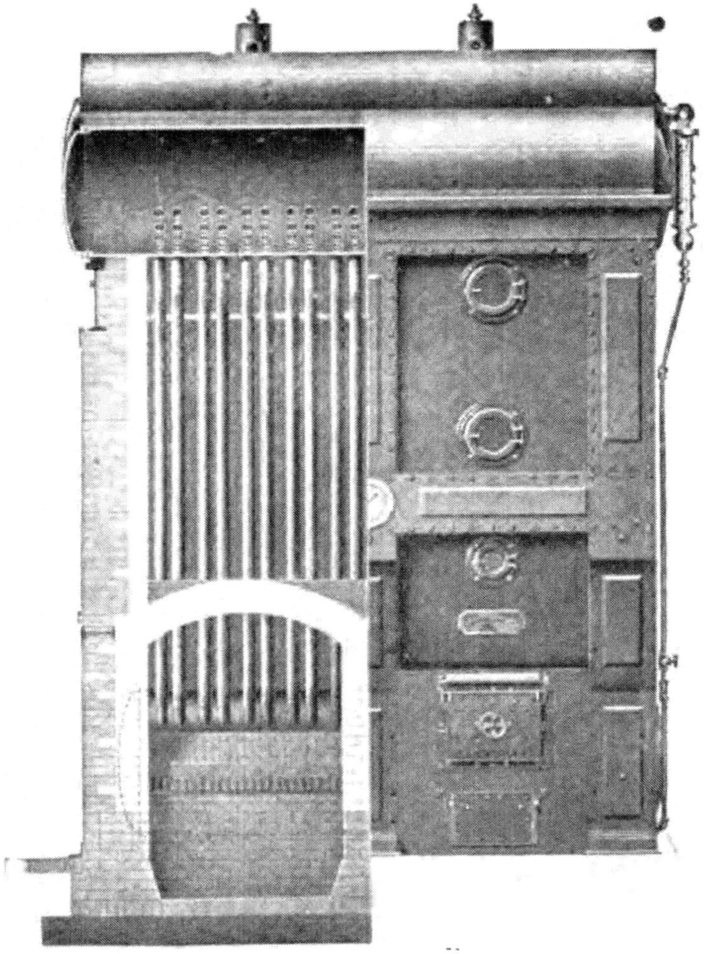

FIG. 12.

pressure to be carried — are so placed that they are not exposed to high temperatures. The drum heads are hydraulically dished to a proper radius; each drum is provided with one manhole, and the manhole plate and arch bars are of wrought steel. Thus four manhole plates give access to the entire interior of the boiler, and expose every tube end, rivet, and joint. The drum interiors are perfectly clear, as there are no baffles, stays, tie-rods, mud pipes, or other obstructions in them. The tubes are of the best lap-welded mild steel. They are slightly curved at the ends to permit them to enter the drums normal to the surface and to provide for free expansion of the boiler. The tubes are expanded directly into reamed holes in the sheets of the drums, and every tube end is visible and accessible.

The furnace is formed by a fire-brick arch sprung over the grates immediately in front of the front bank of tubes. The large space between the boiler front, tubes, and mud drum is available for a combustion chamber and for the installation of sufficient grate surface to meet the requirements of the lowest grades of fuel. The baffles rest directly upon the tubes, and so guide the furnace gases, as explained above, as to bring them into such intimate contact with the heating surface that the heat is quickly and thoroughly extracted from them.

The temperature of the gases in contact with the tubes will evidently be greatest at the bottom of the front bank, and gradually decrease as the gases proceed in their course to the smoke pipe. Obviously, then, the velocity of water circulation and quantity of steam generated will be a maximum in the front bank. In the rear bank there is a slow circulation downward equal to the quantity of water evaporated in the other two banks. There is no constriction in the circulation, as each tube discharges directly into the drums, without the intervention of headers, nipples, or water-legs, and the nearly vertical position of the tubes promotes rapid circulation.

51. The Heine Boiler. — The Heine water-tube boiler may be divided into three main parts, the shell, water-legs, and tubes, (Fig. 13).

The shell is cylindrical in form, varying in diameter from 30 inches to 48 inches, and in length from about 17 feet to 22 feet. The longitudinal seams are of the double-strap butt-joint type while the circumferential seams are all lapped with single or double riveting. The heads of the shells are dished to a radius equal to the diameter of the shell so as to require no internal staying. At the top of the shell near the front end is cut the main steam outlet, a pressed-steel saddle being strongly riveted to the shell for the purpose of attaching the steam tee. The standard form of tee has flanged side and top outlets. Either one of these may be used for the main steam connection, the safety valve being attached to the other. In the bottom of the shell near each end is cut the throat opening for the internal connection to the water-leg. To compensate for the metal cut away, forged steel throat stays bridging these openings are riveted on when the water-legs are attached. Inside and near the bottom of the shell and parallel thereto is fastened a sheet-iron mud drum, which is entirely closed with the exception of a small opening at the top near the front end. The feed pipe which passes through the front head of the shell enters the front end of the mud drum near the bottom, while the blow-off connection which passes through the bottom of the rear head of the shell connects with the back end of the mud drum near the bottom. The theory of the operation of the mud drum will be described later.

Over the throat opening at the front end slanting upwardly to the rear is placed a sheet-iron deflection plate. The deflection plate is closely fitted to the front head and to that portion of the circumference of the shell with which it comes in contact. It extends several feet back of the throat opening and within a

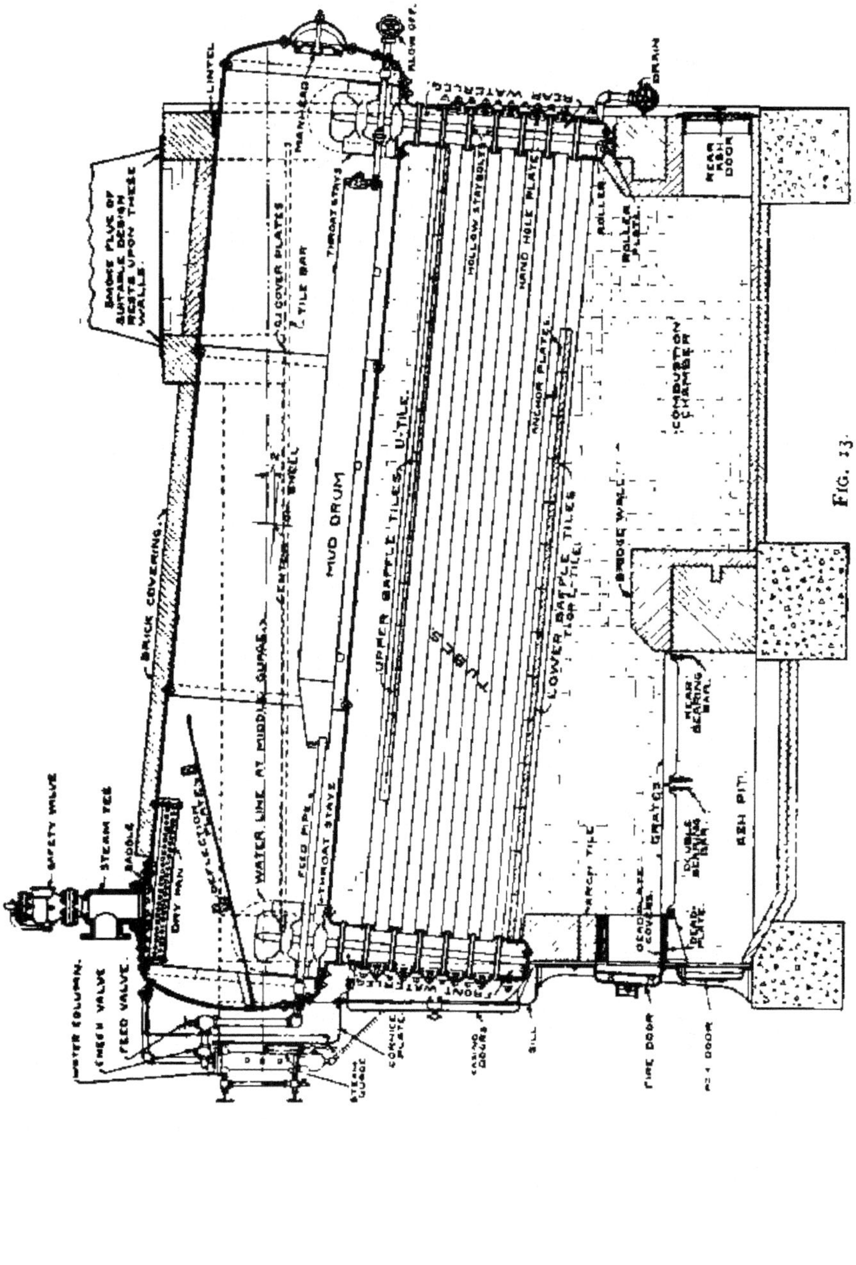
SMOKE FLUE OF SUITABLE DESIGN RESTS UPON THESE WALLS.
LINTEL
MANHEAD
BLOW OFF
DRAIN
REAR ASH DOOR
ROLLER
ROLLER PLATE
HAND HOLE PLATE
HOLLOW STAYBOLT
THROAT STAYS
TILE BAR
MUD DRUM
SHELL
WATER LINE AT MIDDLE GUAGE
UPPER BAFFLE TILES
U-TILE
TUBES
ANCHOR PLATES
LOWER BAFFLE TILES
COMBUSTION CHAMBER
BRIDGE WALL
REAR BEARING BAR
ASH PIT
GRATES
DOUBLE BEARING BAR
ARCH TILE
DEAD PLATE
DEAD PLATE COVERS
BRICK COVERING
SAFETY VALVE
STEAM TEE
SADDLE
DRY PAN
DEFLECTION PLATE
FEED PIPE
WATER COLUMN
CHECK VALVE
FEED VALVE
STEAM GUAGE
CORNICE PLATE
CASING DOORS
SILL
FIRE DOOR
ASH DOOR

FIG. 13.

few inches of the top of the shell. Inside of the shell, just beneath the steam opening, and above the deflection plate, is fastened the dry pan which is a shallow sheet-iron box, in the sides of which are a large number of perforations.

To each side of the exterior of the shell is attached a series of hooks which support the tile bars, the function of which will be described further on. The water-legs are made of two plates, termed respectively the tube sheet and the handhole sheet. These plates are machine flanged, and joined together all around, except at the top, by a butt strap. Being flat surfaces these water-legs require staying to withstand the internal pressure, and for this purpose hollow stay-bolts are used, made of carefully tested mild-steel tubing. These are screwed into tapped holes in the two plates, the projecting ends being carefully upset on the outside. The water-legs are built complete, separately from the shell, and then riveted thereto over the throat openings.

The handholes are closed by means of strong cast-iron or drop-forged steel plates which are inserted from the inside so that the steam pressure tends to make them tighter, and not to loosen them as in the case of plates which are applied from the outside. These plates are held in position by means of yokes and bolts bearing against the outside of the water-leg sheet.

Between the two water-legs extend the tubes which are fastened in position by being expanded with the best type of roller expanders and slightly flared to increase the holding power

The above constitutes the boiler proper, but accompanying it is an artistic front made up of substantial sheet steel and castings, together with grates, buck staves and other parts necessary to properly set the boiler ready for the brickwork, also a steam gauge, safety valve, water column and trimmings, and feed and blow-off valves.

On the lower row of tubes, extending back within three or four feet of the rear end, is placed fire-brick baffle tiling, and likewise on the upper row of tubes extending from the rear to within three or four feet of the front end. These, together with the plates which rest on the tile bars above mentioned, determine the path of the hot gases. Just behind the grates is placed a bridge wall only sufficiently high to hold the coal in place, thus providing ample area for the passage of the gases between the top of the bridge wall and the tubes. The gases of combustion pass over the bridge wall into the large combustion chamber behind it where ample time is given for the complete combustion of the various constituent gases. These then turn upward back of the lower row of tiling into the nest of tubes, thence forward parallel to the tubes, upward again into the space beneath and around the shell, thence backward and upward through the up-take into the breeching. The hot gases are broken up into numberless small streams that completely encircle the tubes, this being due to the very compact arrangement of the tubes, and it is during this passage that the greater part of the heat is absorbed. The gases are further cooled in passing backward under the shell.

The feed water enters the boiler through the front head, passing into the mud drum, which is entirely submerged, the water level being normally at about the center of the shell midway of its length. The water in the boiler when under steam is, of course, at the same temperature as the steam. The feed water when entering is relatively much colder than the water in the boiler, and hence flows along the bottom of the mud drum, being gradually heated by the surrounding hot water to the temperature of this water (this makes it possible actually to force the Heine boiler with feed water of any temperature from 32° up without injury). As this movement is very slow, time is given for the deposition of such substances as may be carried in suspension and also for the precipitation of much of the scale-making impurities. Being entirely without contact with the fire, there is no tendency for this sludge to become baked and hard and it may be blown off through the pipe provided from the rear of the mud drum. The feed water as it becomes hot rises and flows out through the opening in the front end of the mud drum, being carried by the circulation of the water in the boiler to the rear. It will be observed that owing to the position of the boiler there is a much deeper body of water in the shell at the rear than at the front, thus providing, at all times, a solid body of water to keep up the supply to the tubes where the steam is made. The water descends from the shell into the rear water-leg, thence into the various tubes, passing upward toward the front and absorbing in its passage the heat from the gases on the outside of the tubes, bubbles of steam being formed, which pass out of the tubes, together with the unevaporated water, into the front water-leg, thence upward into the shell. The large openings from the shell into the water-leg, or throat openings, while being the most constricted parts in the path of the circulating water, are so large that little or no real obstruction to a free flow is offered.

Owing to the great difference in volume caused by the expansion of the water into steam, the passing out through the water-leg is very rapid and the mixture of steam and water is thrown up with considerable force against the deflection plate, the function of which is to throw down the water, allowing the steam to pass up into the steam space, thence over the upper end of the deflection plate through the holes in the dry pan to the steam outlet.

At the bottom of the rear water-leg is provided a valve for the purpose of draining the boiler. The steam connection of the water column is made at the top of the front head while the water connection is made at the top of the front water-leg. The steam gauge has an independent connection into the steam tee and is fastened in a prominent position in the middle of the ornamental front.

FIG. 14.

The superheater, Fig. 14, consists essentially of a header box of the same type of construction as the water-leg, into one side of which are inserted U-tubes, made of 1½-inch seamless, drawn, mild-steel tubing, expanded into holes provided for them. Opposite the tubes in the other sheet of the header box are a series of handholes closed by inside plates, which give access to the interior of the whole apparatus.

The header box is made entirely of flange steel plate, and is so designed that it is entirely machine made. The hollow stay-bolts, which hold the two sheets of the box parallel, are of the same size and material as those used in the construction of the boiler proper, and, as in the case of the boiler, provide means for introducing the soot blower in order to keep the exterior surfaces of the superheater tubes clean.

The interior of this box is divided into three compartments by means of light sheet-iron diaphragms, which, being nicely fitted, are sufficiently steam-tight to cause the steam to pass through the tubes.

The superheater is located at the side of the shell of the boiler toward the front and just above the last passage of the boiler gases (see Fig. 15), being supported by special castings, which rest upon the boiler tile bar and brick setting. Depending on the capacity and degree of superheat desired, the device may be single and placed only on one side; or in two parts properly connected together, one on each side of the boiler, and above the water line.

The whole is encased in brickwork with a fire-brick roof carried by special T-bars, as seen in Fig. 16.

A small flue (see Fig. 15), built in the side walls of the setting, carries the hot gases direct from the furnace into the superheater chamber, where they make two passes around the superheater tubes. The flow of these gases is controlled by means of a damper at the outlet. When closed the circulation is stopped,

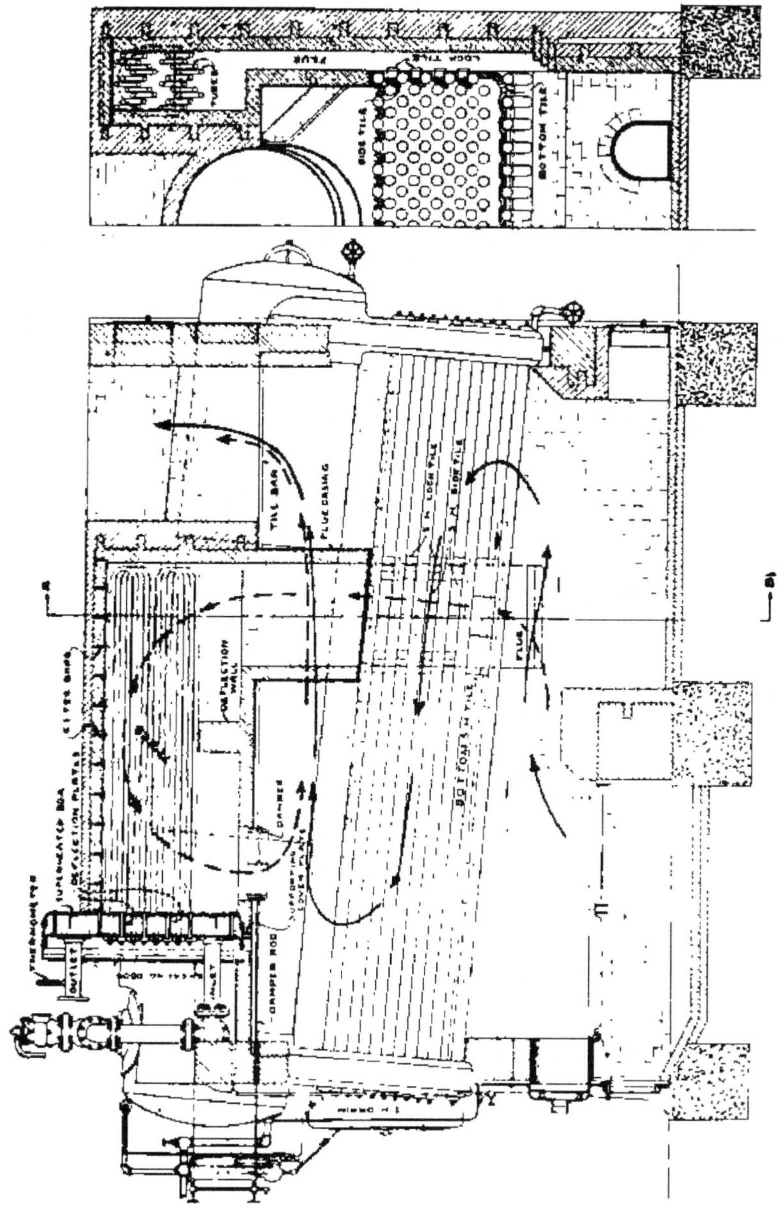

HALF SECTION THROUGH AB

LONGITUDINAL VERTICAL SECTION

FIG. 15

and as soon as the heat from the gases in the superheater is absorbed, only saturated steam will be delivered.

Fig. 16.

By varying the amount of opening of the damper the flow of gases can be regulated so as to give any desired degree of superheat up to the capacity of the apparatus. Since the hot gases do not come into contact with the damper until after passing through the superheater, there is no danger of overheating it.

The usual steam outlet from the boiler proper is connected into the lower opening of the superheater box (see Fig. 15), the steam passing into the tubes of the lower compartment, thence through these tubes out into the middle compartment, whence they go into the second set of tubes connected with this space and through them issuing finally into the third or top compartment, thence out through the opening there into the general piping system. The effect is to mix up thoroughly the steam so that it is of a uniform temperature. Ordinarily it is not deemed necessary to provide a by-pass so as to enable the superheater to be cut out of service entirely, although such an arrangement can easily be provided if desired.

52. The Roberts Boiler. — A type of water-tube boiler shown in Figs. 17 and 18, known as the Roberts Safety Water Tube Boiler, is extensively used. The feed water is divided at the boiler into two streams by a T-connection at *A*, Fig. 17, each stream entering a feed coil *B* on either side of the drum. Passing through the horizontal tubes of these coils it is then delivered into the drum above the water line at *C*, so that any steam that may have formed during the passage through the feed coils may rise to the top of the drum. One of the feed coils delivers into one end of the drum while the other delivers into the opposite end. It is a feature of this boiler that the course of the water through the feed coils is *down*, thus meeting the hot gases as they rise. However much the gases may be cooled by contact with the lower layers of the coils, they meet still cooler layers as they rise, thus establishing the most favorable conditions for the transference of the heat of the gases to the water.

Under the action of gravity the water now flows from both ends of the drum through T-connections into the cross pipes *D* and thence through the downflow pipes *E* into the side pipes *F*, which are on a level with the fire grate. These side pipes are tapped for the upflow coils *G*, through which the water now rises again to the drum. It will be seen, Fig. 18, that the upflow coils rise vertically from the side pipes and then cross in layers under the drum and directly over the fire. The holes

FIG. 17.

in the side pipes for the upflow coils are *staggered* so that the spaces between the cross layers of the coils directly over the fire are only one-half as wide as those between the vertical pipes of the coils. It will be noted too that the upflow coils leading from one side pipe deliver to the drum at *H* on the opposite side. The feed, now arriving in the drum for the second time, is in the condition of damp steam and bubbles through the water in the drum and rises to enter the dry-pipe through perforations all along its top. The dry-pipe extends from one end of the drum to the other and is connected to the heads at the

FIG. 18.

highest points possible. The steam now passes from both ends of the dry-pipe through the pipes *K* into the superheating coils *L* which lie, one on each side, between the boiler and its inclosing casing and above the fire brick *M* of the furnace. The steam passes to the bottom of the superheaters, thence rising through the pipes *N* to a point nearly over the front end of the drum where a T-connection and pipe *O* enables the superheated steam to be led to the engine. Steam of 100° superheat is not uncommon with this boiler when working under the ordinary conditions of practice.

53. **Draft.** — In Chapter III it is shown that air is an essential to combustion, and for a given grate area the quantity of fuel consumed in a given time depends upon the *draft* or flow of air into the furnace. The natural draft ordinarily obtained by means of chimneys or smokestacks depends upon the difference in weight between the column of heated air within the stack and that of the outside air. The weight of the heated gases within the stack is so much less than that of a column of the outside air of the same height that they rise in the stack, and in doing so produce a partial vacuum in the furnace which draws the outside air through the fire. The great disadvantage of natural draft is that its intensity depends directly upon the temperature of the gases within the chimney, and as experience has shown that a temperature between 600° and 700° must be maintained in the chimney in order that the volume of air drawn through the furnace shall be a maximum, it follows that the escape of the furnace gases through the chimney at a temperature so high occasions a very material loss of the heat value of the fuel, a loss in some instances as great as one-fourth. In the effort to save this heat economizers were devised, consisting of a series of pipes placed at the base of the chimney through which the feed water flows on its way to the boiler. By this means it is possible to heat the feed water to a temperature greater than 212°, since the water is under a pressure greater than that of the atmosphere, but in doing so the temperature of the gases may be so reduced as to affect the intensity of the draft.

The efforts to overcome the inherent defects of natural draft led to the development of two systems of mechanical draft known as *induced* and *forced* draft.

54. Induced Draft. — When a fan is placed between the furnace and the chimney, the fan drawing from the furnace and delivering into the chimney, the resulting draft is said to be *induced*. The suction action of the fan produces a partial vacuum in the combustion chamber of the furnace, which causes the air to flow into the ash pit and through the fuel. This system is the one most commonly used in power plants, as it dispenses with the high and expensive smokestack and furnishes the most favorable conditions for the use of economizers. The hot gases in their induced flow pass through the economizer, where they deliver much of their heat to the feed water and are, in consequence, reduced in temperature when they reach the fan for delivery into the chimney.

A different form of induced draft is that of the locomotive boiler. The exhaust steam from the cylinders is discharged through the exhaust nozzle into the smoke pipe at its base. Each puff of steam from the exhaust drives the air ahead of it out of the smoke pipe, producing a partial vacuum in the tubes and smoke box which causes an inrush of air through the furnace. The greater the volume of steam escaping at each puff, and the greater the rapidity of the puffs, the more intense the draft becomes, so that this form of draft virtually controls automatically the fuel consumption and the water evaporation, because at the time the engine is working its hardest and using the most steam, the draft is acting most efficiently.

55. Forced Draft. — When the fan is so placed as to force the air under the furnace grate, the draft is said to be *forced*. The forced draft system is especially adapted to the burning of screenings or low grades of coal which, under other conditions, could not be burned without loss, and to this fact is attributed the economic gain claimed for the system. With forced draft the air is supplied to the furnace in either of two ways: — (1) By making the ash pit practically air-tight and then, by means of a fan, forcing air into it and up through the fuel, thus utilizing the air for the purposes of combustion. (2) By making the fire room practically air-tight and then, by means of a fan, maintaining the necessary pressure therein to supply the required air for efficient combustion.

For stationary boilers the closed ash pit is the only one applicable, while for marine practice both the closed ash pit and closed fire-room systems are employed.

56. Advantages of Mechanical Draft. — The advantages of mechanical draft may be summarized thus:

Its adaptability to both large and small powers; its ability to meet sudden demands for steam by simply increasing the speed of the fan, thereby increasing the intensity of the draft; its entire independence of wind, weather, and atmospheric changes; it furnishes the only conditions under which economizers may profitably be applied; it practically does away with the necessity for high smokestacks; it has increased the rate of fuel combustion from the maximum of 25 pounds of coal per square foot of grate per hour under natural draft to 40 pounds in stationary boiler practice, to 50 pounds in marine practice, and to as much as 120 pounds in locomotive practice; it has increased the efficiency of combustion, and therefore the steaming power of boilers; and it has made possible the profitable use of inferior fuel.

57. The Measurement of Mechanical Draft. — The intensity of mechanical draft is expressed in inches of water, that is, by the pressure that would support a column of water the given number of inches in height. The weight of a cubic foot of fresh water may be taken as 62.4 pounds, so it is evident that for a head of one inch of water the pressure will be $\frac{62.4}{144 \times 12} = 0.036$ pound.

Gauges for measuring draft consist essentially of a U-tube containing water. One leg of the tube is connected with the atmosphere and the other to the space in which the draft is operating. The difference in level of the water in the legs expressed in inches and multiplied by 0.036 gives the force of the draft in pounds per square inch.

58. Stop Valve. — A valve placed in a steam pipe for the purpose of opening or shutting off the supply of steam is called a *stop valve*. It is usually of the globe pattern, as shown in Fig. 19. The bonnet is connected with the body by a beveled-face, ground joint *A*, *N*, and secured to the body by the large hexagonal nut *a*. The beveled-face connection tends to expand the threads, interlocking them firmly by the pressure exerted when wrenching home the hexagonal nut *a*. The valve *V* is of the double-disk construction with two wearing faces, one of which is protected by the disk holder *R* while the other is in use. These faces are reversible and may be reground. The stem *D* is made with forcing collars between which the disk holder slides. The packing is wound around the stem and is forced into the recess by the gland *G*, to which pressure is applied by wrenching down the packing nut *P*. The stem may be packed with this form of stop valve when the valve is wide open, as the faces D_1 and *H* engage, forming a tight joint and not allowing the steam to pass.

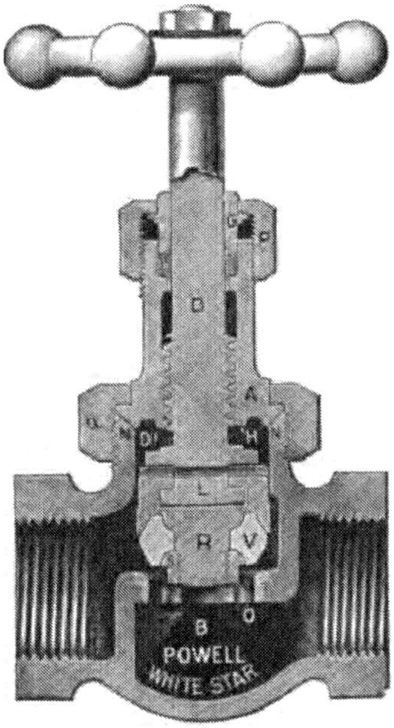

FIG. 19.

59. Gate Valve. — Another form of stop valve, known as a *gate valve*, is shown in Fig. 20. This valve consists of two wedge disks, *V* and *VI*, which are self-adjusting with ball-and-socket backs. The disks are hung on a bushing *C* which travels on the screw stem *D*, enabling the disks to be raised or lowered at will. The backs of the disks are of ball-and-socket construction and as pressure is applied in closing the valve the disks are wedged tight against the seat rings *O*, *O*, the ball and socket admitting an adjustment that insures a tight fit. The bonnet top *A* is secured to the neck of the body *B* by bolts as shown. The screw stem turns in the packing chamber *P*, which is

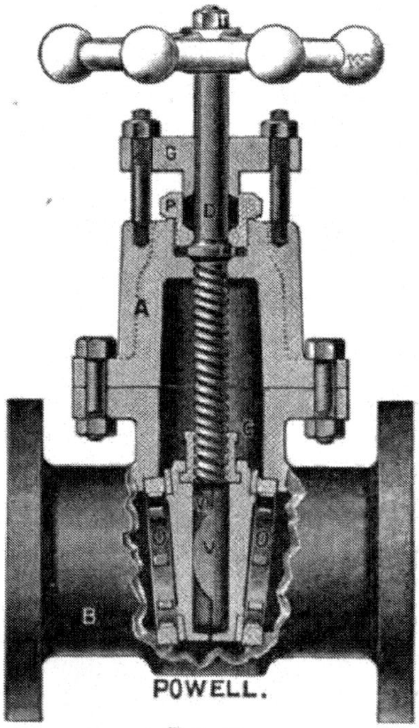

Fig. 20.

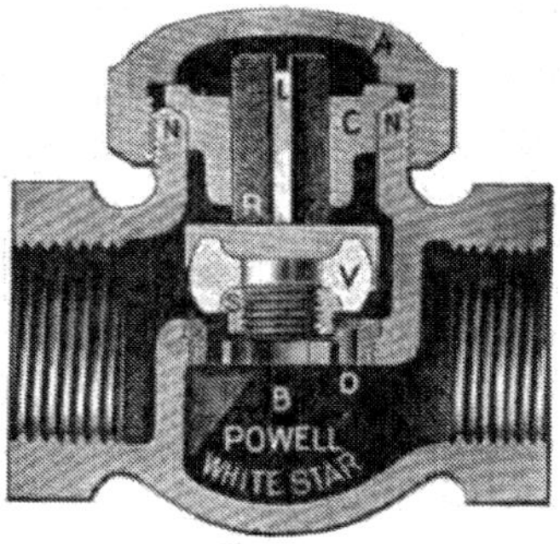

Fig. 21.

securely screwed into the upper part of the bonnet. The gland follower *G* is operated by the two stud bolts shown.

60. Check Valve. — A *check valve* is designed to permit the passage of a fluid in a pipe in one direction only. In the operation of a boiler feed pump or injector such a valve is absolutely necessary to prevent the hot water of the boiler being forced back to the pump when the latter is not working. Fig. 21 shows a section of an approved check valve, its construction being somewhat similar to the stop valve of Fig. 19. The valve *V* is reversible and tightly secured in its holder by the lock nut *S*. The wing guides *L* work freely in the guide collar *C* which is rigidly held in the body by cap *A*. Water from the pump entering the chamber *B* from the right will lift the valve and pass on to the boiler through *O*. Any backward pressure from the boiler will act on top of the valve and promptly close it. There should always be a stop valve in the feed pipe between the check valve and the boiler which can be closed at will, so that should the check valve leak, or fail to act, communication with the boiler may be shut off and the check valve repaired.

61. The Safety Valve. — The object of the safety valve is to relieve the pressure in the boiler when it rises above the working pressure desired. A safety valve should be designed to discharge the steam from the boiler as fast as it is generated when the boiler is forced to its full steaming capacity, and since the lift of a safety valve is quite small, perhaps not exceeding one one-tenth of its diameter, it follows that the area of the valve must be relatively large. Should the quantity of steam generated by the boiler be greater than the demand, the pressure will rise and will, on reaching the height for which the valve is set, open the valve and allow the surplus steam to escape. When the pressure is reduced below that of *blow-off* the valve is closed by its actuating force, that of a weight or of a spring.

There are two kinds of safety valves, known as the *lever* and *spring* types.

62. The Lever Safety Valve. — The illustration, Fig. 22, shows clearly the operation of the lever type of safety valve. The valve remains closed until the pressure in the boiler, acting beneath the valve, exceeds that imposed on top of the valve by the action of the weight and lever, when the valve opens and allows the steam to escape until the pressure is reduced below that of blow-off, the action of the weight and lever then reseating the valve. The lever is graduated so that the position of the weight for any predetermined blow-off pressure is known.

The graduations on the lever are determined by taking moments about the fulcrum. Denote the weights of the weight, the lever, and the valve (including its spindle) by W, W_1, and w respectively. Let a denote the distance of the weight from the fulcrum, b the distance of the center of gravity of the lever from the fulcrum, and let the distance from the fulcrum to the

FIG. 22.

valve spindle be c. If d denotes the diameter of the valve in inches and p the gauge pressure of the steam in pounds per square inch, we shall have $\frac{\pi d^2 p}{4}$ as the total pressure opposed to the weights, and by moments about the fulcrum we shall have

$$\frac{\pi d^2 pc}{4} = Wa + W_1 b + wc.$$

Solving this equation for a will give the position of the weight on the lever for any given pressure p.

63. The Spring or Pop Safety Valve. — The *spring* or *pop* type of safety valve, where the tension of a spring, instead of a weight and lever, is employed to govern the blow-off pressure, is more extensively used than the lever type. It is exclusively used in locomotive and marine practice where the attending jar and motion would interfere with the action of the lever.

The illustrations, Figs. 23, 24, and 25, show the special pop safety valve of the American Steam Gauge Co., in full section, half-section, and quarter-section respectively. The compression of the spring, which governs the blow-off pressure of the valve, is adjusted by the compression screw *G*, which is threaded into the valve bonnet and seated in a follower on top of the spring.

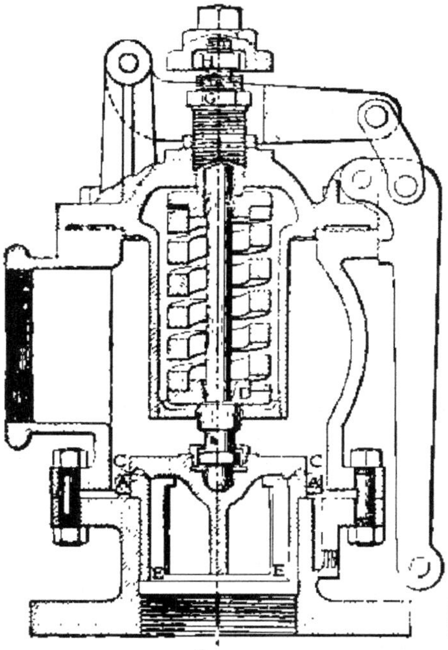

Fig. 23.

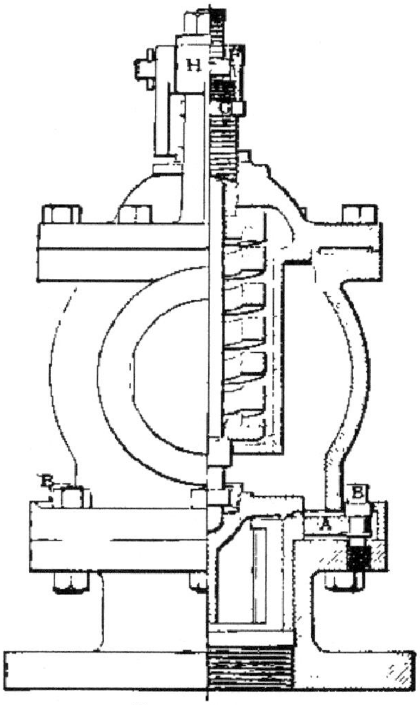

Fig. 24.

The valve spindle passes through the screw *G*, which is bored for the purpose, and has on it a loose flange *D* resting with a spherical bearing on a collar on the spindle; on this flange rests the lower end of the spring, thus distributing evenly the compression load of the spring. It is seen that the spring is incased in the bonnet to prevent injurious contact with the hot steam, the spindle passing through and guided by the bottom of the case. The compression of the spring is transmitted to the back of the valve at a point below the level of its seat, which is for the purpose of insuring the fair and even seating of the valve. In order that the valve may be lifted by hand, which should be done daily, a powerful form of lifting lever has been devised, as is well shown in Fig. 23. The leverage of the long vertical arm is compounded by the short connecting link which becomes a toggle as the long arm approaches the horizontal position and acts with further leverage on the cap on the end of the spindle through the top horizontal lever, as shown in Fig. 25. The valve proper is made of bronze and is guided by perfectly fitting wings in the bushing forming the valve seat.

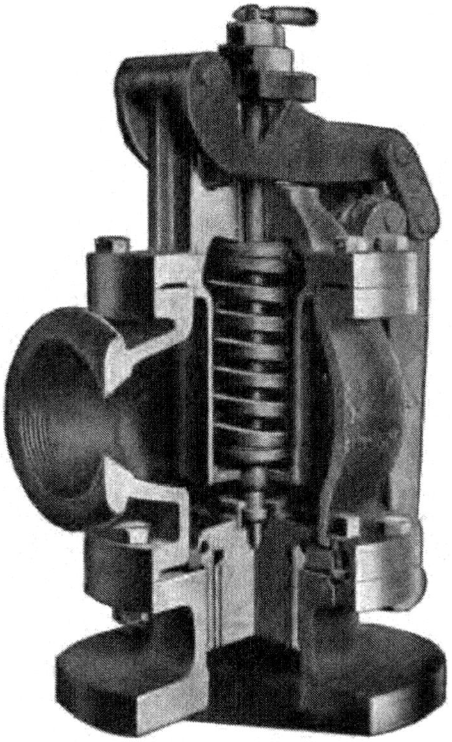

FIG. 25.

It will be observed that the weight to be overcome in lifting a lever safety valve is constant, and that, therefore, when it begins its lift there is no additional pressure required to lift it higher. Such is not the case with the spring type of valve, for the more a spring is compressed the more pressure is required to compress it still further, so that there must be some means of providing additional pressure when a pop safety valve commences to open in order that the varying compression of the spring may be overcome and the valve open to its full lift. This provision is made in the valve under discussion by giving to the valve a grooved lip which overhangs the seat, as seen at *C*. When the valve commences to lift, the rapidly moving escaping steam strikes within the groove and is deflected backward, the reaction on the valve causing additional pressure to continue the lift of the valve. It is desirable that the *blow-down*, that is, the difference between the blow-off and closing pressures, should be from 3 to 5 pounds, and this can be effected only by prolonging the period of blowing off. The ring *A* is for the purpose of regulating the blow-down. It encircles the valve seat and may be raised or lowered by means of the screws *B*, *B*. Lowering the ring lowers the blow-down by giving a free escape to the steam; raising the ring restricts the escape of the steam and raises the blow-down.

In testing a boiler it is subjected to a hydraulic pressure much in excess of its working steam pressure, and in order to protect the spring of the safety valve from the excessive test pressure a testing nut *H* is provided. It screws down on the threaded end of the compression screw *G* until the inner side of the nut top bears upon the collar shown on the valve spindle. This prevents the spindle from rising and holds the valve in its seat without altering the compression in the spring, or without increasing the load on the spring. The test nut must be removed after the test and the regular cap and lock nut replaced, as shown in Fig. 25.

64. Blow-off Valve. — In a pipe leading from the lowest part of a boiler is placed some form of globe valve or conical plug valve, the purpose of which is to blow from the boiler any sediment that may collect, and also for blowing the water out of the boiler when fires are hauled. At the bottom of the boiler there is always a cock which enables the boiler to be drained dry.

65. Water Column. — The height of the water in the boiler is indicated in two ways — by gauge cocks or by a water-glass, and very often by a combination of both. The usual arrangement, Fig. 26, consists of a cast-iron cylinder, called a water column, which is mounted at the front end of the boiler, its upper end connecting with the steam space at the top and to the water space near the bottom, and so placed that the water level in the boiler will be at the middle of the column. To one side of the column is attached three gauge cocks, one at the desired water level, one above, and one below. By opening these cocks at any time the level of the water in the boiler may be determined with certainty by the nature of the discharge. On the front of the water column is the water-glass. It consists of two brass fittings that screw into the water column at its top and bottom. A glass tube connects these fittings, the connections being made by means of stuffing-boxes and rubber packing rings. Each of the fittings has a valve by which the connection with the water column may be shut off when replacing a tube. In addition there is an automatic ball cut-off valve between the valve of each fitting and the water column, shown in section in Fig. 27.

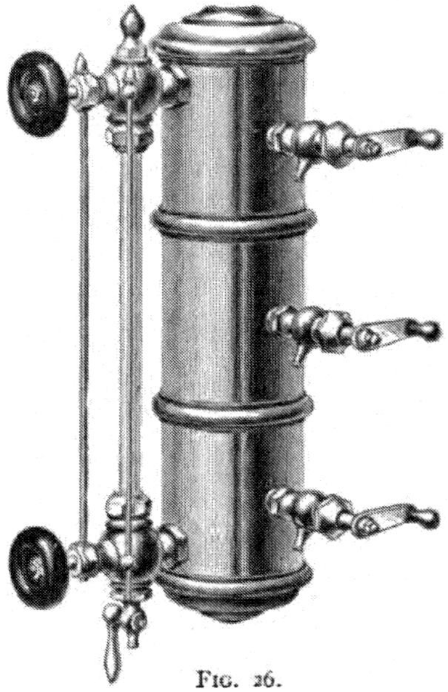

Fig. 26.

Fig. 27.

After the gauge has been placed in position, close the valve by means of the wheel *N*, and in doing so a prolongation of the stem through the valve pushes the ball *L* into the recess prepared for it, and the pressure around the ball becomes equalized at once. The stem *L* must then be withdrawn and the valve left wide open. Should the glass break, the sudden rush of steam or of water would carry the ball home to its seat, thus preventing the emptying of the boiler and all danger to the fireman. Stout wires about the glass guard it from injury. The glass gauge is very convenient, but not so reliable as the gauge cocks. There is danger of the passages to the glass tube becoming choked, to prevent which the drain cock at the bottom should frequently be opened and the glass blown through.

66. Steam Gauge. — The steam gauge of a boiler is in the nature of a safety device, as it indicates the pressure of the steam carried. The principle of the action of the Bourdon tube is universally employed in spring pressure gauges. The single-spring Bourdon pressure gauge is shown in Fig. 28, the graduated face dial being removed.

FIG. 28.

The essential part of the gauge is the seamless-drawn Bourdon tube, elliptical in section, and bent into the arc of a circle. The fixed end of the tube is in connection with a pipe which leads to the steam space in the boiler, while the other end is closed and free to move and is connected to the pointer by means of the links and sector as shown. The teeth of the sector mesh with a small pinion on the shaft at the center, and thus any movement of the free end of the tube is transmitted to the pointer. The small spiral spring shown at the center is attached to the pinion for the purpose of taking up the back-lash of the moving parts when the pressure is lowered. The action of the steam pressure within the tube tends to change its shape from elliptical to circular, causing an outward movement of the free end, and the consequent movement of the pointer over the face of the dial, the graduations on which are made to agree in pounds per square inch above the atmosphere with the indications of a mercury gauge or of a standard spring gauge. If the steam itself were permitted to enter the tube the temperature changes would affect its shape and make the readings inaccurate. To avoid this a siphon bend filled with water is interposed in the pipe between the gauge and the boiler so that the steam pressure is made to act on the tube through the medium of the water.

This gauge is sometimes made with a combination of two Bourdon tubes for the purpose of decreasing its sensitiveness, and thus avoid the vibrations of the pointer due to the jarring motion of a locomotive, but the principle of its action is the same as that of the single-tube gauge.

67. Recording Pressure Gauge. — The uniform maintenance of the boiler pressure required for a power plant is an important factor in the economy of its operation. Wide fluctuations in pressure mean poor economy. Recording pressure gauges are devised to give a graphic representation of the pressure carried in the boiler during the period, day and night, of the plant's operation. By this means the superintending engineer can inform himself as to the degree of safety and efficiency of the prevailing conditions.

An improved recording pressure gauge, manufactured by the Industrial Instrument Co., of Foxboro, Mass., is shown in Fig. 29. The principal parts of the instrument are lettered and described as follows:

The steam connection is made through the socket *F*. From *F* the steam passes through the small tube *E* to the inside of a flat helical tube *H*, the inner end of which is made fast to the cast frame *J*. The shaft *D* has a bearing in the frame at each of its ends, and from its inner end it is connected to the outer or free end of the helical tube *H*. This connection, which is

FIG. 29.

unseen in Fig. 29, is made by means of a thin and flexible metallic strip. On the outer end of the shaft *D* is mounted the pen arm *B*, on the end of which is a V-shaped pen *A*. At *C* is a friction joint in the arm *B*, which enables the pen to be adjusted. A chart attached to the face of the instrument at *G* is revolved by means of clockwork inclosed in the casing *I*. The clock movements are arranged for the following time revolutions of chart: 15 and 30 minutes, 1, 2, 3, 4, 6, 12, and 24 hours.

The operation of the instrument is as follows: The pressure of the steam within the helical tube H causes it to untwist an amount proportional to the pressure, and the movement due to this untwist is communicated directly to the shaft D by means of the flexible connecting strip above referred to. The movement of the shaft in turn is communicated to the pen A through the arm B and the corresponding pressure recorded on the chart.

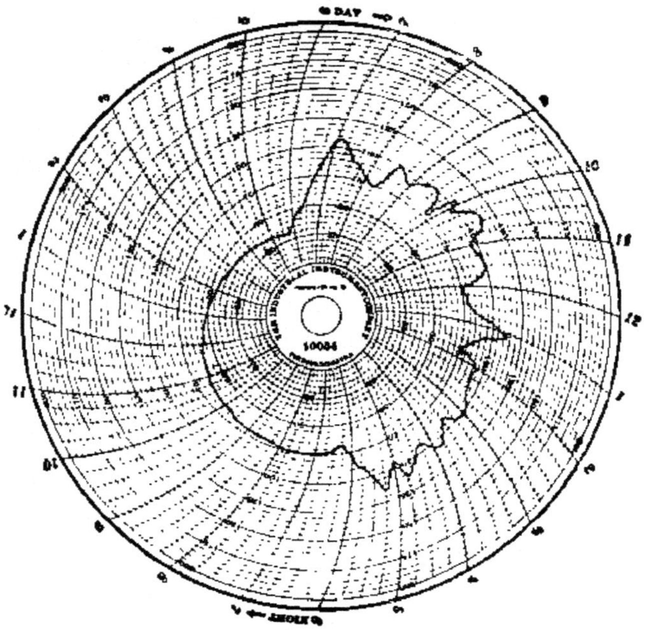

FIG. 30.

In Fig. 30 is shown a reduced reproduction of a chart with a pressure record for 24 hours. This record shows excessive pressure fluctuations resulting from poor and uneconomical firing. The chart shows how day and night portions are made plain by means of skeleton and bold faced type.

For low ranges of pressure — below 10 pounds per square inch — the same company employs an improved diaphragm form of pressure tube, shown in Fig. 31. The pressure tube *H* consists of a series of diaphragms built up in the form of an accordion. The steam entering through the tube *E* causes, by its pressure, an elongation of the tube, which lateral motion is multiplied by means of the restraining coils *I* fastened to one side of the diaphragms. This motion is communicated to the shaft *D*, and thence through the arm *B* to the pen *A*.

The pen, pen arm, and shaft are arranged similarly to those of the helical form, the two movements being interchangeable.

The helical spring movement has the advantage over that of the flexible flat diaphragms in that it provides a much longer length over which the bending effect is distributed.

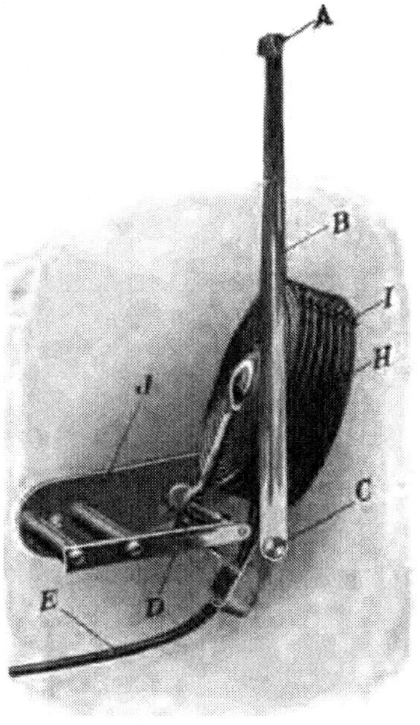

FIG. 31.

68. Vacuum Gauge. — Vacuum gauges are graduated to measure a vacuum in inches of mercury, and since the atmospheric pressure of 14.7 pounds per square inch supports a column of mercury of very nearly 30 inches, it follows that two inches of vacuum is the equivalent, approximately, of one pound pressure. In other words, if one end of a glass tube were connected to a chamber in which there was a perfect vacuum, and the other end dipped in an open cup of mercury, the mercury would rise in the tube to a height of very nearly 30 inches. It is impossible to obtain a perfect vacuum in the condenser of a steam engine, the average vacuum being about 26 inches, so that under such conditions there would be $\frac{30 - 26}{2} = 2$ pounds per square inch back pressure in the condenser opposed to the forward pressure on the engine piston during a stroke.

A gauge with a Bourdon tube similar to the pressure gauge of Fig. 28 is generally used to measure vacuum, the dial being graduated to indicate inches of mercury, the graduations extending from 0 to 30. The action is the reverse of the pressure gauge, the vacuum operating within the tube and the pressure of the atmosphere without.

69. Steam Trap. — The object of the steam trap is to drain from the steam pipe any water that may be carried into it by the steam, or that results from condensation, without permitting the escape of the steam.

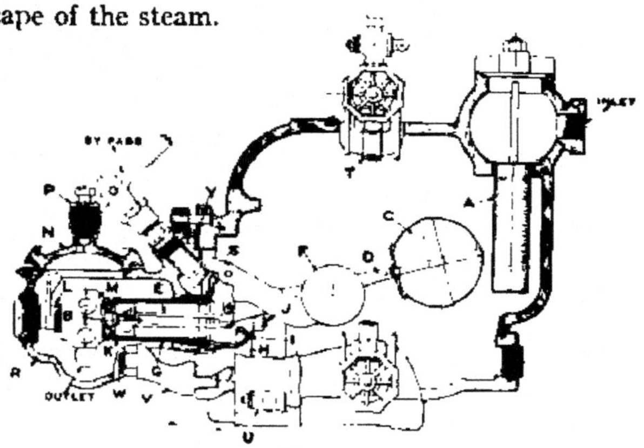

FIG. 32.

A good example of the steam trap is that of the V. D. Anderson Co., shown in Fig. 32. It works continuously and automatically, and its important parts are made of phosphor bronze, though the valve and seat are made of hardened nickel steel when the working conditions are unusually severe.

The water to be drained enters the trap at the opening marked "inlet" and is discharged at the opening marked "outlet."

Any scale or sediment passing in with the steam is caught in the strainer *A*, thus preventing it getting into the working parts of the trap.

The float *C* is made of spun copper, and its lever arm *D* is pivoted at its end as shown. The construction of the end of the lever arm is a mechanical feature of the working of the trap. The end is elongated and pivoted at about its middle, which permits the end to act as a rocker arm as the float rises and falls, the arm of the rocker below the pivot operating the valve *B* through the connection *E*, and the arm above the pivot carrying a brass stud at its end on which rides a projection on the arm of the counterweight *F*. The counterweight is pivoted at the end of its arm and its object is to overcome the effect of the dead weight of the float and its lever arm and thus increase the buoyancy of the float.

When sufficient water has accumulated in the trap to bring its level about 3 inches above the valve *B*, the copper float will begin to rise, the rocker action of the lever *D* drawing the valve *B* from its seat, thus permitting the water to drain through the outlet. As long as the 3-inch water seal is maintained the trap works continuously, permitting no steam to escape. If the water level falls below 3 inches the float falls automatically and closes the valve *B*.

It will be observed that the counterweight constantly imposes a thrust on the brass stud in the upper arm of the lever end as it rides over it, the effect of this thrust being an equal tendency to raise the float as that of the dead weight of the float and its arm is to lower it, thus increasing its buoyancy.

By giving the "by-pass" handle a half turn to the left, the cam *S* is brought in contact with an extension beyond the pivot of the counterweight arm, which has the effect of raising the float through pressure on the brass stud in the upper arm of the elongated head of *D*, and thus open the valve and allow steam to blow through and remove any sediment that may collect about the valve.

The water-glass *T* provides a ready means of noting the water level in the trap, and through the drain cock *U* may be blown any sediment that may accumulate in the trap.

The valve *B* and seat *K* are removable, the seat being held securely in place by the threaded follower *L* and lock nut *M*, and in order to gain access to these parts it is only necessary to remove the head cover *N* held in place by the swinging crab *P*. All working parts of the trap are attached to the head *R*.

70. Reducing Valve or Pressure Regulator. — When it is desired to use steam at a pressure less than that in the boiler, it is passed through some form of reducing valve whose action reduces the pressure to that desired. The Foster reducing valve or pressure regulator is illustrated in Fig. 33. The valve is controlled and operated by the movement of a diaphragm opposed to the action of springs whose tension is adjusted according to the delivery pressure to be maintained, and this pressure is entirely independent of the pressure in the supply pipe.

The steam enters the valve at *A*, and, flowing in the direction indicated by the arrows, passes out at *B*. The regulating valve 4, 5 is of the double-seated balanced type, closing upward, and held in alignment by its hollow stem 3 sliding in the casing above, the stem being grooved for water packing.

In the upper part of the case there is a large circular chamber *D*, inclosed by the diaphragm *d*. This diaphragm is attached to the valve 4, 5 by a connection 2, pivoted at its diaphragm end and held to the valve by a nut at the bottom. This overcomes any slight variation in alignment between the diaphragm and the valve without in any way cramping the motion of the valve. A porthole *E*, running from the delivery side of the valve to the chamber *D*, gives the steam, at delivery pressure, access to the bottom of the diaphragm. The diaphragm is prevented from yielding to this pressure by the combination of levers and springs above. Resting on the jam nut *b* and against the links 8, 8 are the toggle levers *a*, *a*. As the diaphragm is forced upward by the pressure, the ends of these levers are forced outward against the links 8, 8. These links are pivoted at the bottom, and by the action of the toggle levers their upper extremities are pressed against the springs whose tensions are adjusted by the spring bolts *K*. In operation, the steam on

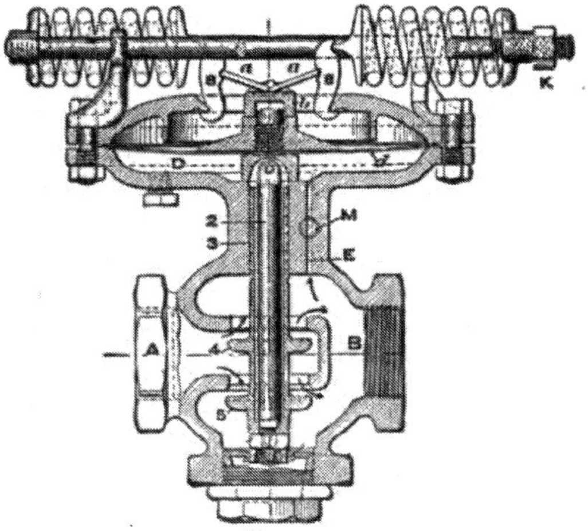

FIG. 33.

the delivery side of the valve passes up through the port *E*, and when it has reached the pressure that overcomes the tension of the springs, the valve is raised, thus restricting the opening for the supply steam and reducing its pressure by wire-drawing to that desired.

If, when delivering steam to an engine, there should be violent pulsations of the diaphragm, close the port *E* partially by turning the screw *M* slowly to the right until only a slight movement is perceptible.

When delivering steam to an engine or pump, locate the regulator far enough away from the steam chest, so that the cubical contents of the pipe between the regulator and the steam chest shall not be less than that of the steam cylinder. This provides a cushion which prevents violent movement and wear of the valve.

There should be a stop valve on the high pressure side of the regulator, which must be wide open when steam is passing through the regulator and closed when no steam is being used.

71. **Back Pressure Valve.** — To avoid the danger incident to an accumulation of pressure in the condenser of a steam engine, a form of safety valve, known as a back pressure valve, is used. Pressure may accumulate by reason of a diminished condensation of the exhaust steam through any failure in the supply of condensing water, or through an increase in the volume of steam exhausted by the engine. To prevent any such accumulation, the back pressure valve acts automatically in releasing the steam to the atmosphere. The area of discharge of a back pressure valve must be much larger than that of a safety valve designed to protect boilers, for the reason that the volume of a pound of steam at atmospheric pressure is much larger than that of the same weight of steam at boiler pressure, about ten times as great as that of a pound of steam at, say, 150 pounds pressure. In addition to this the spouting velocity from back pressure valves is only about one-third that from safety valves, so that a back pressure valve requires nearly 30 times as much area as does a safety valve to pass the same weight of steam.

Very frequently back pressure valves consist either of a single disk held in place by a spring or a weight, or of a cylindrical balanced valve moving over ports in a chest. The disk type of valve is objectionable because of its great size and because of its tendency to hammer; and the cylindrical type, if made loose enough to prevent sticking, is liable to permit a waste of steam through leakage.

The Cochrane multi-port back pressure valve, Fig. 34, is designed to meet the objections alleged against the ordinary forms in use. This valve may be described briefly as consisting of a valve disk *B* upon which are placed a series of small valve disks *D*, each having a spring *E* to hold it closed, a water cushion or dash pot *H* to arrest its motion and prevent hammering, and

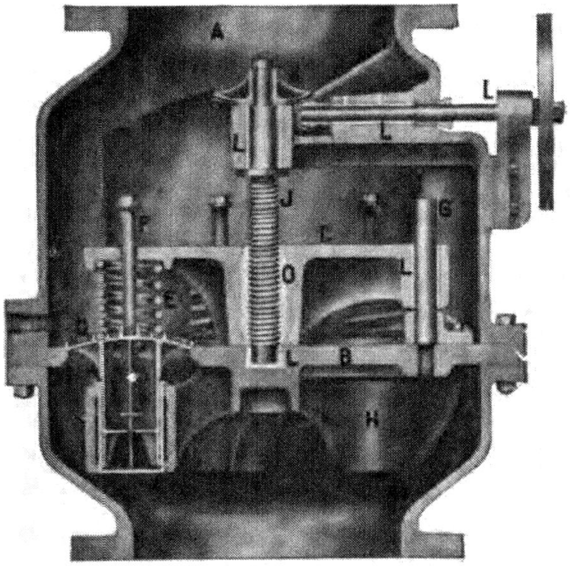

FIG. 34.

a stem *F* by means of which the valve is lifted. The tension of the spring is determined by the position of a pressure plate *C*, which is adjustable by means of a vertical screw *J* that may be turned one way or the other by means of bevel gears *K* connecting with a horizontal stem which passes to the exterior of the casing and carries a handwheel at its outer end. A suitable casing consisting of bottom and top castings, the latter having two or more handhole plates or doors for access to the valve deck, completes the construction.

The valve disks are made of bronze, are small and light and require only a small lift. As these disks are held to their seats by springs instead of weights, the total weight to be lifted when they open is comparatively small, so that they respond quickly to pressure. The seating of the disks is retarded by cushions of water in the dash pots, in which there is always water resulting from condensation. The tension on the springs can be adjusted by raising or lowering the pressure plate.

The multi-port valve, as shown in Fig. 34, is the form particularly adapted to heating and drying systems, to low-pressure turbines, and other apparatus; but when the valve is used with a condenser the springs are lighter and provision is made to protect the vacuum by maintaining a water seal over the disks. This is automatically arranged by means of a float in an independent chamber which controls the inflow of water through a valve, so that the valve deck is submerged as long as the disks are closed. When the disks lift from their seats the water inlet valve is automatically closed by suitable gear.

72. **Steam Separators.** — The object of separators is to remove from a current of steam any water or oil that it may carry, thus rendering the steam dry and free from oil impurities. There are two types of separators, one in which the separation is accomplished by giving to the steam a whirling motion as it passes over a helical path, and the other by bringing the swiftly moving steam in contact with a baffling plate placed approximately at right angles to the direction of flow. In either case the separation is due to the ease with which the direction of motion of a swiftly moving current of steam may be changed, while the heavier particles of water or oil it contains tend to move in their original direction.

73. **The Austin Steam Separator.** — The Austin Figure " E " steam separator for use in a horizontal pipe is shown in Fig. 35. As the rapidly moving steam enters the inlet it comes in contact

with the corrugated angle baffler. The steam readily changes direction, but the heavier particles of water are dashed against the corrugated surface of the baffler and fall by gravity into the receiving chamber below. The steam current, in addition to its tendency to rise, is deflected from the corrugated baffler at an angle that causes it to be thrown against the circular shell at the top of the separator, giving to the steam a centrifugal motion which brings it in contact with the other leg of the angle

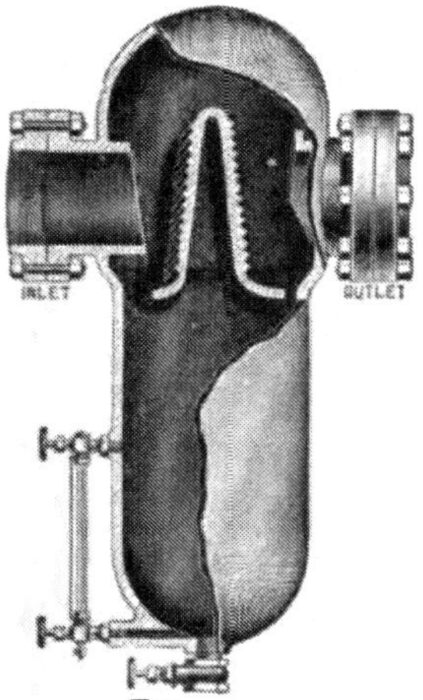

FIG. 35.

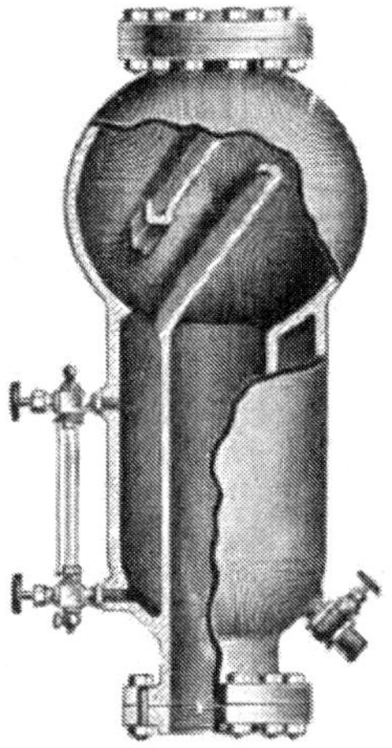

FIG. 36.

baffler to complete the separation before passing through the outlet to the engine. The advantage claimed for the corrugated baffler is that it provides the most effective separation. Any water that may adhere to the inner shell of the separator cannot find exit through the outlet because of the flange or rim around the nozzle of the outlet, around which the water is carried so as to drain into the receiver below. The gauge glass shows the height of the water in the receiving chamber.

An Austin Figure "M" separator for use in a vertical pipe is shown in Fig. 36. As the steam enters at the top it comes in contact with the inclined baffler and is deflected against the spherical inner surface of the shell, thence against the second baffler where it is again deflected to the spherical inner surface and then passes on to the engine. On meeting the first baffler the entrained water in the steam has its progress arrested and, as it collects in the grooved lip of the baffler, flows into the receiving space below. After leaving the first baffler, a further separation of the water from the steam takes place when contact with the second baffler occurs.

74. The Direct Separator. — The vertical type of separator made by the Direct Separator Company is shown in Fig. 37. The current of steam entering the separator impinges upon the conical surface composed of the solid plate *C* and trapping sheet or sieve *B*, through which water may pass freely, but from which it cannot readily escape. Passing through the sieve, the water is deposited on the solid surface of the cone and is conducted by a pipe to the water chamber below. By means of the cone the column of steam is changed into an annular ring, which is comparatively thin. The steam on the outside of this ring comes in contact with the lining *E* of the shell, which is a sieve of the same character as that at *B*. This sieve catches and entrains any water that may be contained in that portion of the current, while the water contained in the inside of the ring is caught in the trough at the lower edge of the cylinder *D*, and thence drained to the water chamber. The current of steam passes through the passages indicated by the white lines and is subjected to a whip-snapping action which will throw off any moisture that has not been caught by the surface over which it has passed. The diaphragm *F* prevents the steam from picking any water out of the water chamber.

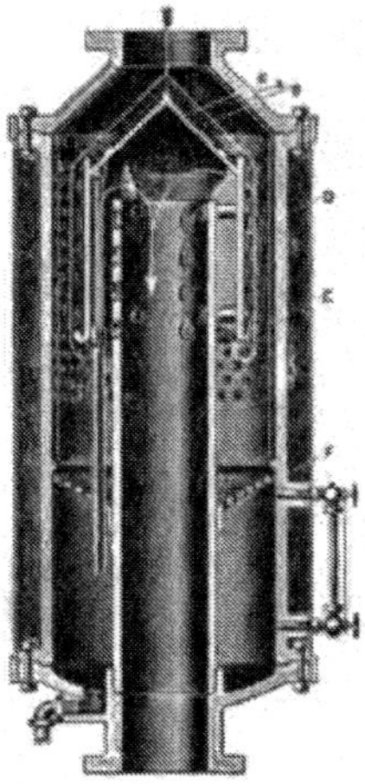

FIG. 37.

Any accumulated moisture on the inside surface of the steam pipe above the separator is caught by the upper edge of the cone *A* and carried down back of the sieve-lining to the water chamber.

The special points of this separator to which attention is called by the makers are: That it is designed to be placed next to the engine, giving no chance for condensation before steam is used; that as soon as a drop of water is separated it does not again come in contact with the steam current, and cannot be picked up and carried to the engine; that the whip-snapping action of the steam in passing downward at a high velocity and then quickly reversing its direction of flow precipitates the moisture, which is heavier than the steam, to the water chamber; that separation takes place largely while the water and steam currents are going the same direction, the steam current assisting gravity; that the upward passage for the steam is larger than that for the downward passage, and the upward current is immediately relieved by the holes shown in the steam pipe, so that there is no tendency to lift any particles of water that have started downward; that the areas are so arranged as not to cause any loss in steam pressure by wire-drawing.

FIG. 38.

The horizontal type of this separator operates in substantially the same manner as the vertical type.

The section, Fig. 38, shows the construction material of this separator. *A* is the surrounding steel jacket, *B* is asbestos lagging to prevent radiation, C is the cast-iron or steel shell, and D is the copper trapping or sieve-lining.

The construction of the oil separators of the Direct Separator Company is the same as that of the steam separators, except that, on account of the low pressure of exhaust steam, they are made lighter, and because the oil prevents rusting, the trapping sheets are made of steel. They are not covered with the non-conductor and jacket as are the steam separators.

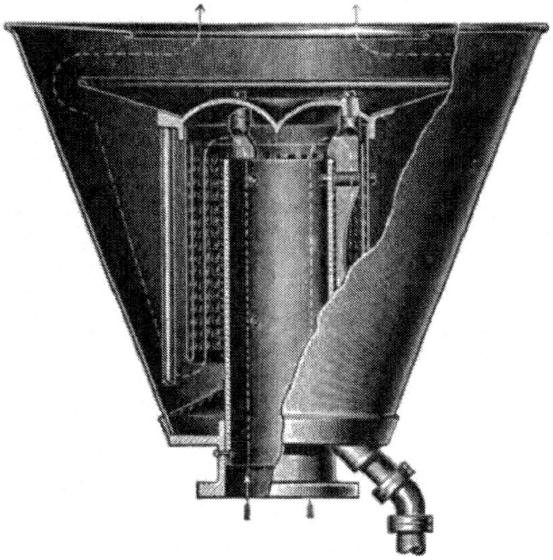

FIG. 39.

75. Exhaust Head. — In cases where the exhaust steam of power plants is released into the atmosphere the exhaust pipe is capped with an *exhaust head*, a device to trap the water and oil and prevent their deposit on roofs and walls in the shape of an unhealthy mixture in summer and ice formations in winter. It also muffles the noise of the escaping steam.

The exhaust head of the Direct Separator Company is shown in Fig. 39. It is built on the principle of the separator, the water and oil being trapped by the perforated lining and throwing it to the bottom of the head by quickly reversing the direction of flow of the current of steam. The course of the steam is shown by the dotted white lines. The entrapped water and oil is carried off by a pipe as shown.

76. Joints. — There are two kinds of steam-tight and water-tight joints in connection with the steam engine and boiler, viz., *fixed* joints and *sliding* joints. Fixed joints are those between the flanges of piping, between the cylinder head and cylinder, between the valve-chest cover and chest, and those between the manhole plates and the boiler. Sliding joints are those which permit, without leakage, rods to enter and leave a cylinder or other vessel under fluid pressure, such, for example, as piston rods, valve stems, and pump rods.

77. Packing. — In making fixed joints there is a great variety of packing used, depending upon the pressure and nature of the fluid. For moderate air and water pressures, and for low steam pressures, hemp and rubber packing may be used. Rubber is never used in its pure state as packing for steam joints, but always in combination with some fibrous material, such as canvas.

The packing when cut to fit the joint is called a *gasket*, and in making the joint it is advisable to coat the gasket with a mixture of black lead and tallow, or oil, a precaution which enables the joint to be remade without destroying the gasket. For high-pressure joints gaskets are made of corrugated sheet copper or of copper wire, the soft copper spreading as the flanges are tightened.

The packing used for sliding joints is either metallic or a mixture of india rubber and canvas. The latter is made in both square and circular section, and consists of a core of rubber wrapped with canvas. Such packing is quickly destroyed when used in connection with superheated steam, or with steam of high pressure. Asbestos has been tried as a material for packing sliding joints, but unsuccessfully, as it wears too rapidly under friction, though practically unaffected by high temperatures. The introduction of metallic packing solved the difficulty in regard to temperature, and its adoption, to as great a degree as any other one thing, has made possible the use of superheated steam in the modern steam engine.

78. Stuffing-box. — The receptacle for the packing in sliding joints is known as a *stuffing-box*, the familiar attachment through which the rod or stem enters steam and water cylinders, valve chests, and the various forms of globe and gate valves.

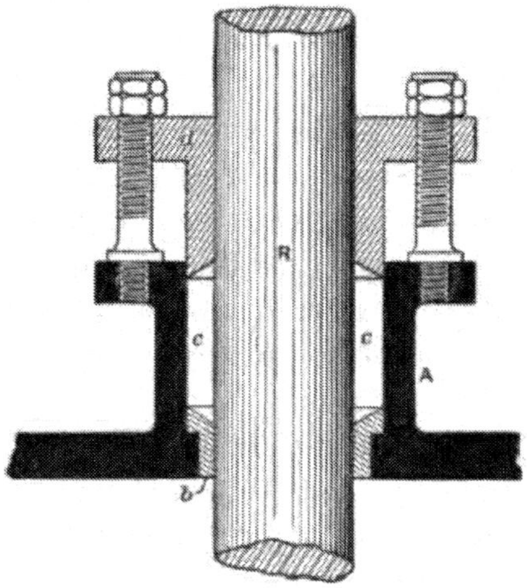

Fig. 40.

In Fig. 40 the stuffing-box A is shown in solid section, bored to a diameter greater than that of the rod. At the bottom of the box is fitted a composition bush b whose bore is just sufficiently large for the passage of the rod R. The packing is cut in strips long enough to encircle the rod and placed around the rod in the annular space c, c. The packing is then compressed by means of the nuts on the stud bolts in forcing the *gland* d against the packing, and thus make the joint tight.

79. Metallic Packing. — Metallic packing is essential to the efficient use of superheated and of high-pressure steam. For stationary and marine engine piston rods and valve stems, the double form of packing produced by the U. S. Metallic Packing Co., shown in Fig. 41, is very extensively and successfully used. The packing proper consists of two sets of babbitt metal rings, 10, 10, 10, the set nearest the cylinder or valve chest being depended upon to do most of the work. These soft metal rings, which come in contact with the rod and do the actual work of the packing, are cut in halves with a small open space between the ends when first applied so that they may readily conform to the surface of the rod when closed up by screwing the gland 9 hard down on the stuffing-box, the ends of the rings coming solidly together.

The two sets of rings are separated by the dividing ring 8, and should any water or steam lead through the first set it is caught by the second set and can be drained off; also should the first set wear out, the second set is already on the rod to take up the work.

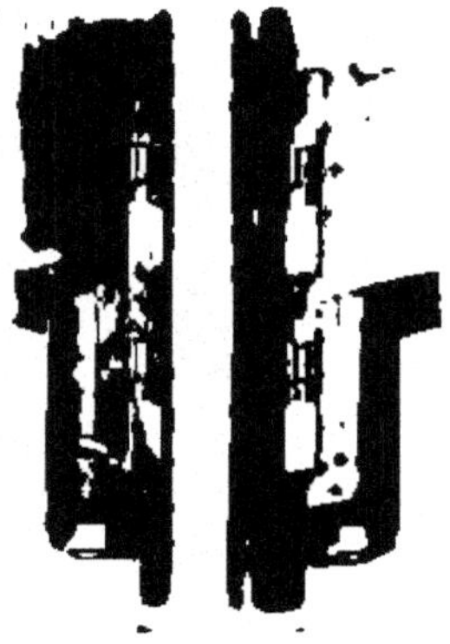

Fig. 41.

The packing rings are conical in shape on their outer surfaces so as to fit into the similarly shaped interior surface of the vibrating cup 6. When the gland is screwed down on the stuffing-box, the packing rings take their adjustment and are held in place by the action of the springs 4 and follower rings 5. The joints between the follower rings 7 and the dividing ring 8 and the gland 9 are each spherical, giving the effect of ball-and-socket joints.

When in operation, the packing rings 10 are forced by the steam pressure into the vibrating cup 6 and against the rod or stem. Flexibility is attained by means of the spherical ball joints, enabling the packing to accommodate itself to rods running out of line without in any way impairing its efficiency or increasing friction.

The particular function of the springs is to hold the packing in place when steam is shut off, the pressure they exert being only about 10 per cent of the total applied to the packing when steam is on.

This packing is automatically adjusted by the steam pressure, works with a minimum of friction, and is durable.

The King type of U. S. metallic packing used for locomotive piston rods is shown in Fig. 42. Its operation will be understood by an inspection of the figure. The gland 3, when screwed

FIG. 42.

home against the stuffing-box, compresses the spring 10, the packing ring 6 being thus adjusted and held in place by the ring 4, the half-pieces 5 and 7, and the retainer 9. The swab holder contains the cotton swabbing, which is kept saturated with oil for lubrication purposes. It should be noted that the bevel face of the packing ring is toward the cylinder.

80. Expansion Joint. — In the piping for conveying steam from the boiler to the engine, or to auxiliaries, provision must be made for expansion, particularly when the steam is superheated and of high pressure. If the pipe were rigidly connected at its ends, the irresistible force of expansion would distort and weaken it, or break the fittings. Bends in small steam pipes between their rigid connections usually make ample provision for expansion because of the curvature of the bends; but in large pipe mains some form of expansion joint must be provided. This takes the form of ordinary stuffing-box and gland in many instances, but is objectionable because of its liability to leak, which, in an exhaust pipe to the condenser, would impair the vacuum. A better arrangement is that of the Wainwright expansion joint, shown in Fig. 43. It consists of a main corrugated tube of soft copper, an inner slip tube of hard copper or composition, external and internal equalizing rings of cast iron, and connecting flanges of cast iron or steel. The corrugated tube withstands the internal pressure, and the equalizing rings insure a uniform movement between the corrugations when expansion takes place.

FIG. 43.

PROBLEMS

1. A boiler having 48 square feet of grate surface, and burning 22 pounds of coal per square foot of grate per hour, consumes 3 pounds of coal and 28 pounds of steam per I.H.P. per hour. The ratio of grate surface to heating surface is 1 to 35, and the weight of contained water in the boiler is 1.5 pounds per square foot of heating surface. How many minutes' cessation of feed would suffice to empty the boiler? *Ans.* 15 minutes.

2. Boilers are operating under a forced draft of 2.5 inches of water. What is the equivalent pressure? *Ans.* 0.09 pound per square inch.

CHAPTER V

THE SLIDE VALVE AND ITS MOTION

81. The Action of the Valve. — In the steam engine steam is admitted to and exhausted from each end of the cylinder by a valve actuated by a part of the engine itself.

With the exception of the drop valves used with beam engines of slow rotational speed and the special cases of rotating valves, such as those used with the Corliss engine, the ordinary locomotive slide valve and its modification known as the piston valve are exclusively used in stationary and marine practice. The inadaptability of the drop and rotating valves to engines of high rotational speed, due to their trip gear, has limited their use.

The smooth and economical working of an engine depends to a large extent upon the proper admission and release of steam to and from the cylinder, and it is therefore very important that study should be given to valves and their design.

In Figs. 44, 45, 46, 47, 48, and 49 the hatched sections represent the valve seat and parts of the cylinder and piston, while the solid shows a longitudinal section of an ordinary slide valve.

In the valve seat there are three passages or ports, two leading from the steam or valve chest to the cylinder, and the other leading either to the condenser, the atmosphere, or, in the case of multiple-expansion engines, to the next cylinder in the order of the expansion. The steam ports leading to each end of the cylinder are marked S and S', and the exhaust port, through which the steam leaves the cylinder, is marked E. The flat face of the valve works steam-tight on the flat seat on the side of the cylinder.

The motion of the slide valve in distributing steam in the cylinder during the stroke of the piston *P* will be understood by reference to the figures, the arrows indicating the directions of motion of the valve and piston.

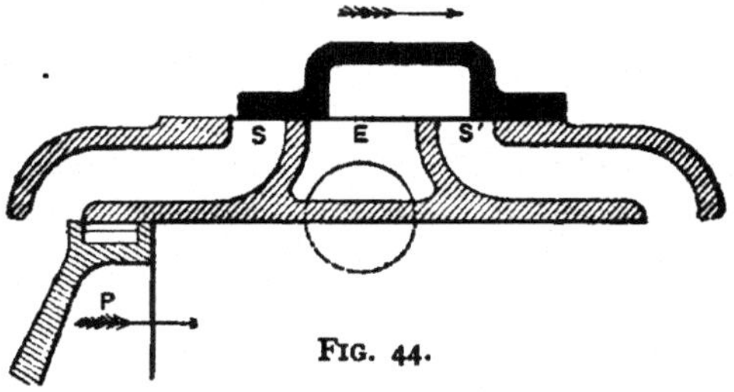

FIG. 44.

In Fig. 44 the piston is just commencing its stroke to the right, the valve showing a slight opening of the port at *S* for the admission of steam into the cylinder. This slight opening of the port for steam when the piston commences its stroke is known as the *lead* of the valve and is always provided for. The entering steam starts the piston on its stroke, the valve moving in

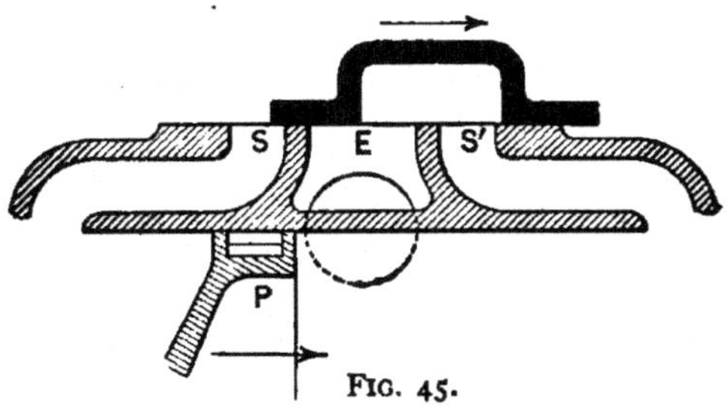

FIG. 45.

the same direction. The port *S'* has been opened by the inside edge of the valve, thus allowing the steam in the cylinder, which had been used on the preceding stroke of the piston to the left, to escape through the exhaust port *E* and exhaust pipe shown in circle. The valve and piston continue their motions in the same direction until the valve reaches the limit of its travel to the right, as shown in Fig. 45, and comes for an instant to rest, the piston continuing its motion. The valve being now at its extreme point of travel to the right, it will be seen that the port S' is wide open to the exhaust, but that the port S is not wide open for the admission of steam. In the design of the valve it is always arranged to have a full opening of the port for exhaust, but a full opening of the port for steam is rarely provided for.

The eccentric, which is the actuating mechanism, now causes the valve to commence its return stroke, the valve and piston moving in opposite directions. When the valve arrives at the

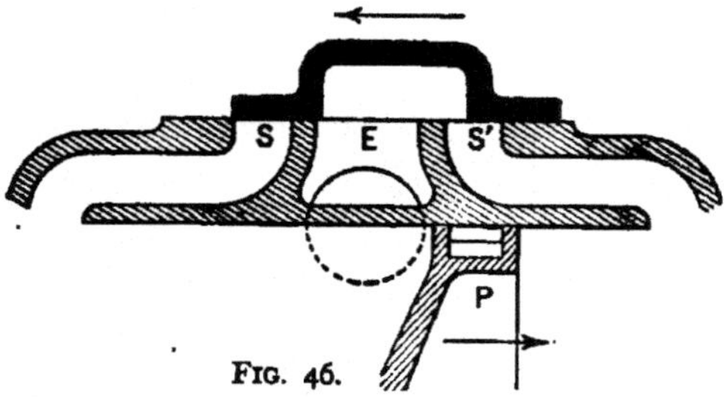

FIG. 46.

position shown in Fig. 46 the important event known as *cut-off* takes place, the steam edge of the valve having just closed the port S, preventing any further admission of steam to the cylinder, the remainder of the stroke of the piston being accomplished by the expansive force of the steam already admitted. The ratio of the final volume of steam found in the cylinder to the volume admitted before cut-off is known as the *ratio of expansion*. The port S', it will be noticed, is still open to exhaust. The valve and piston continue their motions in opposite directions, and Fig. 47 shows the valve in mid position, the piston having nearly completed its stroke, and the exhaust or inside edges of the valve coinciding with the inner edges of the steam ports S and S'. This latter condition results from there being no inside or exhaust lap to the valve, which is rarely the case, as will be seen later. With the valve in this position two events of importance occur.

The closure of the port S' has entrapped within the cylinder some of the exhaust steam of the preceding stroke, and as the piston proceeds this steam is compressed, its pressure gradually increasing as the piston nears the end of its stroke. This *compression* provides an elastic cushion of steam which absorbs the momentum of the reciprocating parts of the engine and brings them to rest without shock. The compressed steam assists the entering steam in starting the piston on the return stroke, and as it fills the clearance space it has an appreciable effect on the amount

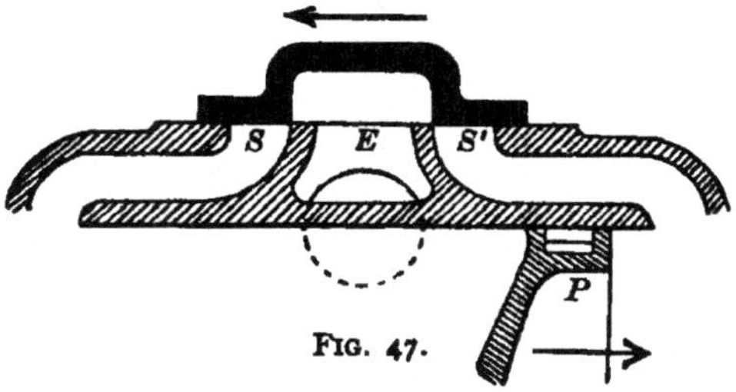

FIG. 47.

expended, since a less quantity need be drawn from the boiler at each stroke.

The left-hand inner edge of the valve having arrived at the inner edge of the port S the other event now occurs. The continued motion of the valve in the direction of the arrow opens the port S and the steam that has been driving the piston is allowed to escape through the exhaust port E, the pressure on the left of the piston being suddenly reduced. This event is known as the *release* of the steam.

The motions in opposite directions of the valve and piston continue, the port S opening wider to exhaust and the pressure of the compressed steam increasing until just before the arrival of the piston at the end of the stroke when the outside or steam edge of the valve reaches the outside edge of the port S', as shown in Fig. 48. Now follows the event of *admission*, and as the port S' uncovers fresh steam is admitted, raising the compressed steam to full pressure which acts in opposition to the piston during the remainder of the stroke and brings it gradually to rest without shock at the end of the stroke.

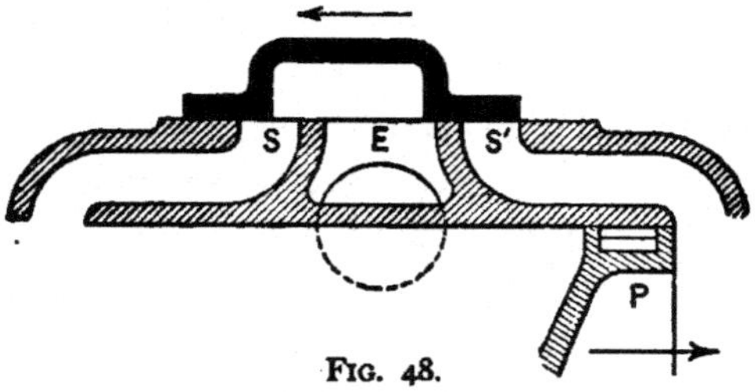

FIG. 48.

Figure 49 shows the position of the valve and piston when the stroke is completed, the port S' being open to the extent of the *lead*. All the events just described are repeated when the piston makes the return stroke, at the completion of which the valve will be found in the position shown in Fig. 44.

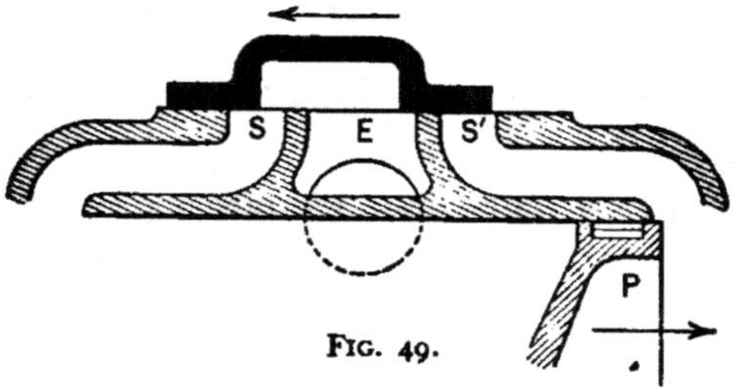

FIG. 49.

In this manner the piston is given a reciprocating motion in the cylinder, which is communicated to the shaft by means of the connecting-rod and crank mechanism, and the shaft revolves as long as steam is supplied to the cylinder.

82. The Eccentric. — Slide valves are usually actuated by an eccentric and rod. The eccentric consists of a disk keyed to the shaft and revolving with it, but has its center eccentric to that of the shaft. In Fig. 50, C is the center of the shaft and E the center of the eccentric. The eccentric disk is usually in two parts, D and D', held together by the strap SS, as shown. As the shaft revolves the eccentric is carried with it, the latter turning within the strap, thus giving to the end *A* of the rod a reciprocating motion which is communicated to the valve. The end *A*, which is guided to move in a straight line pointing to the center of the shaft, is directly connected to the valve stem in engines designed to run in one direction only, but in locomotives and marine engines where the direction of motion is often reversed the reciprocating motion is communicated to the valve through the agency of a link.

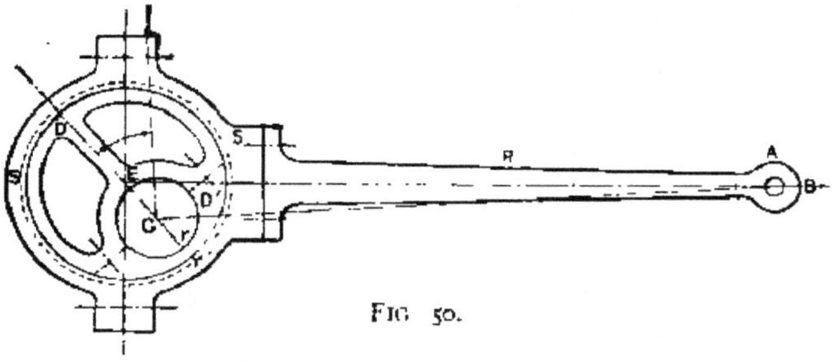

FIG. 50.

An examination of Fig. 50 will show that the motion of the end *A* of the rod along the line *CB* will be twice the distance *CE*, the distance between the centers of the shaft and eccentric. This distance is variously called the *throw of the eccentric*, *eccentric arm*, and *eccentric radius*, and is equal to the half-travel of the valve.

It will readily be seen that the motion given to the point *A* is exactly equivalent to that which would be produced by a crank *CE* and connecting-rod *EA*, from which we conclude that the eccentric and rod is a special case of the connecting-rod and crank mechanism in which the crank-pin is made so large as to include the shaft within its section. In order that this may be, the radius of the eccentric must be greater than the sum of the radii of the crank and of the shaft, that is, $Er' > EC + Cr$ (Fig. 50).

Having considered the action and actuating mechanism of the slide valve in the distribution of the steam in the cylinder, it is next in order to fix the position of the eccentric on the shaft relative to the crank, so that the events illustrated by Figs. 44, 45, 46, 47, 48, 49 shall occur at the proper times.

For this purpose a slide valve with faces of just sufficient width to cover, when in central position, the steam ports S and S', Fig. 51, will be used in illustration.

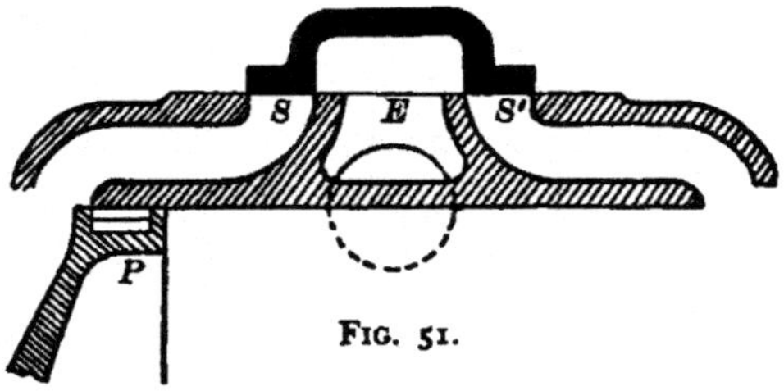

FIG. 51.

In order that alternate strokes of the piston in the cylinder shall be in opposite directions the valve must provide that, at each stroke, steam shall be admitted to the driving side of the piston and that the steam used in the preceding stroke shall be exhausted from the opposite side. The valve in Fig. 51 is in its central position and if its movement were so timed that it would occupy that position simultaneously with the arrival of the piston at the end of either stroke, the provision required of the valve would be fulfilled, provided the movement, or travel, of the valve during one stroke of the piston were equal to twice the width of the steam port, and that the first half of the valve movement were in the same direction as that of the piston and the last half in the contrary direction. Since the throw of the eccentric is equal to the half-travel of the valve, and as the crank moves with the piston, it is at once evident that the desired motion of the valve of Fig. 51 would be obtained by placing the eccentric on the shaft 90° ahead of the crank. We will now consider the distribution of the steam in the cylinder by the valve of Fig. 51, the eccentric being placed on the shaft 90° ahead of the crank. The piston being at the end of the stroke the eccentric will cause the valve to move to the right to admit steam to the cylinder through the port S to drive the piston in the same direction. Simultaneously with the admission of steam into the port S, the right-hand inside edge of the valve will open the port S' to the exhaust, and there will not have been any *release* of the steam before the piston arrived at the end of the preceding stroke, nor had there been any *lead* to the valve to give the full steam pressure on the piston at the moment of commencing the stroke under consideration. After the eccentric arm has revolved through 90°, the valve will have reached its extreme distance to the right, a distance just sufficient to open wide the port S to steam and the port S' to exhaust. The crank and eccentric continue to revolve and the piston proceeds on its stroke; but the instant the eccentric passes the 90° point of its rotation, the valve commences its travel to the left, and after a further angular motion of 90° by the crank and eccentric the piston will have arrived at the end of its stroke simultaneously with the arrival of the valve to its original position. Thus, it is seen, there has been no *cut-off* of the steam before the arrival of the piston at the end of its stroke, and therefore no advantage taken of the expansive force of the steam to complete the stroke; nor has the exhaust closed before the completion of the stroke to provide for that very necessary elastic cushion of compressed steam to bring the piston gradually to rest without shock.

The relative movements of the crank and eccentric, as just explained, will be better understood by a reference to Fig. 52. Let CR and CE be the original positions of the crank and eccentric arms for the position of the valve shown in Fig. 51; and let CR' and CE' be the crank and eccentric positions after the shaft has revolved through 90°, the valve having moved its extreme distance CE' to the right and opened the ports wide to steam and exhaust. A further rotation of 90° by the shaft brings the crank and eccentric to the positions CR'' and CE'' respectively, the valve having returned through a distance $E'C$ to its original position just as the piston completed its stroke; for while the center of the crank-pin has rotated through the semicircle $RR'R''$ its motion of translation has been the diameter RR'' of the crank-pin circle which, of course, is equal to the stroke of the piston. If the shaft were to complete its revolution, as indicated by the dotted semicircles, the crank and eccentric would assume their original positions CR and CE and the valve would have moved its extreme distance CE''' to the left and returned the same distance $E'''C$ to its original position, the piston meanwhile completing its return stroke and the positions of the valve and piston would be as represented in Fig. 51.

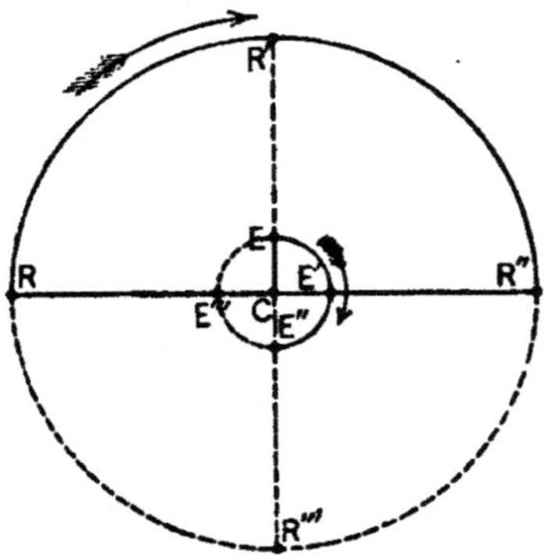

FIG. 52.

The action of this valve is clearly defective and the difficulty results from delaying until the end of the stroke the changes which are necessary for its reversal.

In practice this difficulty is avoided by making the valve longer so that when in its central position its faces more than cover the steam ports, as shown in Fig. 47; and in addition to this lengthening of the valve the eccentric is set on the shaft an angular distance greater than 90° ahead of the crank.

83. Definitions. — *Lap* of the valve — called outside lap or steam lap — is the distance the outer or steam edge of the valve extends beyond, or laps over, the steam edge of the port when the valve is in its mid-position.

Lead of the valve is the distance the valve uncovers the port for the admission of steam when the piston is at the end of its stroke.

It has been pointed out that giving lead to the valve provides that a full pressure of steam shall be exerted on the piston at the commencement of its stroke, and that assistance be rendered the compressed steam in bringing the piston momentarily to rest without shock at the end of the stroke.

84. Angular Advance. — Regarding the valve of Fig. 47 as that of Fig. 51 lengthened to overlap the ports, it will be seen that in order to give it lead when the piston is at the end of its stroke, as shown in Fig. 44, p. 103, we must first advance the eccentric *CE*, Fig. 53, through an angle *ECA*, called the angle of lap, to move the valve a distance *CP* equal to the lap, and then advance it through the angle *ACB*, called the angle of lead, to move the valve a distance *PD*, equal to the desired lead. The sum of these two angles, the lap and lead angles, is known as the ***angular advance*** of the eccentric, and is the number of degrees in excess of a right angle that the eccentric is set on the shaft in advance of the crank.

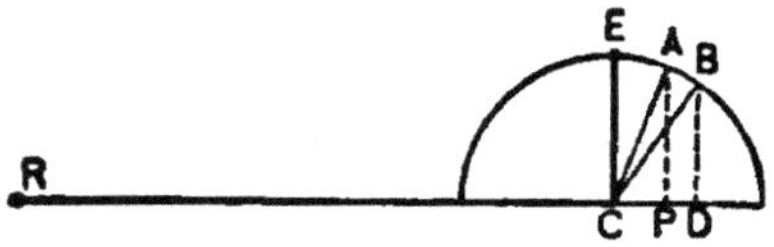

FIG. 53.

85. Observations. — The primary object in giving lap to a valve is to provide means for cutting off the admission of steam into the cylinder before the end of the stroke, so that advantage may be taken of the expansive power of the steam to continue the movement of the piston to the end of the stroke. There is a limit, however, to the amount of lap that should be given a valve. The addition of lap necessitates an increase of the angular advance of the eccentric in order to maintain the lead, and if the maximum port opening is also to be maintained the travel of the valve must be increased. It is always desirable to have a small valve travel so that the work expended in moving the valve shall be a minimum, and it is for this reason that double-ported valves are resorted to, as by their use only one-half the travel of the single-ported valve is necessary for a given port opening. The increase in the angular advance of the eccentric will have the effect of hastening all the operations of the valve, and to whatever extent the opening of the exhaust has been hastened its closure will likewise be hastened and excessive compression may result. With a fixed eccentric the limit of cut-off by means of lap is about five-eighths of the stroke.

If, when the valve is in its mid-position, the inside edges extend over the inside edges of the ports the valve is said to have inside or exhaust lap and its effect is to delay the release of the steam and hasten the compression. When in its mid-position if the inside edges of the valve do not overlap the inside edges of the ports, but show an opening to the exhaust, the valve is then said to have negative exhaust lap, and this is not infrequently the case, particularly with vertical engines where a prompt exhaust of the steam from the cylinder on the up-stroke is desirable. An examination of Fig. 47 will show that with negative exhaust lap there will be a period when both ends of the cylinder will be in communication with the exhaust and with each other. This will occur just before compression and the released steam from the driving side of the piston will flow into the exhaust side, causing an increase in the back pressure which will be shown by a rise in the back pressure line of the indicator diagram just before compression. The period of this communication will be brief, as the motion of the valve is then at its quickest.

The necessity for giving lead to the valve has been shown, its amount being fixed arbitrarily, according to the type of engine and the judgment of the designer. The rotational speed in all engines and the weight of the reciprocating parts compared to piston area in vertical engines are governing features in the determination of the amount of the lead. For slow-running stationary engines the lead varies from $\frac{1}{64}$ to $\frac{1}{16}$ of an inch. For the high-speed stationary engine it is seldom less than $\frac{1}{32}$ of an inch, and for the locomotive it is commonly $\frac{1}{4}$ inch when at full speed and linked up. The weight of the reciprocating parts of a vertical engine acts with the steam on the top or downstroke, and against the steam on the bottom or upstroke, and therefore the necessity for increased lead at the bottom or crank end of the valve to assist compression in bringing the piston to rest without shock and to furnish promptly a full pressure to start the piston on the upstroke. As much as three-eighths of an inch for the bottom lead of a vertical marine engine is not uncommon.

For reasons which will be explained later the angularity of the connecting-rod introduces an inequality in the movement of the piston which occasions a greater piston displacement on the forward (toward the shaft) stroke than on the return stroke for equal angular positions of the crank, and this inequality is greater as the ratio of the length of the crank to the length of the connecting-rod becomes greater. Then, since the cut-off of both strokes is effected by the same eccentric, there will be an inequality in the two cut-offs, later on the forward than on the return stroke. In stationary engines where the crank-connecting-rod ratio is comparatively small this inequality is not of much consequence and a partial compensation is usually provided by giving a trifle less lap on the crank end of the valve than on the head end without seriously disarranging the leads. To provide for an absolute equality in cut-off would require so little lap on the crank end as to make the lead excessive. In vertical marine engines the crank-connecting-rod ratio is comparatively large and the inequality in cut-off more pronounced; but since, as already stated, a large lead is desirable on the bottom or crank end, the lap on that end may be sensibly less than on the top end, which will afford a partial equalization of the cut-off. In addition to this it is common marine practice to make the top cut-off a trifle later than that desired, which will make the bottom cut-off also later and therefore the mean cut-off more nearly that desired.

86. Setting Slide Valves. — The efficient working of the engine depends largely on the proper adjustment of the valve on its stem and the placing of the eccentric in its proper place on the shaft. The performance of these two operations is called *setting the valve.* It is first necessary to locate the dead points of the engine, or put the engine *on its center*, as it is called. When an engine is on its center a line joining the center of the crank-pin and the center of the shaft is in direct line with the axis of the cylinder and the piston is then absolutely at the end of its stroke. To obtain this position the engine is jacked until the crosshead nearly reaches the end of its stroke, when a mark is made on the slide to correspond to one made or existing on the crosshead. At some point on the engine framing near the crank disk, the pulley, or the flywheel, make a center-punch mark and with this mark as a center, and a tram of convenient length as a radius, describe a small arc on the revolving part selected. Then jack the engine past the end of the stroke until the mark on the crosshead returns to the mark made on the slide. From the center-punch mark as a center describe another small arc with the tram on the selected revolving part. The crosshead or the piston is now the same distance from the end of the stroke as it was when the first arc was described on the revolving part, so if the distance between these two arcs be bisected and the point of bisection be marked with the center-punch, then when the engine is jacked so that the tram exactly spans the distance between the two center-punch marks, the engine will be on the center and the piston at the end of the stroke. A like process will determine the other dead center. In putting an engine on its center, and during the whole operation of setting a valve, the engine should be moved in the direction in which it is intended to run in order that all lost motion may be taken up in that direction.

Valves are usually set for equal leads the exceptions being for vertical engines, as already pointed out, and for cases where attempts are made to equalize the cut-off.

To set a slide valve the engine is placed on the center, and if the eccentric be not fixed on the shaft in its proper position it should be approximately so placed, making the angular advance a trifle larger than that required. By means of the nuts on the valve stem the valve is given the proper lead. Turn the engine to the other center, and if the lead shown there is not the same the difference must be corrected — half by altering the length of the valve stem and half by moving the eccentric. With the lead thus equalized the eccentric should be keyed fast on the shaft and the nuts on the valve stem tightened; the operation will then be complete.

Should the valve be designed for unequal leads the method of setting it is exactly the same as just described, the correction being so made as to have the leads as desired.

87. The Link Motion. — With marine engines and locomotives, whose direction of running must often be reversed, the reciprocating motion furnished by the eccentric is communicated to the valve through the agency of a link. The link motion was originally designed as a means of reversing the motion of an engine, but it was afterwards found to furnish a method of materially increasing the range of expansion of the steam in the cylinder.

The action of a link motion in reversing an engine may be understood by referring to Fig. 54, where *CR* represents the crank in some one position. We have already seen that in order to have the crank revolve in the direction indicated by the full-line arrow the eccentric must be fixed on the shaft at some position *CE* ahead of the crank. Now in order to have the crank turn in the direction of the dotted arrow, the eccentric would have to be shifted on the shaft to the position CE' ahead of the crank in the opposite direction. As this is impracticable the difficulty is overcome by having two eccentrics keyed to the shaft, one having its center at E and the other at E'. Each eccentric has its rod connected to one end of a curved link which, as originally designed, is slotted to receive a block directly connected to the valve stem, and by a suitable arrangement of levers the link may be shifted so that the movement of the valve may be under the influence of either the go-ahead or the backing eccentric as desired.

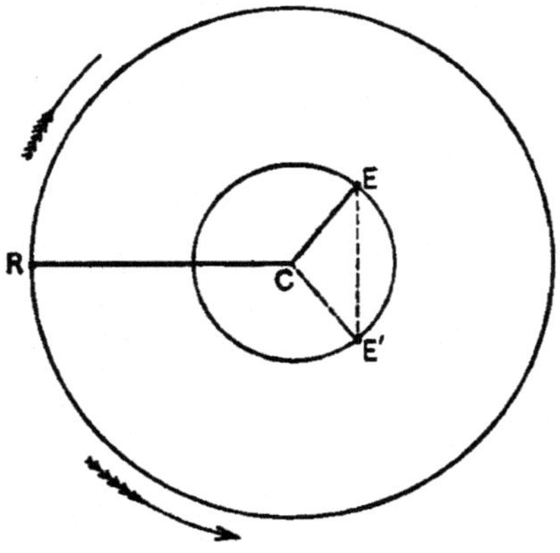

Fig. 54.

Figures 55 and 56 are skeleton sketches of a link motion in which CR is the crank, E and E' the centers of the go-ahead and backing eccentrics respectively, and PQ the link, curved with a radius equal to the length of the eccentric rod EP.

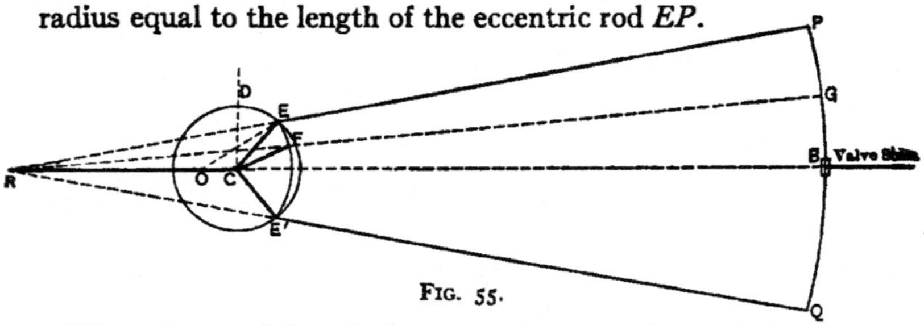

Fig. 55.

When the crank is pointing away from the link and the eccentric rods are joined to the ends of the link nearest to them, as in Fig. 55, the rods are said to be *open*, and if, when the crank is in the same position, the rods are joined to the link as shown in

Fig. 56, the rods are said to be *crossed*. If, as is frequently the case with piston valves, steam is taken at the inside edges, the rods as connected in Fig. 55 would be called crossed, and if as in Fig. 56 would be called open.

When the link is shifted so that the valve block B, Fig. 55, is brought in line with the rod EP it is in full forward gear and the motion of the valve is governed by the eccentric E. If the link be shifted so that the block comes in line with the rod $E'Q$ it is in full backward gear and the eccentric E' governs the motion of the valve. When the block is midway between the two

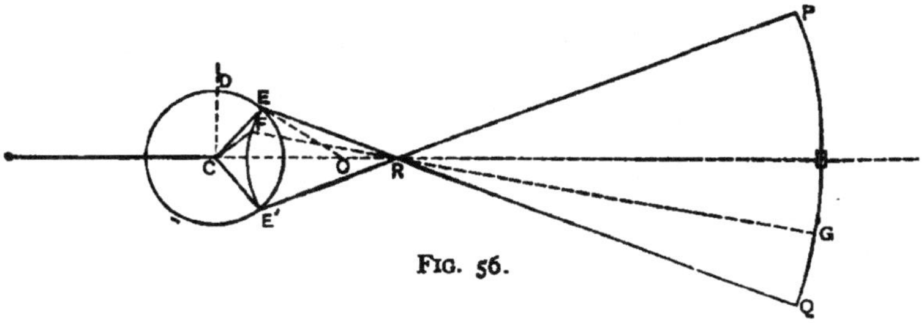

FIG. 56.

extreme positions the link is said to be in mid-gear, and the valve is influenced equally by both eccentrics, with the result that it does not receive sufficient motion to open the ports and the engine remains at rest.

Should the link be shifted so that the valve block is brought to a position intermediate between mid-gear and full forward gear, as at G, the movement of the valve is influenced by both eccentrics, but to such a large extent by eccentric E that the engine will continue to run ahead. The effect, however, of the slight influence of the backing eccentric has been to decrease the travel of the valve, so that all its operations will be earlier than when working in full gear and the cut-off will be, therefore, shorter.

The motion given to the valve block when its position is intermediate between mid and full gear is that due to a *virtual* eccentric of less throw than that of the real eccentric. This question has been investigated by designers, and it has been found that for positions of the valve block intermediate between mid and full gear, the centers of the virtual eccentrics lie in a parabolic curve. The locus of these centers will be represented with sufficient accuracy by an arc of a circle passing through the centers E and E' of the eccentrics and of a radius equal to the product of the eccentric rod and half the distance between the centers of the eccentrics, divided by the distance between the eccentric-rod pins in the link; that is, the required radius is equal to $\frac{EP \times EE'}{2\,PQ}$. This radius is found to be equal to OE, Figs. 55 and 56, and the arc EFE' has been drawn concave to the center of the shaft for open rods, and convex for crossed rods.

To find the virtual eccentric governing the motion of the valve when the block is at G, divide the arc EFE' at F in the same ratio that G divides PQ; then CF is the throw and DCF the angular advance of the eccentric which is virtually actuating the valve. The practical way of determining F is to produce PE and QE' until they intersect at some point R in the center line; then RG divides the arc EE' at F in the same ratio that G divides PQ. The action of the link in providing a variable cut-off may now be understood.

An investigation of the effects of the open and crossed rod connections shows that with open rods the lead increases from full to mid-gear, and decreases with crossed rods; and that the longer the rods are made and the shorter the link the less change the lead will undergo. It is the usual practice to make the open rod connection for stationary engines.

The curvature of the Stephenson link occasions the variation in the lead from full to mid-gear, and this variation becomes greater as the eccentric rods are shortened. Zeuner has shown analytically in his treatise on valve gears that the length of the eccentric rod should be the radius of the arc of the link in order that the lead may be equal. It may be shown too that the motion of the valve becomes more nearly harmonic as the eccentric rods and link are lengthened. Practice, however, has about fixed the length of the rods to about twelve times, and the length of the link to about four times, the throw of the eccentric.

The Stephenson link with open rod connection is well adapted to locomotive practice. The adjustment should be such that when starting the train, with the link thrown in full gear, the cut-off should be as long as from three-fourths to seven-eighths of the stroke in order that, with a partially open throttle, a uniform and moderate pressure be exerted on the piston during a greater part of the stroke to overcome the train friction without causing the driving wheels to slip. This, of course, raises the terminal pressure and is the cause of the noisy exhaust of a locomotive when starting a train.

The cranks of a locomotive being set at right angles, the lead at the start may be very small, but as the speed of the train increases the throttle should be opened gradually, and then the cut-off shortened and the lead and compression increased by raising the link, or by *linking up* as it is called. The locomotive, being a high-speed engine, requires a considerable amount of compression to overcome the inertia of the reciprocating parts.

The crossed rod connection is commonly used for marine engines, for then, with a decreasing lead from full to mid-gear, the engine will stop when the link is put at mid-gear, which it would not necessarily do with the rods open and the engine running with light load.

The slotted Stephenson link is expensive to make and difficult to hang so that the centers of the eccentric rod pins and the center of the valve block shall be in line and the block work easy in the slot. These difficulties have led to the adoption, for marine purposes, of the double-bar link, consisting of two solid bars of rectangular section and of proper curvature. The bars are secured together by distance pieces at the ends. The valve block is between the bars and has flanges which bear upon the tops and bottoms of the bars. The eccentric rods make a forked end connection to pins projecting from the bars near the ends.

88. The Shifting Eccentric and Automatic Cut-off. — As will be seen later on, quick running of an engine is one of the methods of decreasing the losses from condensation and reëvaporation, and it is apparent that the higher the rotational speed consistent with safe piston velocity the greater the efficiency of the steam action in the cylinder, and since piston velocity is a factor of the power, any increase in that direction admits of a decrease in the size of an engine for a given power. Notwithstanding these advantages, the introduction of a high-speed engine was long delayed. The belief was prevalent that as compared with one of lower speed it was more liable to frequent and serious accidents; and a practical objection to its introduction arose from the fact that the limit of rotational speed was soon reached at which the drop cut-off could be applied, and it became essentially necessary to the introduction of the high-speed engine that some contrivance be devised by which the steam should have a positive and variable cut-off, automatically controlled, in order that the speed should be constant under conditions of variable load and pressure.

The vast improvement in the character of the materials of construction; the more perfect understanding of the stresses produced in overcoming the inertia of the reciprocating parts; the skill of the workman — all have made possible the production of an engine of such perfect balance and adjustment that there is no longer a question of safety even when running at the high rotational speed that admits of a direct connection to the armature of a dynamo-electric machine.

An inherent defect in the fly-ball governor precluded its use in the solution of the cut-off problem, so effort was centered in devising a mechanism in which the centrifugal action of revolving weights opposed by the varying tension of a spring should control the action of the valve to meet the exigency of a sudden variation in the load on the engine or in the pressure of the steam. The result was the production of what is known as the *shaft governor*, now almost exclusively used with engines driving dynamo-electric machines in producing electric light and power.

The control of the valve by the shaft governor is either by means of a loose eccentric whose angular advance is made to vary, or by means of a slotted eccentric of variable throw. Examples of each type will be given later on.

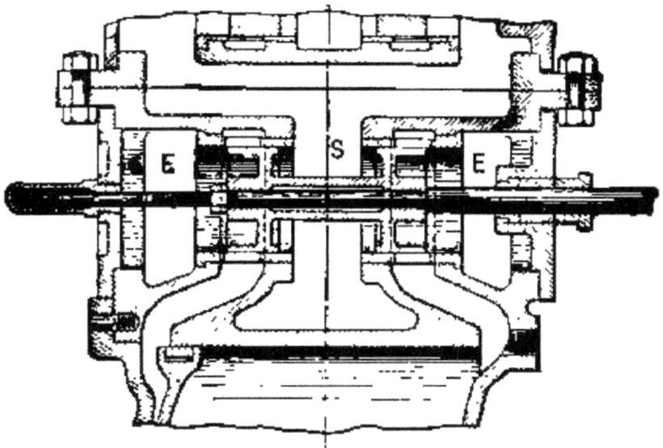

FIG. 57.

89. Piston Valve. — In order that the shaft governor shall be sensitive there must be the minimum of work thrown on it, and this can only be effected by having the least possible resistance offered to the motion of the valve. This clearly points to the use of a balanced valve and the one largely used is a variety of the type known as piston valve, shown in section in Fig. 57. A piston valve may be conceived to be formed from the ordinary slide valve by curving the latter into cylindrical form. It is arranged to take steam at its inside edges, though it could equally well be arranged to do so at the outside edges. It will be seen that the valve is balanced, as the steam completely surrounds it.

When made for large marine engines the pistons of the valve are fitted with springs to keep them steam-tight, and the ports are cast with diagonal bars to keep the rings from springing into the ports, and in order also to afford a continuous guide to the piston so that the rings may pass over the ports without catching on the edges. The chief disadvantage of the piston valve is the difficulty in keeping it steam-tight.

90. Radial Valve Gears. — Efforts have been made from time to time to supersede the link motion and eccentrics in actuating the valves of steam engines, the most successful of which have been the radial valve gears of Marshall and Joy.

In the Marshall gear the link is dispensed with and but one eccentric is used. With the Joy gear the link and both eccentrics are dispensed with, the valve motion being obtained from a point in the connecting-rod.

Radial valve gears when accurately proportioned provide for any degree of expansion with a uniform lead, but their parts are necessarily heavy and difficult of correct adjustment and the most recent marine practice indicates a return to the old but reliable link motion and eccentrics.

PROBLEM

Throw of eccentric, 2 inches; length of eccentric rods, 24 inches; length of chord of link, 15 inches; angular advance of the eccentric, 38°; stroke of engine, 14 inches. Make a skeleton sketch of the link motion, and find the throw of the virtual eccentric when the link is halfway between full and mid-gear, the connection of the rods being open. Scale, 3 inches = 1 foot. *Ans.* Virtual eccentric throw = $1\frac{7}{8}$ inches.

CHAPTER VI

ROTARY MOTION. ACTION OF CRANK AND CONNECTING-ROD

91. Simple Harmonic Motion. — If the point P, Fig. 58, moves in the plane of the paper about C as a center it will describe the circle RPB, and during the period of describing the arc RP the circular motion of P may be considered as compounded of two entirely independent movements at right angles to each other — that of R to Q in the line of the diameter RS and the other of Q to P.

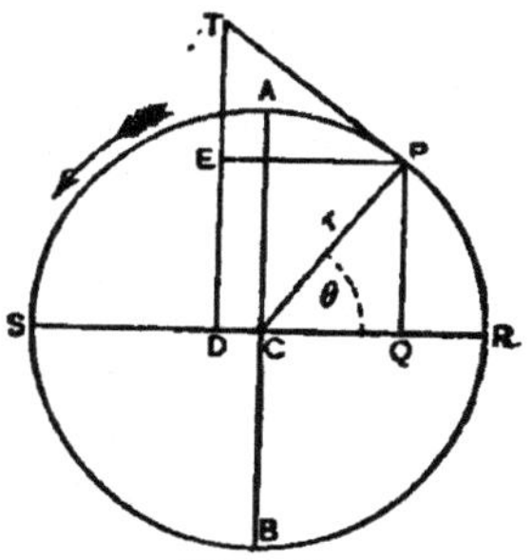

FIG. 58.

Draw the diameter AB perpendicular to RS and let P move with a uniform velocity. If we denote CP by r, and the angle PCR by θ, we may find the relation between the positions of P and Q as follows:

$$RQ = RC - QC = r - r\cos\theta = r(1 - \cos\theta).$$

To find the position of Q when P has described two-thirds of the quadrant RA. Here we have $\theta = 60^\circ$, and $\cos\theta = \frac{1}{2}$, consequently

$$RQ = r\left(1 - \frac{1}{2}\right) = \frac{r}{2}.$$

That is, P describes two-thirds of the distance circumferentially from R to A while Q moves one-half the distance RC, and since the motion of P is uniform it follows that the last half of RC will be passed over by Q in one-half the time it required to pass over the first half. It is thus seen that while the motion of P

is uniform that of Q is variable; and if the relative positions of P and Q were found for values of θ from 0° to 180° it would be found that when P is at R the point Q is at rest, and as P continued its uniform motion the velocity of Q would gradually increase until the value of θ reached 90° when the velocity of Q would be a maximum and equal to that of P; as P moves from A to S, Q would move from C to S with diminishing velocity until its arrival at S when it would again be at rest. During the movements just described Q is said to have a ***simple harmonic motion.***

When a point P moves uniformly in a circle the perpendicular PQ let fall at any instant to a fixed diameter RS intersects the diameter at a point Q whose position changes by a simple harmonic motion.

To ascertain the relation existing between the velocities of Q and P draw the tangent PT, and from T let fall the perpendicular TD upon RS. Suppose the velocity of P to be such that in a given time it moves a distance PT. In the same time Q would move to D, and we shall have

$$\frac{\text{Velocity of } Q}{\text{Velocity of } P} = \frac{QD}{PT} = \frac{PE}{PT} = \frac{PQ}{PC} = \sin\theta.$$

The sine of θ varies in value from 0 to 1, and we can, therefore, find at any period of the motion how much the velocity of Q differs from that of P.

92. Crank-connecting-rod Action. — It has been shown how the valve of a steam engine actuated by an eccentric, or its equivalent, admits steam alternately to the two ends of the cylinder to drive the piston forward and backward along a straight path of definite length called the stroke. This reciprocating motion of the piston is converted into one of continuous rotation of the shaft through the agency of the mechanism of the connecting-rod and crank. In this mechanism the outer end of the piston rod is connected to a crosshead which is constrained by guides to move in line with the axis of the cylinder. To a pin in the crosshead the inner end of the connecting-rod is so attached as to permit an oscillating motion to that end of the rod. In a similar manner the outer end of the connecting-rod is connected to the crank-pin, and if the piston be driven forward or backward the crank will be caused to turn about the axis of the shaft by the alternate push and pull transmitted from the crosshead through the connecting-rod to the crank-pin.

In Fig. 59, let r denote the length CP of the crank and c the length PQ of the connecting-rod. If from R and S we lay off

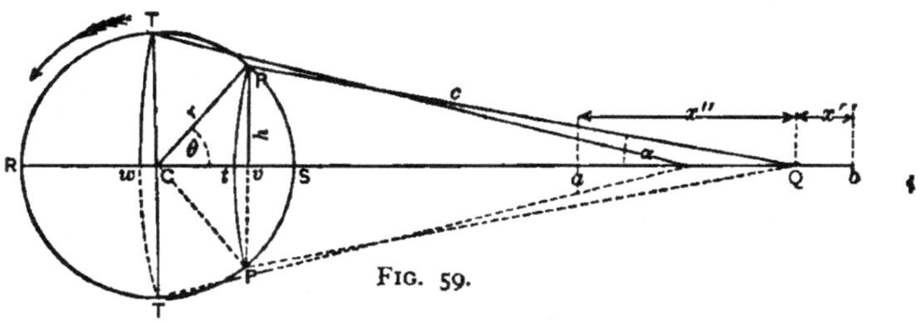

FIG. 59.

Ra and Sb, each equal to c, then ab will be the stroke of the crosshead Q, and for purposes of discussion it is the stroke of the piston, the latter being rigidly attached to the crosshead. Let CP be a position of the crank, making an angle θ with the center line Rb of the engine. Then Q will be the corresponding position of the piston, and is found by striking an arc from P as a center and with a radius c. With Q as a center and the radius c describe the arc Pt. Then tS is evidently equal to QB; and since the diameter RS of the crank-pin circle is equal to the stroke of the piston, RS may equally well represent the piston stroke. Then t conveniently represents the position of the piston corresponding to the crank position CP, and it is at once evident that the movement of the piston is not the result of a simple harmonic motion. The angularity of the connecting-rod, that is, its failure to remain parallel to the axis of the cylinder at all times, introduces the inequality *tw* into the movement of the piston and this inequality gradually increases from its zero value at the ***dead point*** *S* until the crank arrives at the position *CT*, 90° in advance of *CS*, when its value *wC* is a maximum. As the crank continues its rotation the inequality gradually diminishes until, at the dead point *R*, it is again zero.

An examination of Fig. 59 shows that when the crank is at the mid-point of its revolution from *S* to *R* the piston has traveled a distance *Cw* beyond its mid-stroke; and it is seen that during the first quarter of the revolution on the forward stroke from *b* to *a* the piston travels a greater distance than during the second quarter.

This inequality can be avoided only by giving to the piston a simple harmonic motion, but as that could be obtained only by making the connecting-rod infinitely long the inequality must always exist with the crank-connecting-rod mechanism. The shorter the connecting-rod the greater will be the inequality in the movement of the piston, and for this reason, if for no other, the crank-connecting-rod ratio is made as small as possible. In marine practice, where space limits the length of the connecting-rod, the ratio is from 1 to 4 to 1 to 5, while for stationary engines it is not uncommonly as small as 1 to 7.

A further examination of Fig. 59 shows that during the return stroke of the piston conditions exactly the reverse of those of the forward stroke obtain, as shown by the dotted lines.

It will be observed that the use of the connecting-rod of finite length causes the cut-off to be considerably later on the forward stroke than on the return stroke and the reference to this question on page 113 can now be understood.

The inequality in piston displacement for the same crank position of the two strokes results in an unequal distribution of steam in the cylinder, and therefore an irregularity in the driving power of the engine. The angularity of the connecting-rod is a disturbing element in the consideration of every important dynamic question of the steam engine, rendering more or less difficult the solution of problems which would be otherwise simple.

With the aid of Fig. 59 we may find an expression for the relative positions of the crank and piston at any instant.

Let α denote the angle made by the connecting-rod with the center line of the engine, and denote by x' the distance the piston has traveled on the forward stroke when the crank has the position CP.

$$x' = Cb - (Cv + vQ) = r + c - (r\cos\theta + c\cos\alpha)$$
$$= r(1 - \cos\theta) + c - c\cos\alpha.$$

We have $\quad h = c\sin\alpha = r\sin\theta,\;$ whence $\sin\alpha = \dfrac{r\sin\theta}{c},$

and $\quad \cos\alpha = \sqrt{1 - \dfrac{r^2\sin^2\theta}{c^2}}.$

Therefore $\quad x' = r(1 - \cos\theta) + c\left[1 - \sqrt{1 - \dfrac{r^2\sin^2\theta}{c^2}}\right].$

If x'' denotes the distance of the piston from the crank end of the stroke for the same crank position, we shall have

$$x'' = ab - Qb = 2r - x'$$
$$= 2r - \left\{r(1 - \cos\theta) + c\left[1 - \sqrt{1 - \frac{r^2\sin^2\theta}{c^2}}\right]\right\}$$
$$= 2r - r + r\cos\theta - c\left[1 - \sqrt{1 - \frac{r^2\sin^2\theta}{c^2}}\right]$$
$$= r(1 + \cos\theta) - c\left[1 - \sqrt{1 - \frac{r^2\sin^2\theta}{c^2}}\right].$$

In general terms we shall then have

$$x = r(1 \mp \cos\theta) \pm c\left[1 - \sqrt{1 - \frac{r^2\sin^2\theta}{c^2}}\right], \qquad (1)$$

where the top sign must be taken if x is measured from the head end of the stroke, and the lower sign if measured from the crank end.

The position of the piston for any given crank position, or conversely the crank angle for a given position of the piston, can be found by equation (1).

The ratio $\frac{r}{c}$ will become very small if raised to a power above the square, so if the expression $\left[1 - \frac{r^2 \sin^2 \theta}{c^2}\right]^{\frac{1}{2}}$ be expanded by the binomial theorem, and the terms containing higher powers of c than the square be rejected no appreciable error will result and the formula will be more convenient for use.

Thus

$$x = r(1 \mp \cos\theta) \pm c\left\{1 - \left[1 - \frac{r^2 \sin^2 \theta}{c^2}\right]^{\frac{1}{2}}\right\}$$

$$= r(1 \mp \cos\theta) \pm c\left\{1 - \left[1 - \frac{r^2 \sin^2 \theta}{2c^2}\right]\right\}$$

$$= r(1 \mp \cos\theta) \pm \frac{r^2 \sin^2 \theta}{2c}. \qquad (2)$$

Example. — Let $c = 40$ inches, $r = 10$ inches, $\theta = 210°$. Find the position of the piston, measured from either end of the stroke.

Solution. — Here we have $\frac{r^2}{2c} = \frac{5}{4}$, $\cos 210° = - \cos 30° = - 0.866$, and $\sin 210° = - \sin 30° = - \frac{1}{2}$.

Substituting in (2) we have

$$x' = 10(1 + 0.866) + \tfrac{5}{16} = 18.66 + 0.3125 = 18.9725 \text{ inches},$$

when measured from the head end of the stroke; and

$$x'' = 10(1 - 0.866) - \tfrac{5}{16} = 1.34 - 0.3125 = 1.0275 \text{ inches},$$

when measured from the crank end of the stroke.

Equation (2) is of use in investigating the effect of the obliquity of the connecting-rod on the motion of the piston. For practical purposes the relative positions of the crank and piston may be obtained graphically.

It was shown on page 123 that $r(1 - \cos\theta)$ would be the value of x if the movement of the piston were the result of a simple harmonic motion, but as that could be obtained only with a connecting-rod of infinite length the term $\frac{r^2 \sin^2\theta}{2c}$ in equation (2) is the measure of the error in piston displacement due to the angularity of the connecting-rod of finite length.

93. Forces in Crank-connecting-rod Motion. — Let *CF*, Fig. 60, denote the crank position on the forward stroke when

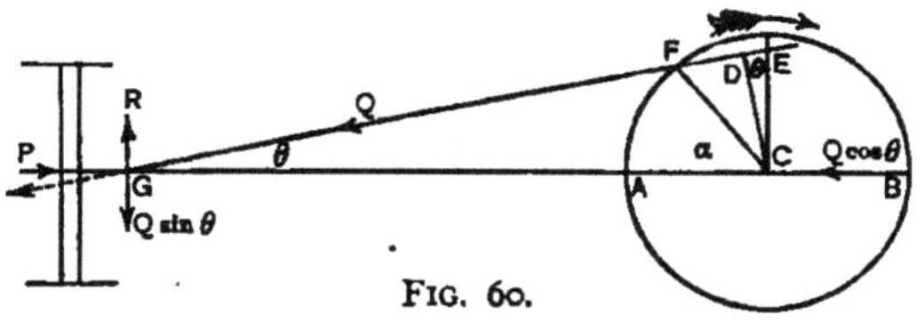

FIG. 60.

the connecting-rod *GF* makes an angle θ with the axial line of the cylinder, the crank making an angle α with the same line.

The total pressure *P* on the piston is opposed by the resistance of the crank to being pushed about the center *C* of the shaft, producing a compressive force *Q* in the connecting-rod. The resultant of the forces *P* and *Q* at the point *G* is the pressure, $Q \sin\theta$, on the guide, which is balanced by the reaction *R*. At

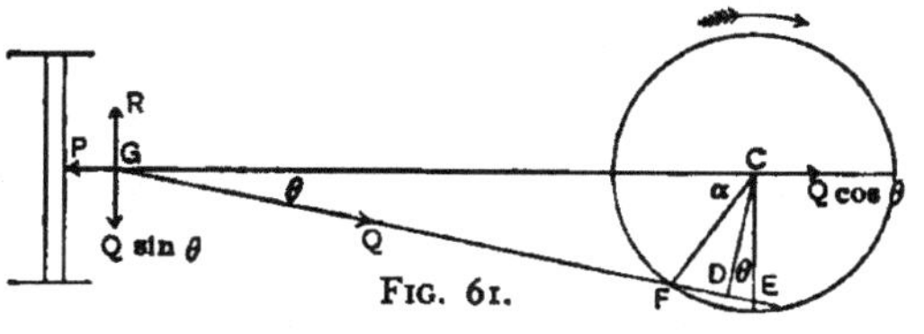

FIG. 61.

the corresponding crank position on the return stroke, Fig. 61, the force in the connecting-rod has changed from compression to tension, caused by the resistance of the crank to being pulled about the center C. The pressure on the guide is always $Q \sin \theta$, and for the three forces P, Q, and R acting at the joint G, we have for each stroke the static equations

$$P = Q \cos \theta \quad \text{and} \quad R = Q \sin \theta,$$

whence

$$Q^2 = P^2 + R^2.$$

The direction of rotation in Figs. 60 and 61 is that for the ahead motion of an engine, but should the direction of rotation be reversed, as is frequently the case with marine engines and

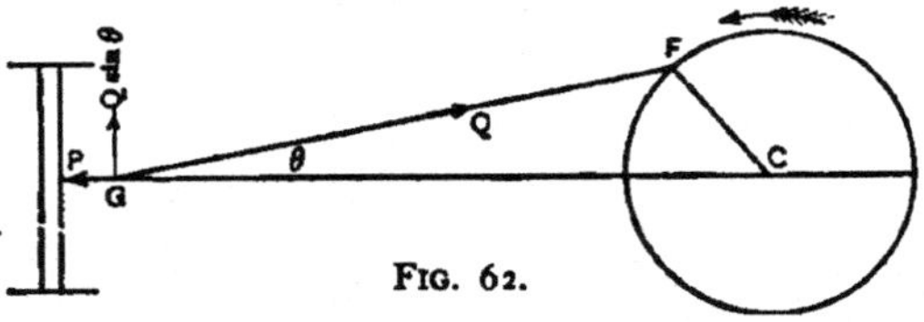

FIG. 62.

locomotives, the resultant of the forces P and Q, Fig. 62, will not press down on the guide, but will act in the opposite direction, necessitating an upper guide.

94. Twisting Moment. — In the operation of converting the reciprocating motion of the piston of a steam engine into one of continuous rotation of the shaft, through the medium of the connecting-rod and crank, the force that either pushes or pulls the crank about the axis of the shaft is applied to the crank-pin and results in an effort to twist the shaft while it is being held in a horizontal position in its bearings. The measure of this effort is the product of the force and the perpendicular distance of its line of action from the center of the shaft, and is known as the *twisting moment*.

If from the indicator diagram of an engine we ascertain the effective pressure p in pounds per square inch acting on the piston at any point of the stroke and multiply it by the area A of the piston in square inches, the product pA will be the total pressure P acting on the piston at the point. Produce the line of action of the force Q, Fig. 60, to its intersection E with the vertical through C. Draw CD perpendicular to GE. Then the twisting moment is $Q \times CD$. Resolving Q horizontally and vertically we have $Q \cos \theta = P$, whence $Q = \frac{P}{\cos \theta}$. In the right triangle DCE we have $CD = CE \cos \theta$. Hence

$$\text{Twisting moment} = Q \times CD = \frac{P}{\cos \theta} \times CE \cos \theta = P \times CE.$$

That is, for any angle of the crank, the twisting moment is equal to the total pressure on the piston multiplied by the length of that part of the perpendicular to the center line of the engine through C intercepted by the center line of the connecting-rod, measured in inches to the same scale that CF represents the length of the crank. Assuming the pressure on the piston to be constant, it is evident that the twisting moment will vary with every position of the crank as the length of CE varies, increasing from zero at the dead center A, Fig. 60, when the piston is at the end of its stroke and the crank and connecting-rod in line with the axis of the cylinder, to the maximum on the forward stroke when the connecting-rod and crank are at right angles and before half stroke is completed, then gradually decreasing until the dead point B is reached when it again becomes zero. On the return stroke the twisting moment increases from zero at B to the maximum beyond half stroke where the connecting-rod again becomes tangent to the crank circle, and then gradually decreases to zero at A.

The irregularity in the twisting moment, or turning moment, becomes more pronounced under the conditions of actual practice where the pressure on the piston is constantly changing, the moment becoming rapidly less after expansion begins. The effect of this irregularity in the twisting moment occasions intermittent stresses in the shaft that are unduly large in comparison with the power transmitted, which is proportional only to the mean twisting moment. The earlier the cut-off and the shorter the connecting-rod relative to the crank, the greater the irregularity in the twisting moment and in the smoothness in the running of the engine. The unevenness in twisting moment furnishes the reason for the use of flywheels with simple engines, the energy stored in the wheel when the twisting moment is greater than the mean being given back when the moment falls below the mean and thus carry the engine over the dead centers. In all stage-expansion engines, where the several cranks are set on the shaft at various angles with each other, the irregularities in twisting moment are greatly reduced and a greater evenness in the running of the engine secured.

95. Inertia of the Reciprocating Parts. — In what has preceded concerning the crank-connecting-rod forces the effect of overcoming the inertia of the reciprocating parts in modifying the effective pressure transmitted to the crank-pin has been neglected.

During each stroke of an engine its reciprocating parts, consisting of the piston, piston rod, crosshead, and one-half the connecting-rod (only one-half the weight of the connecting-rod is assumed to have a reciprocating, or back-and-forth, movement), are moved from a state of rest at the commencement, attain their maximum velocity near mid-stroke, and are then gradually brought again to a state of rest at the completion of the stroke. The force required to overcome the inertia of the reciprocating parts and accelerate them during the early part of the stroke is obtained at the expense of the steam pressure on the piston, and during the latter part the retarding force required to bring them gradually to rest occasions a virtual addition to the steam pressure. In order to obtain from the indicator diagram the effective pressure on the piston at any point of the stroke this alternately accelerating and retarding force must be expressed in pounds per square inch of piston area and then subtracted from or added to the diagram pressure at the point, as the case may be.

Suppose, for the purposes of this problem, the weight of the reciprocating parts to be concentrated at the crank-pin and revolving with it. It may be shown that when a moving body describes a circle of radius r, with a uniform velocity v, the body is acted on by a radial force toward the center, the acceleration of which is $\frac{v^2}{r}$. Then, since the measurement of force is the product of mass and acceleration, we shall have for the centripetal force of the problem

$$F = \frac{Mv^2}{r} = \frac{Wv^2}{gr},$$

in which W is the weight of the reciprocating parts in pounds, v the velocity of the crank-pin in feet per second, r the length of the crank in feet, and g the accelerating force of 32.2 feet per second due to gravity.

Let F, Fig. 63, denote the centripetal force for any position of the crank making an angle α with the center line of the cylinder. The horizontal component $F \cos \alpha$ produces the acceleration of the reciprocating parts in a horizontal direction, the vertical component $F \sin \alpha$ producing only pressure on the bearings.

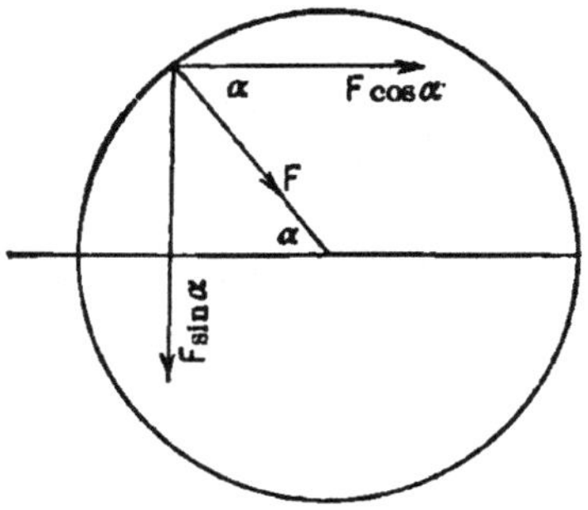

FIG. 63.

Neglecting for the moment the angularity of the connecting-rod, or assuming it to be infinitely long, at the commencement of the stroke, when $\alpha = 0$ and $\cos \alpha = 1$, the accelerating force $F \cos \alpha = \frac{Wv^2}{gr} \cos \alpha$ becomes $\frac{Wv^2}{gr}$, showing that the whole of the centripetal force is then acting horizontally and the acceleration then a maximum. At mid-stroke, where $\alpha = 90°$ and $\cos \alpha = 0$, the accelerating force becomes zero, showing that there is no acceleration and the velocity of the reciprocating parts therefore a maximum. Beyond the 90° crank position the cosine becomes negative, showing the retarding action of the force in gradually bringing the reciprocating parts to rest at the end of the stroke, where $\alpha = 180°$ and $F \cos \alpha = -\frac{Wv^2}{gr}$.

The action of the force throughout the stroke, the connecting-rod being assumed infinitely long, is illustrated in Fig. 64, OO' representing the stroke of the piston.

Expressing the accelerating force in pounds per square inch of piston area, we shall have at the commencement of the forward stroke

$$F = \frac{Wv^2}{Agr},$$

and at the end

$$F = -\frac{Wv^2}{Agr},$$

in which A is the area of the piston in square inches. From O, the beginning of the stroke, set off OC perpendicular to OO' and equal in length to $\frac{Wv^2}{Agr}$ expressed in pounds to the same scale as that of the indicator diagram. From O' set off $O'D$ perpendicular to OO' and equal in length to $-\frac{Wv^2}{Agr}$. The line joining C and D will of course pass through the middle point E of the stroke. The ordinates of the triangle OCE show the gradual decrease in the accelerating force from its maximum OC at the commencement to its zero value at mid-stroke E. The area of the triangle OCE represents the work done by the steam on the piston in accelerating the reciprocating parts from commencement to mid-stroke, and the ordinates of this triangle measure the pressures in pounds to be deducted from the corresponding ordinates of the indicator diagram to get the effective pressures on the piston that are transmitted to the crank-pin. The area of the triangle *OCE* also represents the kinetic energy stored in the reciprocating parts due to the work done by the piston in accelerating them from zero velocity at commencement to

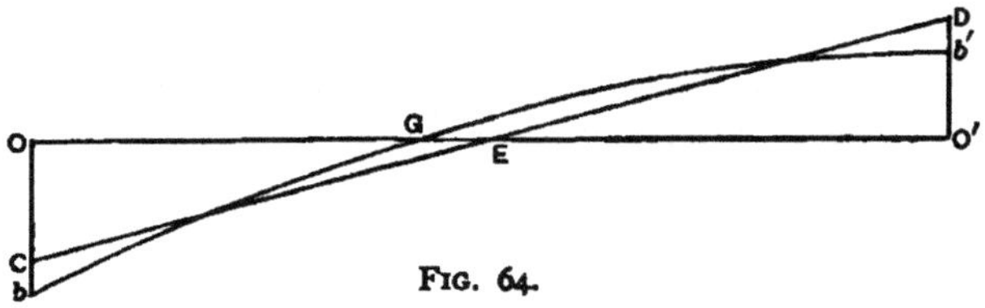

FIG. 64.

maximum velocity at mid-stroke. During the second half of the stroke the reciprocating parts give up their stored energy in the performance of work on the crank-pin in retarding their own velocity from the maximum at mid-stroke to zero at the end of the stroke, the triangle *O'DE* representing this work. The ordinates of the triangle *O'DE* represent the pressures that are to be added to the corresponding ordinates of the indicator diagram to get the effective pressures on the piston that are transmitted to the crank-pin during the second half of the stroke.

It is thus seen that the solution of the problem of the inertia of the reciprocating parts would be easy of solution were it possible to use connecting-rods of infinite length; but since rods of finite length alone are possible, the disturbing element of their obliquity enters the problem and renders its exact solution difficult.

96. The Effect of the Finite Rod. — With a connecting-rod of finite length the line of inertia is no longer the straight line *CD* of Fig. 64, because of the fact that the maximum velocity of the piston and the zero of acceleration do not occur at mid-stroke, as may be seen by the construction of a crank-pin-piston velocity diagram.

Let CF, Fig. 65, be the position of the crank when the connecting-rod GF makes an angle θ with the center line of the cylinder. Produce the center line of the connecting-rod to its intersection E with the vertical through the center of the crank-pin circle.

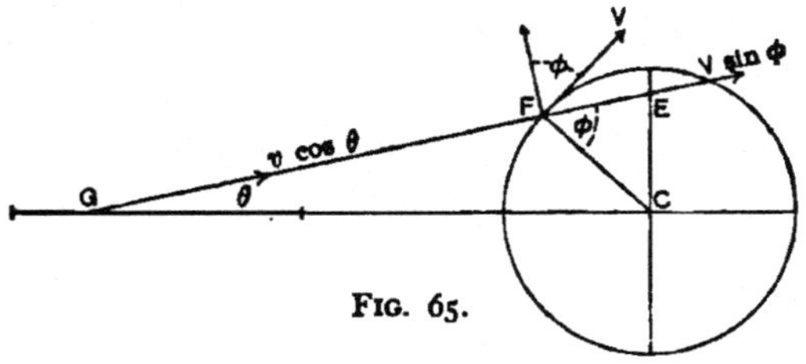

FIG. 65.

The direction of the velocity of the crank-pin F at any point of its path is the tangent to the crank-pin circle at the point. The crosshead G may equally well represent the piston, so if we denote the velocity of the crank-pin by V, and that of the piston by v, we shall have the resolved velocities of F and G, or of the crank-pin and piston, along the connecting-rod GF equal, since GF is a rigid rod of constant length. Hence

$$V \sin \phi = v \cos \theta.$$

That is,

$$\frac{\text{Velocity of piston}}{\text{Velocity of crank-pin}} = \frac{v}{V} = \frac{\sin \phi}{\cos \theta} = \frac{\sin \phi}{\sin GEC} = \frac{CE}{CF}.$$

Now if to any scale we let CF denote the assumed constant velocity of the crank-pin, then CE, to the same scale, will denote the velocity of the piston. From this the curve of piston velocities may be constructed, as follows:

Let CF and FG, Fig. 66, be the positions, respectively, of the crank and connecting-rod corresponding to the piston position G. Produce GF to its intersection E with the vertical through C. With C as a center and CE as a radius describe the arc Ee, intersecting CF at e; then e is a point of the curve. The locus of e is the polar curve of piston velocities, the radius vector in line of the crank representing the piston velocity for that crank position to the same scale that the length of the crank represents the velocity of the crank-pin If the connecting-rod were of infinite length, the locus of e would be two circles described on the vertical positions of the crank as diameters.

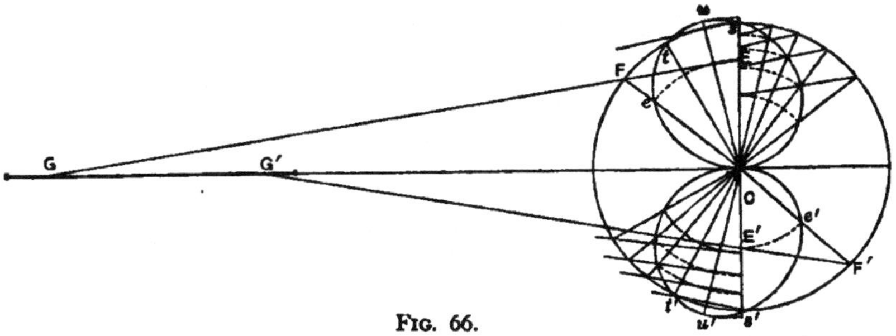

FIG. 66.

The velocities of the crank-pin and piston are seen from the diagram to be equal when the crank is in its vertical positions, and again at crank positions Ct and Ct' where the connecting-rod produced passes through s and s'. Between the points s and t and s' and t' the velocity of the piston is greater than that of the crank-pin, and at all other points it is less. Within the limits of the crank-connecting-rod ratios of actual practice the piston has its maximum velocity when the crank and connecting-rod are about at right angles, or when the connecting-rod is tangent to the crank circle.

The position of the piston when at its maximum velocity may readily be found by construction, thus locating the zero points of acceleration in the inertia curve, as follows:

On the center line of the cylinder take AB, Fig. 67, equal in length to the stroke to some convenient scale. Taking the connecting-rod as 4 cranks in length, we shall have AO as the length of the connecting-rod and $OO' = AB$ as the path of the crosshead, or of the piston. Take CD at right angles to AB as the 90° position of the crank on the forward stroke. From D draw DH perpendicular to CD and equal in length to the connecting-rod AO; DH will then be tangent to the crank circle at D. Assuming CD and DH to be rigidly connected at D, revolve the right angle CDH about C as a center until the point H falls into the path of the piston at G. The point D will then be found at

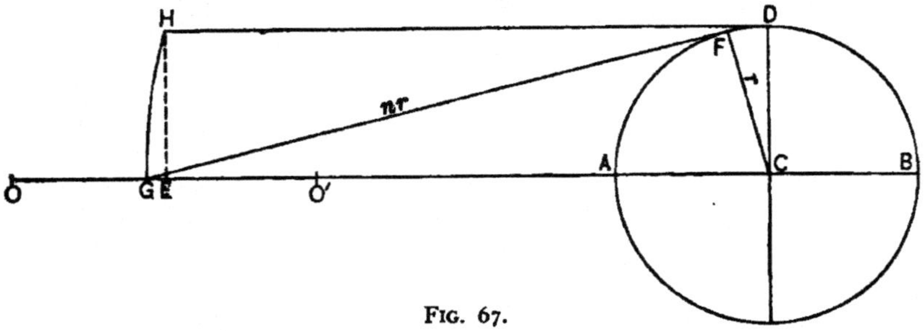

FIG. 67.

F and GF will be the position of the connecting-rod when at right angles to the crank; therefore GE will measure the distance from mid-stroke when the piston on its forward stroke has its maximum velocity, and G will be the zero point of acceleration. If we denote the ratio $\frac{\text{connecting-rod}}{\text{crank}}$ by n and the length of the crank by r, then $HD = GF = nr$, and we shall have

$$GE = GC - EC = \sqrt{n^2r^2 + r^2} - nr,$$

which locates the point G, the point E being at mid-stroke.

With a connecting-rod infinitely long the magnitude of the forces accelerating the reciprocating parts at the ends of the stroke have been found to be $\pm\frac{Wv^2}{Agr}$ in pounds per square inch of piston area. With a connecting-rod of finite length the accelerating forces are affected by the centrifugal action of the connecting-rod in its tendency to revolve about its crosshead end as a center, in addition to its movement of translation. The acceleration of this centrifugal force is $\frac{v^2}{nr}$, and its magnitude is $\frac{Wv^2}{Agnr}$ in pounds per square inch of piston area. At the near dead center this force acts towards the center of the shaft and with the centripetal force of the reciprocating weights, and at the far dead center it acts away from the shaft center and against the centripetal force. The effective accelerating force at the near dead center with the finite rod is then

$$F = \frac{Wv^2}{Agr} + \frac{Wv^2}{Agnr} = \frac{Wv^2}{Agr}\left(1 + \frac{1}{n}\right),$$

and for the far dead center we shall have

$$F = -\frac{Wv^2}{Agr}\left(1 - \frac{1}{n}\right),$$

in pounds per square inch of piston area. If N denotes the revolutions of the engine per minute, a convenient expression for F is

$$F = \frac{Wv^2}{Agr}\left(1 \pm \frac{1}{n}\right) = \frac{W\,(2\pi rN)^2}{Ar \times 32.2 \times 60 \times 60}\left(1 \pm \frac{1}{n}\right)$$

$$= \frac{0.00034\,WrN^2}{A}\left(1 \pm \frac{1}{n}\right).$$

Having now the magnitude of the force accelerating the reciprocating parts at the dead centers and the point of zero acceleration, the inertia curve may be drawn with sufficient accuracy for practical purposes.

Thus, in Fig. 64, we have $GE = \sqrt{n^2r^2 + r^2} - nr$, which fixes the point G. Then, to the same scale of pounds as that of the indicator diagram, make Ob equal in length to $\frac{Wv^2}{Agr}\left(1 + \frac{1}{n}\right)$, and make $O'b'$ equal in length to $\frac{Wv^2}{Agr}\left(1 - \frac{1}{n}\right)$, thus establishing the points b and b'. Through the three points thus found draw the flat curve bGb' as the curve of inertia. The ordinates of the area ObG are to be subtracted from, and those of the area $O'b'G$ are to be added to, the corresponding ordinates of the indicator diagram to get the effective pressures transmitted to the crank-pin for the forward stroke. For the return stroke the ordinates of the area *O'b'G* are to be subtracted from, and those of area *ObG* be added to, the corresponding ordinates of the indicator diagram.

The indicator diagrams of Fig. 68 are from an engine whose cylinder diameter is 9 inches; stroke, 10 inches; revolutions per minute, 350; weight of reciprocating parts, 110 pounds; ratio of crank to connecting-rod, 1 : 3; and scale of indicator spring, $\frac{1}{60}$.

The effective pressures on the piston shown by the indicator, that is, the difference in the pressures on *opposite* sides of the piston, must be determined before correcting the diagrams for the inertia of the reciprocating parts. This can be done only by getting diagrams from the two ends of the cylinder as in Fig. 68. During the forward stroke, *abc* is the line indicating the pressure forcing the piston ahead against the simultaneous back pressure on the opposite side of the piston indicated by the line *def*, so that the ordinate included between these two lines at any point of the forward stroke is the effective pressure on the piston at the point; in like manner the ordinate included between *ghi* and *jkl* is the effective pressure during the return stroke.

Take the length of the diagram to represent the stroke of 10 inches and divide it into ten equal parts, erecting ordinates midway between the divisions. The dotted lines of Fig. 69 were determined by taking the lengths of the ordinates included between the forward-pressure line of each stroke and the back-pressure line of the opposite stroke. It will be noticed that for a short distance toward the end of each stroke the pressure is negative. The dotted lines of effective indicated pressures must now be corrected for inertia of the reciprocating parts in order to get the real pressures that are transmitted through the connecting-rod to the crank-pin.

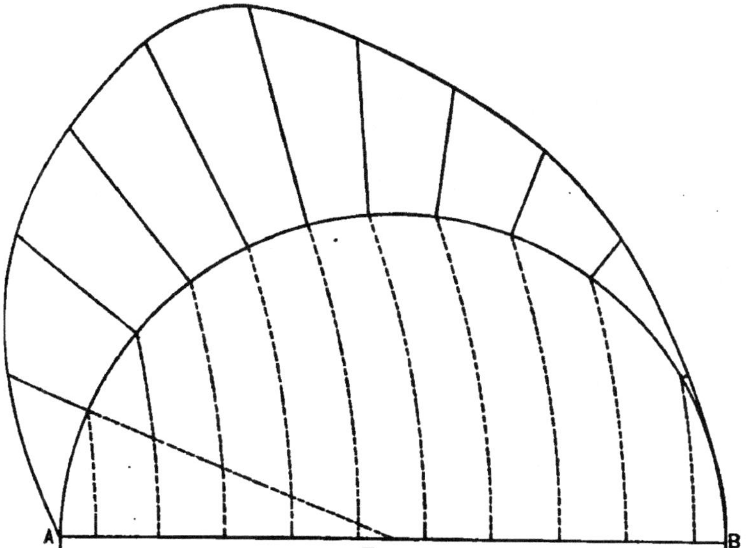

FIG. 70.

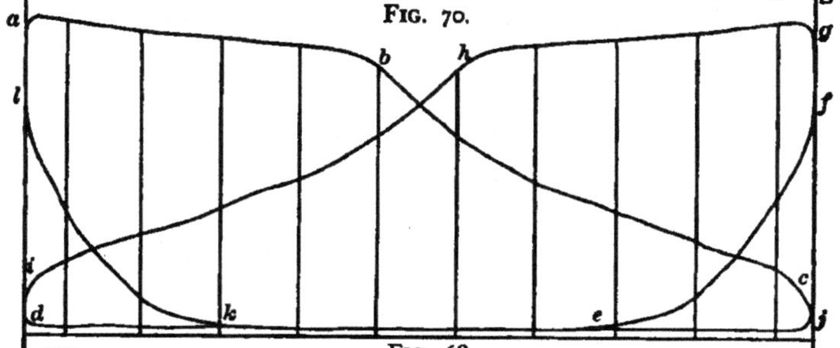

FIG. 68.

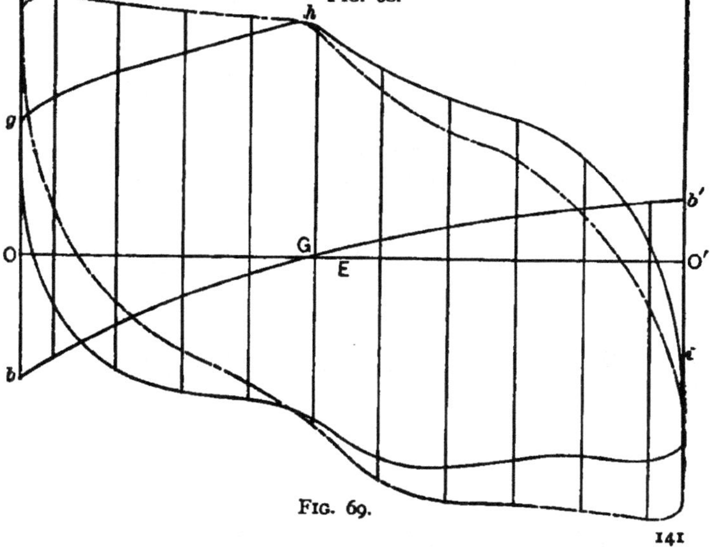

FIG. 69.

We first determine the point of zero acceleration, or the point of the stroke where the velocity of the piston is a maximum, by substitution in

$$GE = \sqrt{n^2r^2 + r^2} - nr = \sqrt{(3)^2 \times (5)^2 + (5)^2} - 3 \times 5 = 0.81 \text{ inch.}$$

That is, the zero acceleration of the reciprocating parts occurs when the piston on its forward stroke is 0.81 inch from, and before, mid-stroke, which fixes the point G of the inertia curve.

For the accelerating force at the beginning and end of the stroke, we have

$$F = \frac{0.00034\,WrN^2}{A}\left(1 \pm \frac{1}{n}\right)$$

$$= \frac{0.00034 \times 110 \times 5 \times 350 \times 350}{81 \times 0.7854 \times 12}\left(1 \pm \frac{1}{3}\right) = 40 \text{ and } 20$$

pounds per square inch of piston area.

To the scale of the indicator spring, $\frac{1}{60}$ inch to the pound, set off Ob and $O'b'$ equal in length respectively to $\frac{40}{60}$ and $\frac{20}{60}$ inch. Then bGb' is the inertia curve. Deducting the lengths of the ordinates of the triangular figure ObG from, and adding those of the triangular figure $O'b'G$ to, the ordinates of the dotted line of effective pressures, we get the full-line curve $ghiO'$ of the true effective pressures transmitted to the crank-pin. Where the pressure falls below OO' it is negative, which occurs at the end of the stroke, showing insufficient pressure either to accelerate the reciprocating parts or to carry the engine over the dead center, the stored energy in the flywheel supplying the deficiency. The full-line curve of the true effective pressures for the return stroke is shown in the lower half of Fig. 69.

We have now sufficient data to construct a curve of twisting moments or of tangential efforts on the crank-pin, which is the driving force of the engine.

On a diameter AB, Fig. 70, of the same length as the indicator diagram, describe a semicircle to represent the half of the crank circle for the forward stroke of the piston. By projections we get on AB the piston positions corresponding to the ordinates of the indicator diagrams of Fig. 68, and then, with a radius equal to the length of the connecting-rod, obtain the respective crank positions. From the upper part of Fig. 69 the effective pressures transmitted to the crank-pin are obtained, and these pressures, multiplied by the area of the piston in square inches, give the total effective pressures, P, on the piston for each crank position. Then, $P \times CE$, Fig. 60, gives the twisting moment. The twisting moments have been calculated for the engine of the example and plotted outside the crank circle on the prolongations of the respective radial crank positions, Fig. 70, to the scale of 1 inch = 10 tons-inches. This twisting-moment diagram illustrates the lack of uniformity in the turning force of the engine during the stroke, there being absolutely no force to turn the engine either at the beginning or at the end of the stroke. In single-cylinder engines the stored energy in the flywheel is relied upon to carry the engine over the dead centers, and such engines are very largely used in those industries where they are required to run in one direction only.

In locomotive and marine practice where there is much starting, stopping, and reversing, the two-cylinder type of engine with cranks at right angles on the shaft is used, for the reason that if the engine stops with one crank on the center, the other crank is at its best position for starting, and, with the engine running, when one crank passes a dead center with no turning moment the other crank is operating at about its maximum turning moment. This arrangement obviates the necessity of a flywheel. The employment in marine practice of the stage-expansion engine with two or more cylinders greatly facilitates starting, stopping, and reversing, and produces a greater uniformity in twisting moment, which would be shown by plotting a polar curve similar to that of Fig. 70. In constructing such a curve the indicator diagram pressures must be reduced to a common scale, or expressed in equivalent pressures on one or the other of the pistons, and the curve for each cylinder must commence on the crank circle at the angular position corresponding to the angular position of its crank on the shaft. The sum of the vectors of these curves at any radial crank position will be the total turning effort at that position.

In engines of the vertical type account must be taken of the dead weight of the reciprocating parts which acts with the steam pressure on the down stroke and against it on the upstroke. This will cause a modification of the inertia curve, which may be made conveniently by first constructing the curve as if for a horizontal engine and then shifting its base line for each stroke to compensate for the influence of the dead weight of the recip-

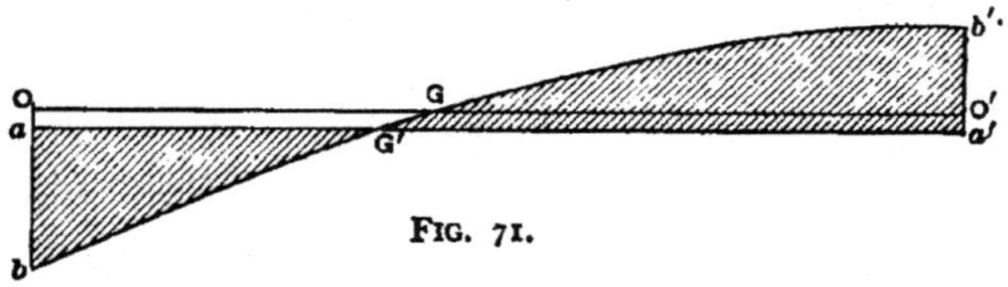

FIG. 71.

rocating parts. Suppose *bGb'*, Fig. 71, to be the inertia curve found for the engine under the supposition that it is of the horizontal type. Let *w* denote the weight of the reciprocating parts expressed in pounds per square inch of piston area. At the beginning of the down stroke the action of *w* is with the steam pressure so that the accelerating force required is diminished by the extent of *w* pounds, and at the end of the stroke the energy expended in retarding the reciprocating parts is increased by the amount of *w* pounds. To the same scale that *Ob* represents the accelerating force of the horizontal engine at the beginning of the forward (down) stroke set off *Oa* equal to *w* pounds. Draw *aa'* parallel to *OO'*. Then the ordinates of the triangular figure *abG'* represents the pressures that are to be deducted from the net pressures of the indicator diagram, and the ordinates of the triangular figure $a'b'G'$ represent the pressures to be added to the net indicator diagram pressures. For the upstroke the conditions are exactly reversed.

An inspection of the inertia curve of Fig. 64 shows the important effect the action of the reciprocating parts has on the running of the engine. In single-cylinder expansion engines of high rotational speed the action of the reciprocating parts is conducive to smooth running, from the fact that the force required to accelerate the parts relieves the crank-pin of the otherwise high pressure to which it would be subjected at the beginning of the stroke; and at the end of the stroke, where the steam pressure has become low from expansion, the energy restored by the reciprocating parts supplies pressure on the crank-pin, the net effect of the action being to promote the equalization of pressure throughout the stroke. If the speed of rotation becomes excessive the required accelerating force at the beginning of the stroke may be too great for the steam pressure to supply, in which case the crank will actually pull the piston until the steam pressure begins to act, at which instant the force in the connecting-rod changes from tension to compression with a resulting injurious knock in the brasses. In instances of this kind one of three remedies, or a combination of the three, must be applied, viz., decrease the revolutions of the engine, increase the initial pressure, or decrease the weight of the reciprocating parts.

PROBLEMS

1. Length of connecting-rod, 66 inches; stroke, 33 inches. Find the position of the piston, measured from both ends of the stroke, when the crank has passed through an angle of 135° on the forward stroke.

Ans. $x' = 29.1984''$. $x'' = 3.8016''$.

2. Diameter of cylinder, 12 inches; length of connecting-rod, 28 inches; stroke, 14 inches. Find the pressure on the guides and the compressive

force in the connecting-rod when the crank is at the 110° position of the forward stroke, the effective pressure on the piston at that point being 26 pounds per square inch. *Ans.* 940 pounds; 4115 pounds.

3. Find the twisting moment in the shaft of the engine of problem 2.
Ans. 24,000 pounds-inches.

4. Find by calculation the effect of inertia at the ends of the stroke, expressed in pounds per square inch of piston area, of the engine of problem 2 when making 250 revolutions per minute, the weight of the reciprocating parts being 260 pounds. Find also the distance from the commencement of the stroke where the piston is at its maximum velocity.
Ans. 26.17 pounds, 15.7 pounds, 6.14 inches.

CHAPTER VII

THE RECIPROCATING STEAM ENGINE. TYPES AND DETAILS

97. Action of the Reciprocating Steam Engine. — A perspective view of a reciprocating steam engine is shown in Fig. 72. The cover of the valve or steam chest is removed in part, and the valve chest, valve, and cylinder are shown in section. The valve, the section of which is shown in solid black, is of the ordinary D-slide type. The piston, which is at about one-third its stroke in the cylinder, is moving from left to right under the action of the pressure of the steam entering the cylinder from the steam chest through the port S, the motion of the eccentric on the shaft having moved the valve so as to uncover the port S. The steam in the right end of the cylinder, which was used on the preceding stroke of the piston from right to left, is escaping through the port S' into the exhaust port E, and thence to the atmosphere in the case of a non-condensing engine or to the condenser in the case of the condensing type of engine. The motion of the piston through the piston rod is communicated to the crosshead, and thence through the connecting-rod to the crank-pin, the reciprocating, or back-and-forth, motion of the piston thus causing the crank to turn about the axis of the shaft and give to the shaft a motion of continuous rotation, as explained in Chapter VI. In this simple type of engine the crank is usually mounted in a disk having a counterbalancing weight diametrically opposite the crank to give evenness to the rotation. The illustration shows how the eccentric, revolving with the shaft, gives a sliding motion back and forth to the valve

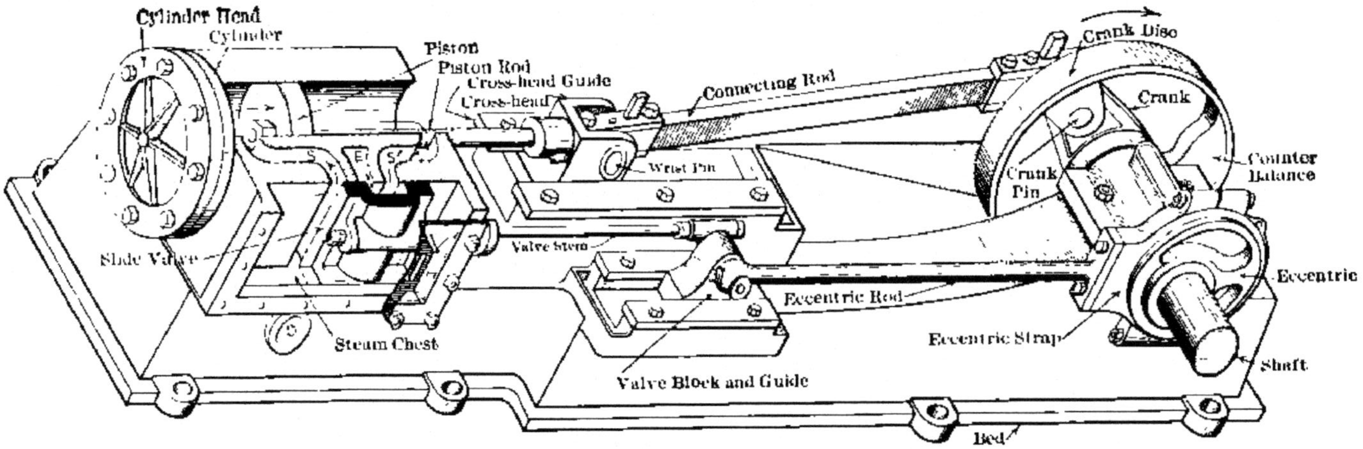

Cylinder Head
Cylinder
Piston
Piston Rod
Cross-head Guide
Cross-head
Connecting Rod
Crank Disc
Crank
Counter Balance
Crank Pin
Wrist Pin
Valve Stem
Slide Valve
Steam Chest
Valve Block and Guide
Eccentric Rod
Eccentric Strap
Eccentric
Shaft
Bed

FIG. 72

through the agency of the eccentric rod and valve stem, as explained in Chapter VI.

98. The Operation of the Link Motion. — The engine illustrated in Fig. 72 is designed to run in but one direction. Should it be desired to reverse its motion a second eccentric, as explained in Chapter V, would have to be employed and the reversal made through the medium of the link motion.

The action of the link motion as ordinarily applied to a locomotive, whose motion has frequently to be reversed, is shown in Fig. 73. The link is shown in its lowest position, that for going ahead, and the motion of the link block, and therefore that of the valve, is influenced by the *ahead* eccentric. The valve does not get its motion direct from the link block, as indicated in Fig. 72, page 148, but indirectly through the intervention of a *rocker shaft* and its arms. For this reason the eccentric, instead of being 90° plus the angular advance ahead of the crank, is set on the shaft at an angle of 90° minus the angle of advance behind the crank. It is seen in Fig. 73 that the valve is open to the extent of the lead and that the piston is about to commence its stroke to the left, the shaft revolving in the direction shown by the arrow. If, by means of the reversing lever and rod, the link were raised to its highest position so that the motion of the link block be influenced by the backing eccentric, the revolution of the shaft would be in the opposite direction and the motion of the engine therefore reversed. An inspection of Fig. 73 shows clearly that, with the link raised and the backing eccentric influencing the motion of the link block, the direction of rotation of the shaft must be contrary to that indicated by the arrow in order that the valve may move from right to left and admit steam to the cylinder. If the rotation were in the direction indicated by the arrow the influence of the backing eccentric would move the valve from left to right and no steam could enter the cylinder.

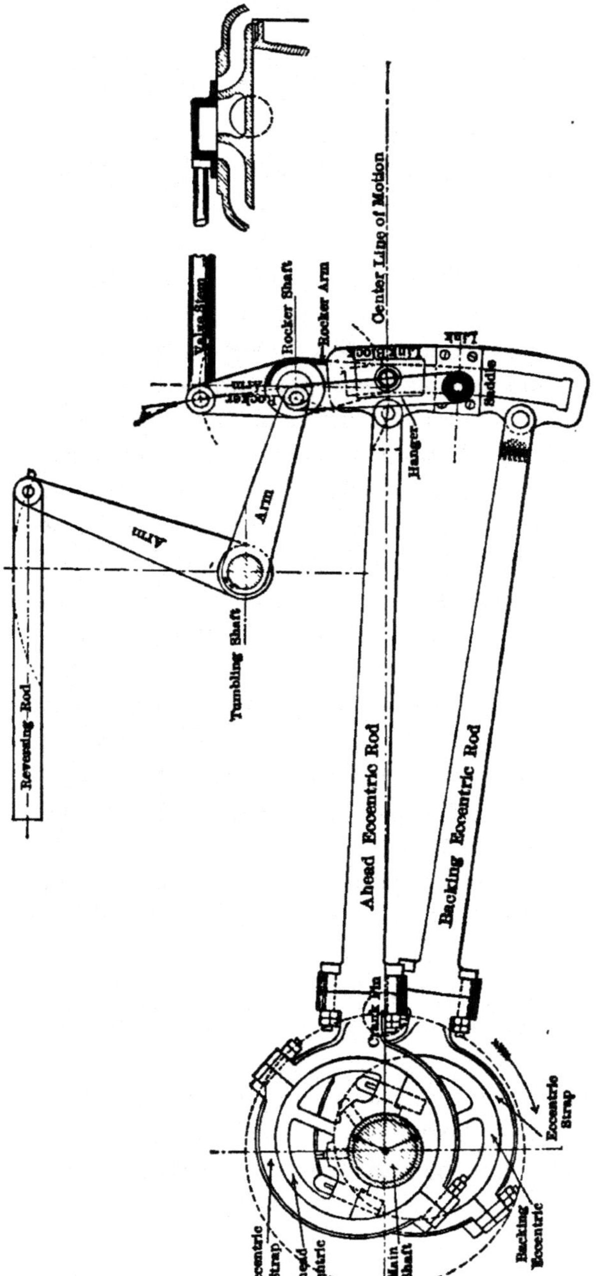
Center Line of Motion
Valve Stem
Rocker Shaft
Rocker Arm
Rocker Arm
Link
Saddle
Hanger
Arm
Arm
Tumbling Shaft
Reversing Rod
Ahead Eccentric Rod
Backing Eccentric Rod
Crank Pin
Eccentric Strap
Eccentric Strap
Ahead Eccentric
Main Shaft
Backing Eccentric

FIG. 73.

99. Governing the Steam Engine. — A very important feature of the operation of the steam engine is that of governing its speed of rotation. The flyball governor as originally introduced by Watt, and still extensively used in an improved form, is not capable of close regulation at high speeds and is being displaced by the more reliable type of shaft governor referred to in Art. 88, page 120.

The action of the flyball governor is shown in Fig. 74. The object of the governor is to maintain as nearly as possible a constant speed of rotation of the engine regardless of changes of load or in steam pressure. Two iron balls A, A have their arms pivoted at the upper end of a vertical spindle B. The arms of the balls are connected by arms C, C to a sleeve D which is free to slide on the spindle B. The whole is supported and guided by the column E. At the lower end of the spindle is a bevel gear which engages with a similar gear on the horizontal shaft F. The shaft F carries a pulley G which is belted to another pulley on the engine shaft, the ratio of the pulleys being such as to give to the balls a speed of revolution proportional to that of the shaft. Through this means the spindle is made to turn when the engine is in operation, the balls revolving about it as an axis. The bell crank K has one arm attached to the sleeve D, its other arm operating a balanced throttle valve situated in the steam pipe between the stop valve and valve chest. With the engine at rest the governor balls are at their lowest position

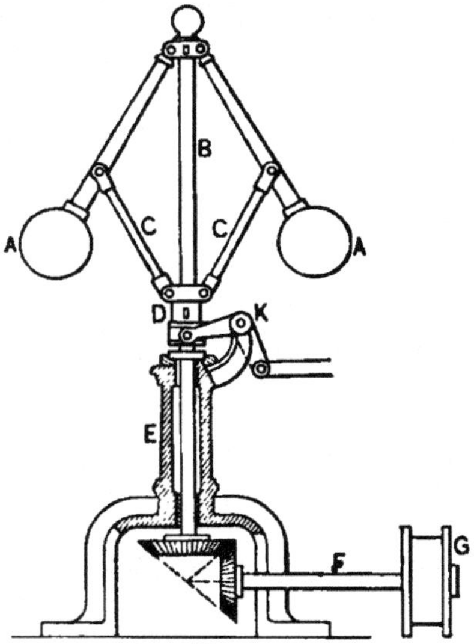

FIG. 74.

and the throttle valve wide open. By opening the stop valve steam is admitted to the cylinder and the engine starts, the balls of the governor flying out under the action of centrifugal force until the height of the cone of revolution to which the governor is adjusted is reached. This action of the balls has raised the sleeve *D* and, through the bell crank, has closed the throttle valve to a degree which admits to the cylinder only sufficient steam to operate the engine at the desired speed. Any decrease in the load on the engine or increase in the steam pressure will cause the balls to fly out a greater distance and raise the sleeve *D*, resulting in a further closing of the throttle, which stops the racing of the engine and restores as nearly as possible the normal speed. An increase in the load or a decrease in the pressure results in an action exactly the reverse. In certain types of the flyball governor there is a spring attachment by which the governor may be adjusted to act at any desired speed.

It will be observed that the action of the flyball governor as just described, controls the speed by means of throttling the supply of steam to the cylinder of the engine, the point of cut-off and ratio of expansion remaining constant. The effect of throttling is to wire-draw the steam and decrease the initial pressure, both processes being conducive to inefficiency. In some instances, notably in the case of the Corliss valve gear, the flyball governor performs its office of speed règulation by changing the point of cut-off and of the degree of expansion of the steam.

100. Stage-expansion Engines. — The reciprocating engine thus far described is known as a simple expansive engine, having but one cylinder, and making use of the expansive force of the steam by cutting off its admission to the cylinder before the piston completes its stroke. In Chapter XI it is shown that in order to utilize the expansive force of steam of the high pressures that have come into vogue it is necessary to compound the cylinders; that is, distribute the expansion between two, three or four cylinders, thus producing the double-expansion, the triple-expansion, and the quadruple-expansion engines. In these stage-expansion engines the steam at boiler pressure is admitted to the first cylinder and after a partial expansion, and consequent reduction in pressure, it is admitted into the next and succeeding cylinders for further expansion in the performance of work, until, in the last cylinder, the expansion is completed and the steam exhausted into a condenser or into the atmosphere. In order that the work shall be evenly divided between the cylinders, each cylinder in the expansion is made larger in section area than the one that immediately precedes it in the same proportion as the steam pressure is reduced. The first cylinder of the system is called the *high-pressure* cylinder and the last the *low-pressure* cylinder, the cylinders intervening being called *intermediate* cylinders.

101. **The Buckeye Shaft Governor.** — The action of the shaft governor in securing speed regulation by automatic cut-off of the steam supply to the cylinder has already been referred to.

The Buckeye shaft governor, Fig. 75, gives a remarkably close speed regulation to the well-known engine of that name. The mechanism of the governor is mounted within a pulley keyed to the crank shaft of the engine, so that its rotation is the same as that of the shaft. The action of the governor depends upon the change in the centrifugal force of revolving weights, by which the eccentric is made to operate an independent valve so as to vary the point of cut-off and thus regulate the speed of the engine by changing the ratio of expansion, but maintaining the initial steam pressure.

The levers *a*, *a*, one on each side of the shaft, carry the weights *A*, *A*, whose positions on the levers are adjustable. The outer ends of the levers are pivoted at *b*, *b* and their inner ends are linked to the eccentric *C* by the links *B*, *B*, the eccentric riding loose on the shaft. With the weights resting on their stops *f*, *f*, the eccentric is in the position for the longest cut-off, about five-eighths stroke; and when the weights are at the outer limit of their movement the earliest cut-off takes place, at about one inch from commencement of stroke, the eccentric moving on the shaft through an angle of about 90° for the two extreme positions of the weights. The main springs *F*, *F* are anchored

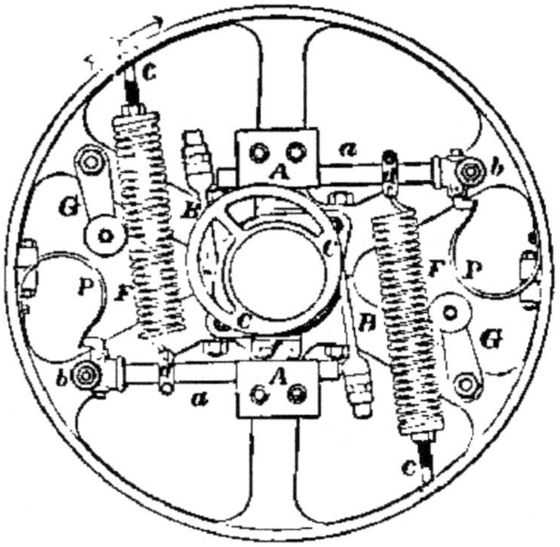

Fig. 75.

to the rim of the wheel, their inner ends being secured to the levers *a*, *a* by means of the adjustable clips *d*, *d*. The tensions of the springs may be varied by means of the tension screws *c*, *c*. The auxiliary springs *P*, *P* are introduced to aid the centrifugal force during the early part of the outward movement of the weights, or until the tension of the main springs is sufficient to give close regulation for light and varying loads.

The position of the weights and the tensions of the springs being adjusted for a given load and speed, the regulating action of the governor is as follows:

Opening the steam stop valve, the engine starts and centrifugal force quickly moves the weights outward to their adjusted position for the load, the combined action of the links *B*, *B* fixing the angular advance of the eccentric to effect the cut-off at the proper point to maintain the desired speed. Any increase of the load or decrease of the pressure would momentarily decrease the speed, but it would also decrease the centrifugal force of the weights, which would cause them to fall toward the center, thus decreasing the angular advance of the eccentric so as to effect a later cut-off and restore the speed without affecting the admission or compression, both of which events are controlled by a separate eccentric operating the main valve. For a decrease in load or an increase in pressure the action would be reversed and an earlier cut-off effected.

102. The Fleming-Harrisburg Governor. — The shaft governor of the well-known Fleming-Harrisburg engine depends for its action on a combination of centrifugal and inertia forces. It is shown in Fig. 76. Its mechanism is mounted within a pulley keyed on the crank shaft, but its slotted eccentric, instead of being keyed to the shaft, is suspended from a point *a* in the pulley, so that, as it swings on its point of suspension as a pivot, its center moves across the shaft, thus altering the throw of the eccentric and varying the travel of the valve. Two arms *A*, *A* of equal weight are pivoted at *b*, *b*, their pivoted points and centers of gravity *g*, *g* being diametrically opposite. Each of these arms is connected by a link *d* with the eccentric, the links pivoting on studs on the arms and on the eccentric. The arms have weight pockets at their ends, the pocket nearest the center of gravity being the larger. In adjusting the governor to a desired speed of the engine, small weights are placed in these pockets as may be found necessary.

The inner ends of the springs are attached to the arms as shown, the outer ends being secured to blocks in curved slots near the rim of the wheel. This admits of changing the position of the outer ends of the springs in adjusting the sensitiveness of the governor. In Fig. 76 the arms are shown resting on the hub stops *c*, *c*, the engine being at rest, and the eccentric at its maximum throw.

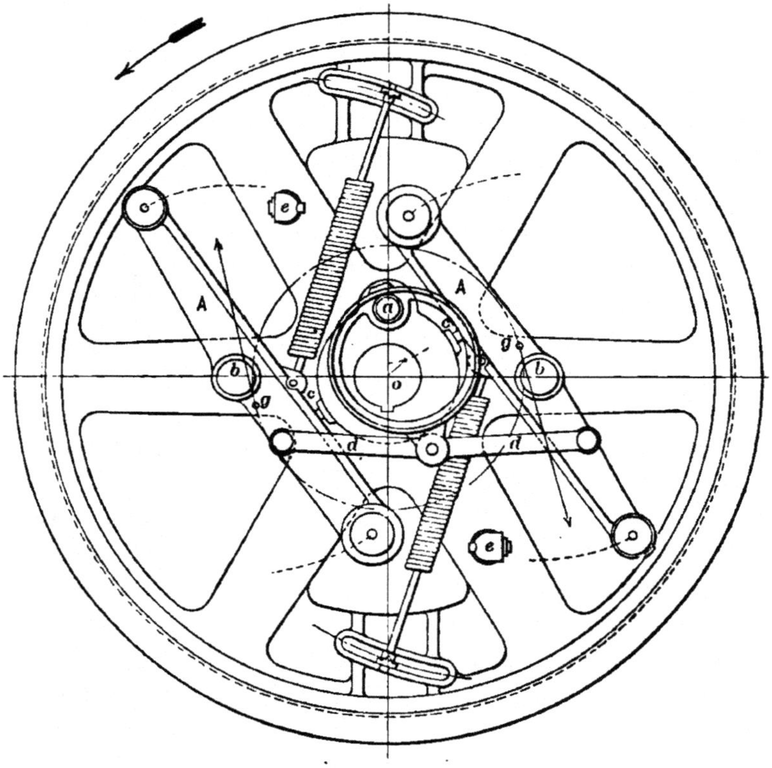

FIG. 76.

The governor being adjusted for a given load and speed, its action in maintaining that speed within narrow limits under the conditions of varying load or pressure is as follows:

As the engine starts, the arms, under the action of centrifugal force and the restraining action of the springs, move about their points of suspension, their extremities moving in the dotted arcs, until they reach their previously adjusted positions for the load and speed. This movement of the arms, through the links *d*, *d* moves the center of the eccentric in its arc of vibration about *a* as a center, decreasing the throw of the eccentric, and therefore the travel of the valve, so as to make the valve cut off the steam at the proper point to maintain the speed. Any increase in the load or decrease in the pressure would immediately and momentarily decrease the speed, but at the same time would decrease the centrifugal force of the arms, causing them to fall back toward the hub stops and thus increase the throw of the eccentric so as to effect a later cut-off and restore the speed. A decrease in the load or an increase in the pressure would cause an opposite action and effect an earlier cut-off.

While centrifugal force is largely the steadying influence of the governor and the force which holds the arms in any position they assume due to the load on the engine, it cannot be depended upon for quick action under changes of load, for in order to admit more steam to meet an increase in load the speed must first be reduced to diminish the centrifugal force sufficiently to move the arm toward the center of the shaft. The term "angular acceleration" is applied to that tendency of the double-ended lever arms to rotate about their pivotal points when changes in velocity occur in the wheel. If these lever arms were not restrained by the springs and other governor mechanism, and were simply pivoted loosely on their pins, they would assume a certain position in the wheel and remain in that position as long as the velocity remained unchanged.

Assume, for illustration, a sudden retardation of the rim velocity due to an increase in the load. The lever arms immediately tend to rotate about their pivotal points, due to their inertia and the energy stored in them, the rotation tendency being in the same direction as that of the wheel. This "angular acceleration" is taken advantage of to shift quickly the eccentric so as to increase its throw and admit more steam to the engine before the reduction in centrifugal force is felt. For a decrease in the load and an acceleration of rim velocity the reverse action will take place and the cut-off be shortened.

There is another force, that of tangential inertia, which has a steadying influence on the lever arms, preventing violent slamming and changes of position under sudden variations of load. The tangent to the circle described by the centers of gravity *g*, *g*, drawn in the direction opposite to that of rotation, is the direction of action of the force of tangential inertia, and as the construction of the governor is such that these tangents can never pass outside of the lever pivots, it follows that the force of tangential inertia is opposed to, and restrains, the violent action of centrifugal force and angular acceleration in shifting the arms to their outer positions if the load is suddenly removed.

Should one of the springs break the engine could not run away, because the arms in flying out would bring up against the stops *e*, *e*, bringing the center of the eccentric in the line *ao* joining its point of suspension and the center of the shaft, the position for the minimum travel of the valve, admitting no steam to the cylinder and causing the engine to stop.

The regulation by this governor is so close that a variation in speed of less than 2 per cent accompanies a shift from full load to no load.

103. The Defects of the Reciprocating Engine. — The operation of the reciprocating engine with the simple slide valve controlling the admission and the exhaust of the steam to and from the cylinder was early recognized as uneconomical and inefficient. The range of cut-off was limited, and the operation of cut-off could only be effected by a gradual closing of the valve with the consequent wire-drawing of the steam. The effort to hold the engine to a fixed speed under the varying conditions of load and pressure through the agency of the flyball governor was ineffective and inefficient, as the governor lacked sensitiveness, and its action in throttling the steam was another source of the evil of wire-drawing. The valve was necessarily placed at some distance from the ends of the cylinder, necessitating long ports and consequent large clearance spaces, the wasteful effect of which will be considered later.

An immense step in the improvement of the steam engine was made in 1849 when Geo. H. Corliss, an American inventor, introduced his system of semi-rotating valves for the admission and exhaust of steam to and from the cylinder, and to this day the type of engine bearing his name, within the limits of its possible rotational speed, is regarded the world over as the highest form of reciprocating engine construction.

104. The Corliss Engine. — The distinctive features of the Corliss engine are:

1. Four semi-rotary valves, two for steam and two for exhaust, and so placed at the ends of the cylinder as to reduce clearance to a minimum. With the cylinder horizontal, which is usually the case, the steam valves are placed at the top and the exhaust valves at the bottom of the cylinder, so that all water from condensation may be drained in the exhaust.

2. The employment of a *wrist-plate* which receives an oscillating motion from an eccentric on the main shaft. Pivoted to the wrist-plate are four arms, one connecting to each of the four valves, and so adjusted as to obtain rapid opening and closing of the valves, and to maintain a full opening of port during the period of admission and of exhaust.

3. A simple and effective means of releasing the steam valve from the driving mechanism, so that the steam shall be cut off at the desired point of the stroke and the valve seated quickly by the suction of a vacuum dash pot.

4. A governor, driven by a belt from the main shaft and entirely independent of the valve gear, its purpose being that of adjusting the power of the engine to the work to be done by determining the point at which the valve shall be released to effect the cut-off with the minimum variation in speed in any change of load.

The arms radiating from the wrist-plate to the valves are provided with right- and left-hand screws by means of which their lengths may be adjusted independently, so that the lead, the point of exhaust, and the point of compression may be varied at will, whatever the point of cut-off. The bell-crank connections of the valve gear insure a quick opening of the valve to steam, and the suction of the dash pot effects its instant closure at cut-off, so that wire-drawing is reduced to a minimum and the full expansive force of the steam realized in the cylinder.

It will be observed that a combination of the special features of the Corliss valve gear realizes all the features that were lacking in the slide valve.

The operation of the steam valve gear of the Corliss engine is shown in Fig. 77. The wrist-plate (not shown) receives an oscillating motion from the eccentric on the main shaft. The valve stem *A* leads to the valve through a sleeve projecting from the steam chest. To the end of the valve stem is keyed the arm *B*, at the end of which, on the farther side, is a projecting block *b*. To *B* is also attached the rod *d* leading to the dash pot. The bell crank CC' is mounted on the sleeve encasing the valve stem. To the end of the arm *C* is attached a rod leading to the wrist-plate, and at the end of the arm C' is a projecting stud on which is pivoted a claw having the arms *D* and *E*. The arm *F*, riding loosely on a boss of the arm *B*, is connected to the governor by the rod *h*, and on its boss *S* is a projecting block *e*. The arm *F* is free to move, and its position, and therefore that of the projecting block *e*, is determined by the governor according to the load on the engine. The arm *E* of the claw is made to press against the boss *S* of the arm *F* by the action of the spring *k*. In the position shown in the figure the catch *f* engages the block *b*, and as the arm *C* is moving from right to left the arm *B* is rising and turning the valve so as to open the port for the admission of steam into the cylinder. The upward movement continues to open the valve to steam until the arm *E* strikes the block *e*, which action forces the arm *D* outward, releasing the catch *f* from the block *b*, the arm *B* then falling under the suction action of the dash pot, thus closing the valve to steam and effecting a quick cut-off. On the return stroke of the engine the valve gear at the other end of the cylinder operates in a similar manner while the arm *C* moves from left to right, causing the catch *f* to engage again with the block *b* in preparation for the succeeding stroke. In case of a variation of the load on the engine the governor changes the position of *e* so that cut-off will be sooner in case the load is decreased or later in case of a heavier load and in this manner maintain a nearly constant speed of the engine. In case the driving belt of the governor should break the governor balls would fall to their lowest position, moving the arm *F* so far to the right that the arm *E* would not come in contact with *e* and there would be no cut-off of the steam and the engine would run away.

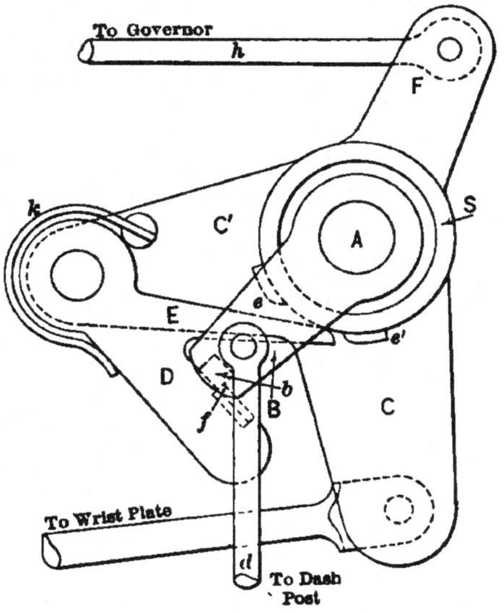

Fig. 77.

To prevent this the block e' is provided, which, in the emergency just mentioned, would be overridden by the arm E and thus prevent the catch f from engaging with the block b, causing the engine to stop.

In Fig. 78 is shown the wrist-plate and valve mechanism on the side of the cylinder, the steam valves being at the top

FIG. 78.

and the exhaust valves at the bottom. The illustration shows how the oscillation of the wrist-plate gives a back-and-forth rotational movement to the valves. The part section shows that the exhaust chamber is separated from the cylinder in order that the walls of the cylinder shall not be chilled by the exhaust steam. Both the steam and exhaust valves are double ported, as shown in the section views, which is for the purpose of giving the necessary port opening with a reduced movement of the valve. The dash pot used in connection with the steam valve at one end is shown. It has a double piston, made in one casting, the lower and smaller one working in a cylinder from which the air is excluded, while the upper piston works in a larger cylinder which has small openings to the atmosphere near its bottom. When the valve arm *B*, Fig. 77, is raised the double piston lifts, forming a vacuum in the lower cylinder and admitting air into the upper one. When the arm *B* is released its fall is hastened by the suction of the vacuum, giving to the valve a quick cut-off motion. The air holes in the upper cylinder are so proportioned as to restrict the escape of the air as the pistons fall so as to provide a cushion to bring them to rest.

It will be observed that the eccentric center, as with the slide valve, will be in the line of dead points, and the valve will have its maximum displacement when the crank has turned through an angle of 90° minus the angular advance, or before the piston has reached half stroke; so that the arm *B* will, if not released before the valve has its maximum displacement, remain under the control of the wrist-plate linkage and, in consequence, cut-off will not take place until near the end of the stroke, as determined by the angular advance. From this it is seen that the range of cut-off extends from the beginning to less than half stroke. A longer cut-off might be provided for by giving negative angular advance to the eccentric, that is, by setting the eccentric on the shaft less than 90° ahead of the crank. Such an arrangement with a single eccentric, however, would interfere with release and compression. When a cut-off later than half stroke is required, as is generally the case in the L.P. cylinder of a compound engine, there must be two wrist-pins, one for the steam valves and the other for the exhaust valves, and worked by separate eccentrics. The steam valve eccentric may then be given negative angular advance and provision made for a cut-off later than half stroke.

The disadvantages of the Corliss engine lie in the facts that its flyball governor does not give a close regulation of the speed under varying conditions of load and pressure, and that its trip cut-off gear with dash-pot attachment limits its rotational speed to little, if anything, in excess of 100 revolutions per minute, which necessitates a long piston stroke with its attending cylinder condensation losses.

105. The Four-valve Engine. — The four-valve type of engine, automatically controlled by a shaft governor, is now becoming extensively used. Its steam and exhaust valves are operated by separate eccentrics, and the cut-off is effected without the aid of trip gear or dash pot. It possesses all the advantages of the Corliss engine, except that the closing of the valve at cut-off is not quite so prompt; but it has the additional advantages of closer speed regulation, higher rotational speed, and valve gear of simpler construction.

Figure 79 is a general view of the Fleming-Harrisburg four-valve engine. The cut-off is controlled by the centrifugal-inertia type of shaft governor and slotted eccentric illustrated in Art. 102. The eccentric operating the steam valves is connected to a rocker arm on the side of the engine frame. From this rocker arm a connecting-rod transmits the motion of the eccentric to bell cranks mounted on the side of the cylinder, and thence, through links, the motion is conveyed to the valve arms. This combination of levers is designed to give to the rotating valve an accelerated motion at commencement of stroke and at cut-off. The eccentric operating the exhaust valves is fixed on the shaft, its rod also connecting with a rocker arm on the engine frame. From this rocker arm a connecting-rod transmits the motion of the eccentric direct to the exhaust valve arms.

This engine operates very efficiently at rotational speeds as high as 250 r.p.m., and may be termed a moderately high-speed engine. It is built in both the simple and compound forms.

106. The Buckeye Engine. — A vertical section through the cylinder and valve chest of the well-known Buckeye engine is shown in Fig. 80. It is an engine of moderately high rotational speed, governed automatically by the shaft governor described in Art. 101. The main valve *B* is of the cylindrical balanced type, the steam surrounding it and pressing equally in all directions, and is driven by a fixed eccentric. The cylindrical cut-off

Fig. 80.

valve *C* rides within the main valve and is operated by a loose eccentric whose angular advance is controlled by the action of the governor, as explained in Art. 101. The riding of the cut-off valve within the main valve is made possible by having the stem of the cut-off valve pass through the hollow stem of the main valve. The valve movement is so adjusted that cut-off occurs, whether early or late, when the motion of the cut-off valve is fastest. As shown in Fig. 80, the piston is about to start on its stroke from left to right, the steam valve being open to the extent of the lead, and the steam of the preceding stroke escaping to the exhaust at the right end.

A cross-compound type of the Buckeye engine is shown in Fig. 81.

107. The High-speed Engine. — One of the factors in estimating the power of an engine is the distance through which the piston moves in a given time. This factor is known as the *piston speed,* and is expressed in feet per minute. The piston speed is then the product of the number of strokes made by the engine in a minute and the length of the stroke in feet. Since the piston makes two strokes while the crank makes one revolution, we shall have $2LN$ as an expression for the piston speed, in which L is the length of the stroke in feet and N the number of revolutions per minute. From this it is seen that, for a given piston speed, as the length of the stroke is decreased the number of revolutions may be increased. The piston speeds for different types of engines are quite well determined from practical considerations, and are not likely to be much exceeded; so that, for a given power, the stroke of an engine may be decreased as its rotational speed is increased, with the resulting decrease in cost and in space occupied.

It has been seen that the releasing-gear valve-motion of the Corliss engine limited the rotational speed to about 100 r.p.m., necessitating a long stroke to that type of engine to obtain the required piston speed. Such engines are termed *low-speed* engines, the speed referring to the number of revolutions per minute and not to the piston speed. With the four-valve engine and other types that have discarded the releasing gear and substituted a positive motion from the eccentric to the cut-off gear, the rotational speed increased to about 250 r.p.m., which classed them as engines of moderately high speed.

The production of the rotational speed of more than 400 r.p.m. that now characterizes the *high-speed* engine resulted in advan-

FIG. 81.

tages other than a reduction in first cost and in space occupied. It was realized that engines of such rotational speed would be suitable for direct connection to electric generators, and it is in such service that the high-speed stationary engine has its field of widest application, though they are extensively used in almost every other service in units not exceeding 800 I.H.P. The quick running of such engines reduces the time during which the steam remains in the cylinder per stroke, and thus lessens the loss from cylinder condensation. Owing to its high rotational speed the high-speed engine is susceptible to close speed regulation, its cut-off being automatically controlled by a shaft governor.

The high-speed engine, as usually constructed, having a balanced valve, generally of the piston type, and fitted with self-oiling devices, is a marvel of mechanical skill, operating through protracted service with a degree of efficiency little, if at all, excelled by the low-speed type of engine. Its high rotational speed has made possible a compromise reduction in the size of the cylinder involving both stroke and diameter, and also a reduction in the diameter of the flywheel, thus decreasing the dead weight on the engine bearings and permitting a smaller diameter of shaft.

108. The Fleming-Harrisburg High-speed Engine. — A good example of the high-speed engine is that of the Fleming-Harrisburg single-valve, self-oiling engine, and its details will be given as typical of good practice.

It is built in units developing from 12 to 800 I.H.P., the revolutions per minute varying from 500 to 120, and the piston speed from 500 to 720 feet per minute. The valve is of the balanced piston type, its point of cut-off being automatically controlled by the centrifugal-inertia shaft governor described in Art. 102.

A horizontal sectional view through the cylinder and valve chest is shown in Fig. 82. The piston is shown in the position for the commencement of the forward stroke, the piston valve, taking steam at its inner edge, showing an opening for the admission of steam equal to the lead, and showing a full exhaust opening at the other end.

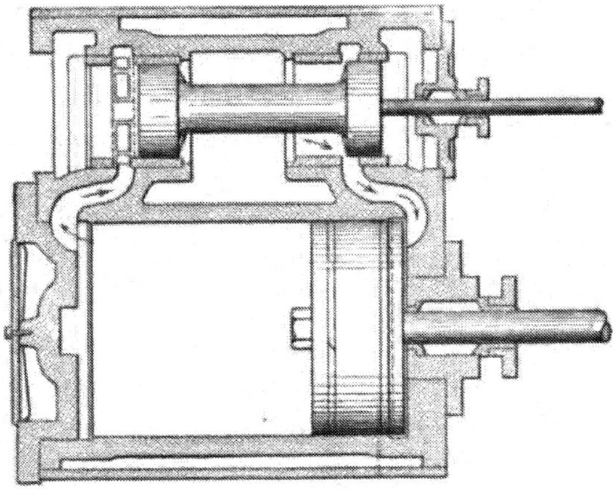

FIG. 82.

In Fig. 83 is shown the piston valve and the cage in which it works. The piston type of valve is the simplest form of the slide valve, is balanced under all operating conditions, reduces

FIG. 83.

friction to a minimum, and affords a greater port opening than any other type of single-ported valve. The valve is designed to work in bushings, or cages, so that repairs may be made promptly and inexpensively in case of wear. The valve is made of chilled cast iron, and instead of being a one-piece casting, the heads are separate and may be adjusted relatively to each other by means of bronze lock nuts. The valve and cage are ground true to gauge, and the steam ports in the cages are accurately milled.

The valve gear is shown in Fig. 84. The eccentric and strap are of selected, close-grained cast iron, are strong but light, and have ample bearing surface. The strap is lined with the special grade of babbitt metal employed in all the other bearings of the engine. The eccen-tric rod is a one-piece forging of open-hearth steel, while its box connection with the pin on the valve block, or ram, is made of phosphor bronze. The ram between the ends of the eccentric rod and valve stem is circular in form, and is accurately ground to gauge in its guide.

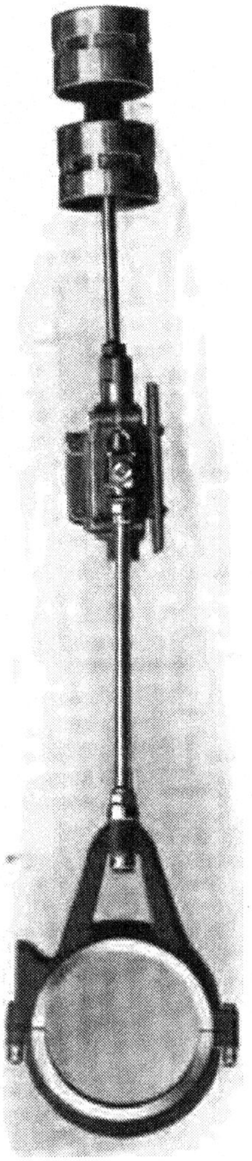

FIG. 84.

The piston, piston rod and piston packing rings are shown in Fig. 85. The piston is of cast iron, made as light as possible, consistent with strength. Instead of the ordinary piston packing ring with the diag-onally cut joint, the Fleming-Harrisburg ring is L-shape in section and has a square cut joint covered with a brass plate, so that no steam can pass it. These rings are

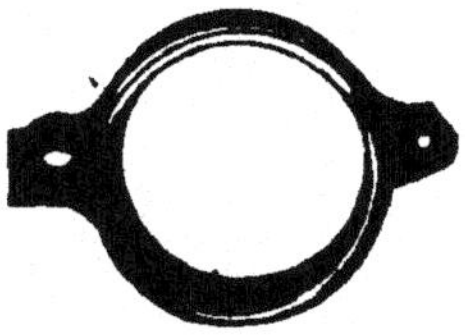

made of the finest quality of hard close-grained charcoal iron, and when sprung in position they fit accurately into grooves in the piston, the vertical stem of the L fitting snugly in its groove, while a space is provided under the horizontal stem to admit steam under it so that it is pressed out against the cylinder

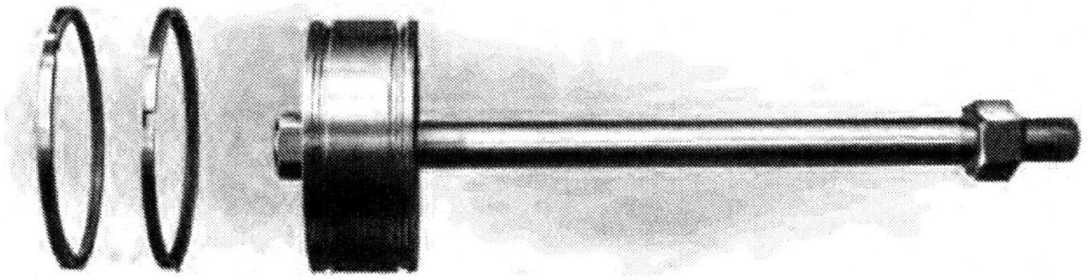

FIG. 85.

wall to make a lasting steam-tight joint. The piston rod is made of the finest rolled steel containing a percentage of nickel. The rod is turned to a taper at the piston end and is driven home to a shoulder and securely held by a cottered nut. The other end is screwed in the crosshead and prevented from turning by a jam nut.

FIG. 86.

The crosshead is shown in Fig. 86. It is a steel casting, the upper and lower surfaces of which are accurately planed to receive the liners and shoes. These shoes are of phosphor bronze and are firmly bolted to the crosshead. They are finished after attachment, thus insuring alignment. The wrist-pin is an open-hearth steel forging, turned, hardened, and ground accurately to gauge. The design of this pin is a taper of the same angle toward both ends, permitting its secure fastening by means of lock nuts. Access to the wrist-pin is provided by means of an opening in the engine frame, through which it may be removed without disturbing the crosshead.

The connecting-rod is shown in Fig. 87. It is a rod of the marine type, made of an open-hearth steel forging direct from the billet, the crank end being a steel casting with cap and bolts. The cap is locked on by means of nuts, and, as an addi-

FIG. 87.

tional safeguard, the ends of the bolts are drilled and provided with cotter pins. The casting and its cap have carefully scraped babbitt liners of high-grade anti-friction metal, in which the crank-pin works. The crosshead end of the connecting-rod is forged solid, drilled and slotted for the crosshead-pin brasses, which are made in halves. The brasses are lined with phosphor bronze and are made adjustable by means of a wedge and bolt.

The crank shaft and overhung crank is shown in Fig. 88. The shaft is a one-piece, hammered, open-hearth steel forging. It is accurately turned throughout its length and, for a direct-connected generating set, is closely finished to size to carry the armature or rotor of the generator for which it is designed. The crank-disk and crank-pin, shown in Fig. 89, are of cast steel and in one casting, making possible the use of a pin of larger diameter than could be employed where the pin is made separately and forced into the disk. By making the pin and disk in one piece, the pin serves as a substantial reinforce-

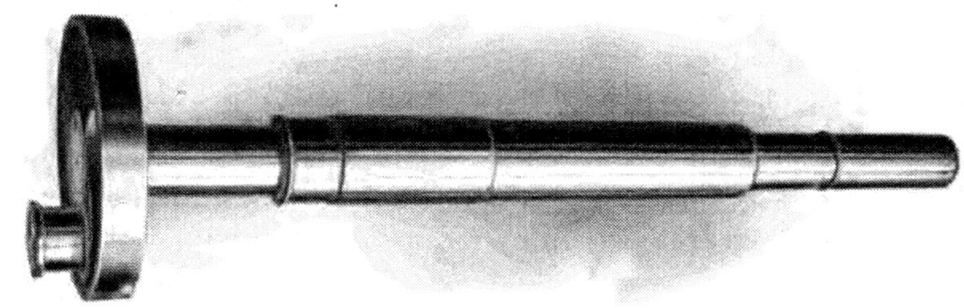

Fig. 88.

ment of the disk, instead of weakening it as is the case where the usual practice of boring out and pressing the pin into place is followed. The disk is closely fitted to the crank shaft, besides

Fig. 89.

being keyed to it. To obtain the best possible balance the crank-disk is nicely proportioned to counteract the inertia of the reciprocating parts.

The method of lubrication which characterizes the Fleming-Harrisburg high-speed engine as "self-oiling" is shown in Figs. 90 and 91. The main supply of oil is concentrated in the crank-

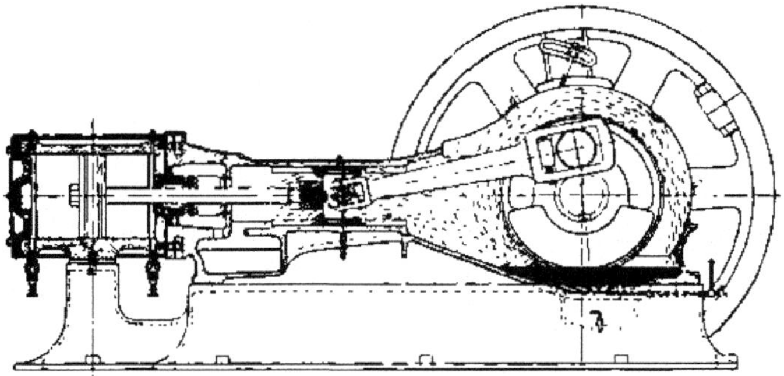

FIG. 90.

case oil well, or crank pit, whose hood is removable. An attached sight glass shows at a glance the quantity of oil in the pit. The rim of the circular crank-disk is constantly in contact with the oil, Fig. 90, and carries with it, as it revolves, a copious supply of oil which is thrown by centrifugal force back upon the crosshead and guides. The crank-disk also throws a continuous spray into a trough, placed across the inside face of the oil hood, which becomes filled as soon as the engine turns over. A short tube of ample bore conveys this oil to a point over the main bearing, Fig. 91, to which it flows in a constant stream that is always visible to the attendant. This stream of oil enters

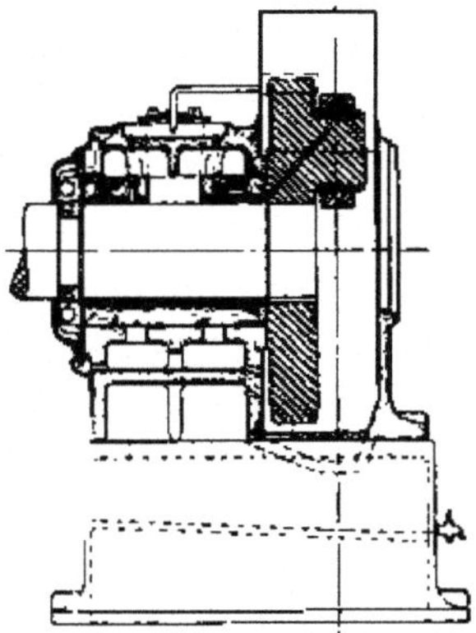

FIG. 91.

and flows through the main shaft bearing, from which it passes inwardly into an annular eccentric groove cut in the outside face of the crank-disk. From the most eccentric point of this groove angular holes are drilled through the disks to the middle of the crank-pin surface, through which a part of the oil discharged to the bearing is forced by centrifugal pumpage to the crank-pin bearing. This lubricating device is automatic in its action, and starts and stops with the engine.

All the details given of the Fleming-Harrisburg high-speed engine may be considered as examples of good engineering practice. Those of other engines differ, but the main objects sought are identical.

109. Lubrication. — In the operation of the steam engine there is power lost in overcoming the friction between sliding surfaces, and it is the object of lubrication to reduce this loss to a minimum. If bearing surfaces were permitted to rub each other without lubrication they would soon become damaged from heating and would render the engine inoperative. The lubricant most commonly used is mineral oil, and its introduction between the surfaces in moving contact, in the shape of a thin film, greatly reduces the friction and permits a safe and easy movement. Some form of sight-feed drip cup, filled and operated by hand, is extensively used in lubricating the bearing surfaces of engines, though a preferable method is that of having a system of pipes lead to the bearings from a central tank, the oil being fed either by gravity or by pressure. The lubrication of the moving parts of an engine that are in contact with the steam, such as the valve and piston, is effected mechanically, or hydrostatically through the medium of the steam. When mechanical means are employed, such as a hand force pump, oil is injected into the steam pipe close to the valve chest, or into the valve chest itself, and is there atomized and carried by the steam to the wearing surfaces. The steam sight-feed oiling devices are operated by the hydrostatic pressure of a head of water formed in the device by steam condensation, and by the displacing power of water over oil.

110. Sight-feed Lubricator. — Sight-feed lubricators for introducing oil into the valve chest and cylinder of an engine are of two kinds, viz., *single-connection* lubricators, which have but one connection with the steam pipe, the oil passing through the same passage that admits the steam, and *double-connection* lubricators which have two openings into the steam pipe, one for the admission of steam and the other for the passage of oil.

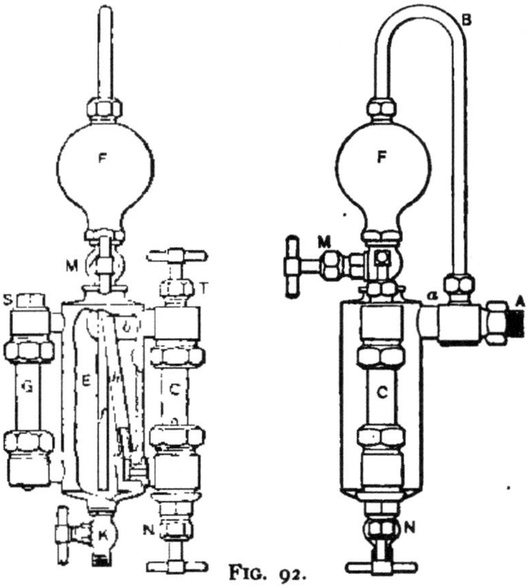

Fig. 92.

Figure 92 shows a sight-feed lubricator of the single-connection type. The connection with the steam pipe is made at A, the steam flowing through the pipe B into the condenser F. The steam also flows through the connection a and cored passage b into the sight-feed glass C. The steam is condensed by radiation in both the condenser and sight-feed glass, the water in the condenser flowing into the oil reservoir E through the pipe f. The reservoir being filled with oil, the water which enters through pipe f, by its displacing power, lifts the body of oil in E, causing it to escape through the pipe *h* into a nozzle at the bottom of the sight-feed glass; thence, because of its lesser specific gravity, it rises through the water in the sight-feed glass, and then through *a* and the cored passage *b* it enters the steam pipe and is carried by the steam into the valve chest and cylinder.

As the pressure on top of the water in the condenser and that on top of the water in the sight-feed glass is the same, there is no tendency for the oil to flow due to the steam pressure, the displacing force being that due to the difference in height between the column of water in the condenser and that in the sight-feed glass.

The oil reservoir is filled through the plug opening at *S*, and the rate of flow of the oil is governed by the needle valve *N*. The water from the condenser can be shut off by the valve *M*, and the water may be drained from the reservoir by means of the cock *K*. The gauge glass *G* shows the height of oil in the reservoir. In case of injury to the sight-feed glass, steam can be shut off by means of the valve *T*. The oil flows upward in drops in the sight-feed glass, giving the lubricator what is known as an *upfeed*.

In Fig. 93 is shown an arrangement for oiling the crank-pin, the eccentric and shaft bearing. In the case of the crank-pin, the oil drips from an ordinary sight-feed drip cup into the oil receiver, whose position is in the center line of the shaft. From the receiver the oil is carried outward by centrifugal force through the revolving pipe arm, and thence through the pipe connection *A* to the crank-pin, and then, through small passages, circulates between the surfaces of the crank-pin and brasses. The oil for the shaft bearing is fed by gravity from a sight-feed cup, from the standard of which projects an arm which carries a similar cup which feeds into a trough on the eccentric strap.

111. **Grease Cup.** — A heavy grease product of mineral oil is often used for lubrication. In some instances the grease flows

from the cup by gravity as it is melted by the heat of the bearing, but more frequently the grease is forced between the bearing

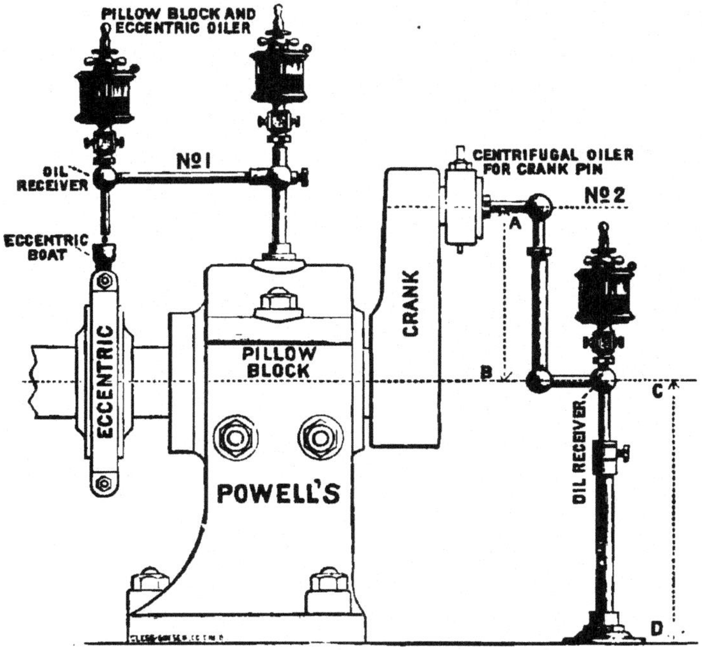

FIG. 93.

FIG. 94.

surfaces by the compression of a spring. Fig. 94 shows a compression grease cup in section. The filling of the cup and its action is as follows: By turning the jam nut *N* to the right the spring *O* is compressed and the plunger *K* drawn to the top, exposing the full length of the stem *H*. Then unscrew the cap *C* and fill the chamber *A* with grease, leaving space at the top to receive the plunger *K*. Replace cap *C* and apply pressure by releasing the jam nut *N*, running it to the top of the stem against the button J. The quantity of grease fed is controlled by the cut-off R in the shank B. To stop the feed at any time, run jam nut N down to the cap C, thus taking the tension off the spring. The jam nut is provided with a spring stop D which engages the flat on stem H and is of sufficient tension to hold it at any desired point.

112. Classification of Engines. — Engines are classified according to the service for which they are intended and may be considered as of three classes, viz., stationary, locomotive and marine.

The stationary engine delivers its power to the shaft, from which it is transmitted directly, or by belt or rope, to its point of application. It has countless applications in the industries and in the generation of electric light and power.

The locomotive and marine engines deliver their power directly, the former to the driving wheels and the latter to the propeller.

Of the two forms of engines, the *horizontal* and *vertical*, the former is very generally used in plants of small and moderate powers. It is cheaper in construction and much more accessible for inspection and repair than the vertical engine, though it occupies more floor space. The main objection to horizontal engines of large power is the difficulty of supporting the weight of the piston and preventing it from cutting the cylinder and wearing it out of true when traveling at the high piston speeds required in modern practice. It is because of this objection that the vertical engine is used in plants of large power. The shaft of a vertical engine rests in bearings which are a part of the bed plate that rests on the foundation, the cylinder being held in vertical framing, thus throwing the weight of the piston on the bearings. The vertical engine runs more smoothly and with a greater mechanical efficiency than the horizontal engine, due to the fact that the unbalanced forces inseparable from the operation of the steam engine are more easily counteracted.

Inaccessibility for repairs and the excessive head room required for the vertical engine are its chief disadvantages.

113. **Direction of Running.** — A horizontal engine is said to run *over* when the crank-pin describes the upper half of its circle of revolution while the piston makes its *forward* stroke, that is, the stroke towards the shaft. For engines that run in one direction only, that is, not subject to reversal, the condition that they run *over* is sufficient to insure that the resultant pressure on the crosshead will be *down*, and that, therefore, only a single slide will be necessary to guide the crosshead. Should the engine run *under*, or should its motion be reversed, the resultant pressure on the crosshead would be *up*, necessitating a top and bottom slide to guide the crosshead and prevent the deflection of the end of the piston rod from its horizontal motion. Double crosshead slides are required for vertical engines and for locomotive and marine engines, whose direction of running is frequently changed.

114. **The Choice of an Engine.** — The choice of an engine for any particular service presents a problem of so many conflicting conditions that a great degree of skill is required for its satisfactory solution. The cost of operation of a power plant, which includes the fixed charge of interest on the first cost and the cost of fuel and attendance, is always a controlling influence in the choice of an engine suitable for the power and service required. The most economical engines in the use of fuel are usually of high initial cost, which increases the fixed charge of interest, but such engines require the least boiler power to operate them, so that the cost of the engine alone is not the only thing to be considered, but rather the cost of the entire plant. If the service required of an engine be continuous and the load constant the first consideration is then the cost of the fuel, and the most efficient engine should be chosen regardless of first cost. For light powers under these conditions, the

simple high-speed single-valve engine would be suitable; for intermediate powers, the engine of moderately high speed, whether simple or compound, should be the choice; and for large powers, the low-speed compound engine of either the horizontal or vertical type is applicable. For continuous service and variable load, some form of condensing engine should be chosen, because of its wider field in the expansive use of steam without materially affecting its efficiency.

The choice of an engine for intermittent service, such as hoisting engines and engines furnishing power for the industries that operate for only a portion of the year, is attended with so many conflicting circumstances that no definite rule applies. Engines of widely different types are found operating under such conditions.

Whether or not a plant should be operated by a condensing or a non-condensing engine is often determined by the local conditions as to the supply, cost and purity of condensing water. The condensing engine is not only more economical in the expenditure of fuel than the non-condensing engine, but can be made smaller for equal powers and is therefore cheaper in first cost. With an available supply of cooling water it is greatly in the interest of economy to operate the engine with a condenser, particularly for the larger powers, and under such conditions the compound types of engines have their greatest field of usefulness. The choice of the non-condensing engine is governed by the use it is proposed to make of the exhaust steam as, for example, in heating and drying systems, heating feed water, and in other uses to which it is applicable.

CHAPTER VIII

CONDENSERS. FEED-WATER HEATERS. PUMPS

115. Condensers. — The economical advantage in operating a steam engine in connection with a condenser is explained in Chapter XIII. Local conditions as to the supply of condensing water, or as to the extent of the plant, are usually the determining influences governing the employment of condensers. The situation of a power plant may be remote from a natural water source, such as a lake or a stream, necessitating the purchase of city water, but if the power to be developed is large it will be in the interest of economy to use a condenser since by so doing a saving of about 25 per cent in steam consumption may be effected. A saving in steam consumption means not only a saving in fuel, but means also that the power of an existing non-condensing plant may be increased one-fourth by the employment of a condenser.

Suppose, for example, that the buildings and equipment of an existing non-condensing plant of 400 H.P. cost $125 per H.P. If this plant were operated with a condenser its H.P. would be increased to 500, and the cost per H.P. reduced to

$\frac{400 \times 125}{500} = \$100$, a saving of $25 per H.P., which is vastly greater than the cost of the most efficient condenser.

The condenser may be of the *jet* or *surface* type. In the jet condenser the exhaust steam from the engine enters the condensing chamber where it is brought in contact with a jet of cold water, resulting in the condensation of the steam and the production of a vacuum. The mixture of the water from the condensed steam and the condensing water falls to the bottom of the condenser and must be pumped out regularly by the air pump; otherwise the air contained in the condensing water would accumulate and destroy the vacuum.

It will be seen that with the jet condenser, the feed water is a mixture of the condensing water and the water resulting from the condensation of the steam, about 25 pounds of the former to 1 pound of the latter, so that in case the condensing water contains salts, acids and scale-producing impurities, the feed water resulting from the mixture is capable of doing great injury to the boiler.

With the surface condenser the exhaust steam on entrance into the condenser comes in contact with the cold metallic surfaces of tubes through which condensing water is being forced by a circulating pump. In this process the steam is at once condensed and a vacuum formed. The water resulting from the condensation of the steam falls to the bottom of the condenser, whence it is drawn, together with any air or uncondensed vapor that may be present, by the air pump.

With the surface condenser there is no mixture of the condensed steam and the condensing water, and therefore a purity of feed water is produced that prevents an accumulation of dirt in the boiler and the formation of scale on the heating surface.

The choice of a type of condenser is governed by local conditions. If the condensing water is suitable for feed water and its supply plentiful, the jet condenser is to be preferred, owing to its simplicity, its comparative small bulk and cost, and its requirement of only one pump. With the surface condenser the cooling water and its contained air does not mix with the pure water resulting from the condensation of the steam, and in consequence the air pump maintains a better vacuum and has less work to perform than with the jet condenser. The surface condenser is the more bulky and expensive of the two types, but its use is imperative in marine practice where salt water is used for cooling, and on land where good cooling water is scarce or expensive.

116. Cooling Ponds and Towers. — In some cases where cooling water is scarce or expensive resort is made to cooling ponds and cooling towers, devices which admit of cooling the condensing water after its discharge from the condenser, thus permitting its repeated use. The cooling pond, which is of artificial construction, is of sufficient capacity to supply the condenser with cooling water. This water, after passing through the condenser, is discharged into the pond, where by evaporation and radiation into the air its temperature is lowered to an extent that permits its re-use in the condenser.

Cooling towers are cylindrical in shape and of steel construction. The cooling or condensing water after passing through the condenser is discharged at a temperature of from 90° to 100° into the tower at its top. The interior of the tower is provided with partitions of tile checker work or of galvanized iron screens, and as the water falls through the checker work or through the meshes of the screens it is broken up and exposed to the cooling action of an ascending current of air from a fan at the bottom.

117. The Blake Automatic Jet Condenser. — The sectional view, Fig. 95, shows clearly the action of this condenser. The adjustable cone is for the purpose of regulating the thickness of the spray and amount of injection water supplied to the condenser. It is also a convenient means of flushing out the condenser in case it becomes clogged with grass, or other obstructions that may enter with the injection water. The exhaust from the engine enters at the top of the condenser and is condensed immediately by the injection spray. The gate valve shown on the injection inlet is for the purpose of shutting off the water supply, as the cone should not be used for this purpose. The gate valve, however, should be left wide open when the condenser is in operation and the adjustment of the water supply made with the cone.

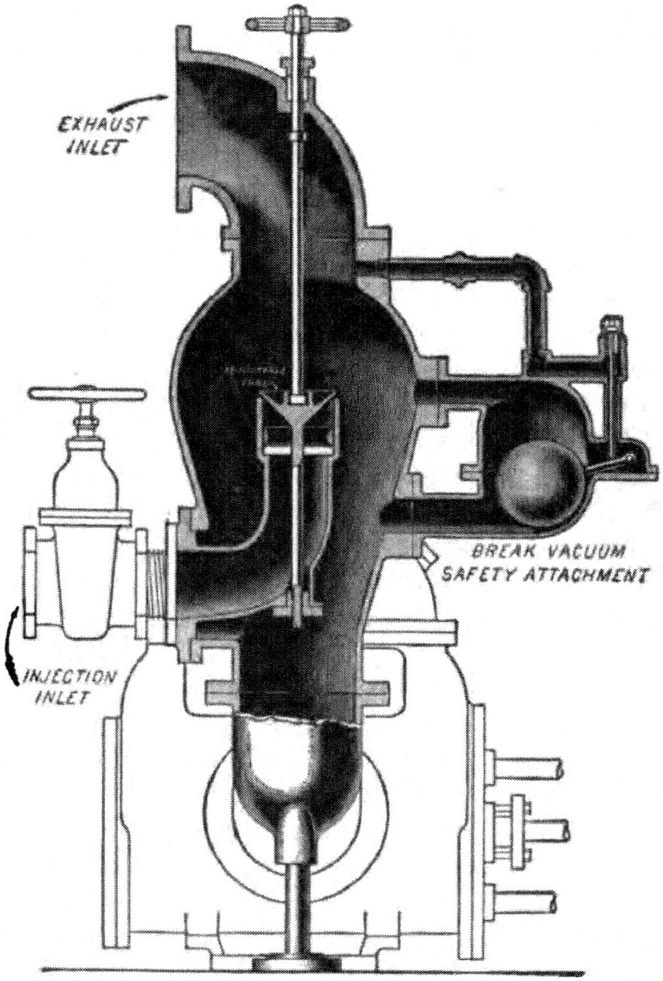

Fig. 95.

The Automatic Break Vacuum Safety Attachment is an important feature of this condenser. Its purpose is to prevent the water from rising any higher in the condenser than the proper level. If, by accident, the air pump should stop, the water will rise in the condenser and, when it has reached a certain level, lifts the float which in turn raises the relief valve from its seat, thus admitting air into the condenser and destroying the vacuum. As soon as the pump is again started the float drops to its normal position, the air relief valve closes, and the work of condensation is promptly resumed. In this way the possibility of flooding the engine, with all its attendant dangers, is avoided.

FIG. 96.

118. The Blake Admiralty Surface Condenser. — The cylindrical form of this condenser is shown in Fig. 96, the bonnet at the circulating pump end being removed. The circulating water enters the lower chamber at the right end, passes through the lower nest of tubes, returns through the upper nest of tubes into the upper chamber at the right end, and thence into the discharge pipe through the flanged orifice shown. The exhaust steam enters the condenser at about the middle of its top, and circulating about the cool tubes is condensed and falls to the bottom, whence it is removed by the air pump.

The tube heads are independent plates and the tubes of seamless drawn brass, thoroughly tinned inside and out to protect them from corrosion. The tubes pass through stuffing-boxes in the tube heads which have threaded brass ferrules, each with

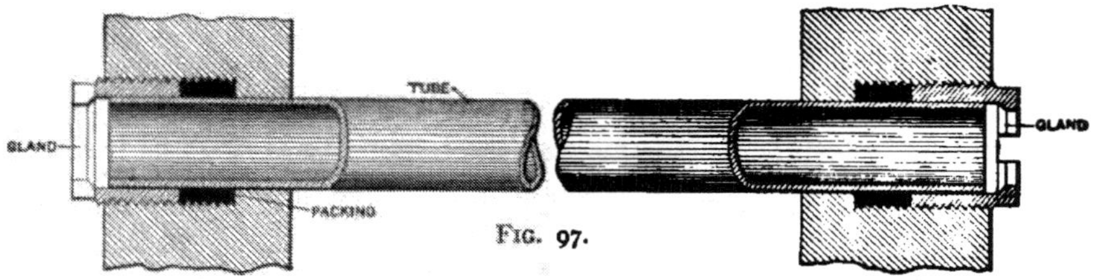

FIG. 97.

a lip on the inside, as shown in Fig. 97. This lip prevents the tube from creeping, and at the same time permits expansion and contraction. The tubes are readily removed and the stuffing-boxes easily repacked.

FIG. 98.

This form of condenser is shown mounted over a combined air and circulating pump in Fig. 98. The three cylinders are placed in tandem, the steam cylinder in the middle, the piston of which operates the two water pistons. This arrangement saves space, is simple and efficient, and is largely used in marine practice.

119. Oil Separators. — Exhaust steam from engines and pumps is more or less saturated with oil that it has picked up during its passage through the valve chests and cylinders and it is very important that this oil be removed before the steam is admitted to the condenser or into an open feed-water heater. The introduction of oil into the boiler through the medium of the feed water is attended with serious results, as it forms a non-conductive deposit on the heating surface, with the consequent loss in boiler efficiency and great danger of overheating the metal. In order to avoid these ill effects separators are used to deprive the exhaust steam of its oil.

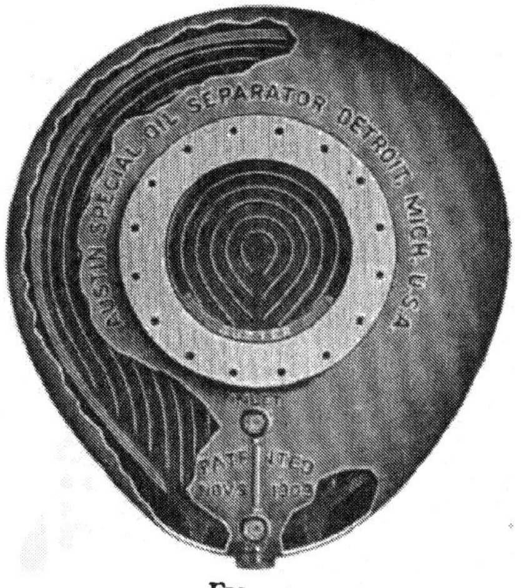

Fig. 99.

Fig. 100.

120. The Austin Oil Separator. — One form of the Austin oil separator is shown in part sectional side view in Fig. 99, and in end outside view in Fig. 100. The flat oval shape of the separator is designed to give a large internal area so as to reduce the velocity of the steam. The baffle plate is deeply corrugated with circular grooves, all leading to the receiver part of the separator. The baffle plate has a high flange at its outside edge to prevent oil or water from rebounding over the side. The exhaust steam enters the flanged opening at the center, and coming in contact with the baffler the oil trickles down the grooves of the corrugations to the receiver, the steam escaping over the edge of the baffle plate and continuing its course to the condenser. As a precaution to effect a complete separation, there is a ring flange around the outlet orifice to catch any oil that the steam may deposit on the sides of the separator. A large percentage of the oil in exhaust steam gathers upon the lower inner side of the pipe along which it flows under pressure of the steam, and it is to catch this oil that the gutter at the inlet orifice is provided, from which it drains to the receiver.

121. Feed-water Heaters. — It is the purpose of feed-water heaters to utilize the heat of the exhaust steam of the engine in raising the temperature of the water fed to the boiler, and thus effect an economy in the expenditure of fuel. No direct saving in fuel can result from heating feed water unless the heat is obtained from some source of waste, such as the exhaust steam of the engine. Live-steam feed-water heaters, in which the feed water may be heated to boiler temperature if desired, are feed-water purifiers as well, in that they free the water of scale-making impurities by causing their precipitation before the water is discharged to the boiler, and as such precipitation takes place only at temperatures considerably higher than 212°, the use of live steam in them is a necessity. Their saving in fuel occurs indirectly from the fact that they very much reduce the accumulation of scale on the heating surface of the boiler, and thereby maintain a higher rate of heat transmission from the products of combustion to the water. The high temperature of the feed water from a live steam heater undoubtedly prolongs the life of a boiler by lessening the expansion and contraction in the metal incident to the use of feed water of lower temperature.

Exhaust-steam feed-water heaters are of two types, the *open* and the *closed*.

The open type consists essentially of an open chamber in which the exhaust steam and the water to be heated are brought into intimate contact by spraying the water through the steam. The pressure in such heaters is necessarily that of the exhaust steam, about atmospheric pressure. The water resulting from the condensation of the steam mixes with and becomes a part of the feed water, which is a bad arrangement unless a separator is used to deprive the steam of its entrained oil. If the supply of steam be ample the temperature of the feed water is usually raised to about 210° F.

In the closed feed-water heater the steam and water are separated by metallic walls, the water being usually circulated through tubes and the exhaust steam flowing in the spaces between the tubes. In some instances the arrangement is the reverse of this, the steam flowing through tubes that are surrounded by the water to be heated. The transmission of heat from the steam to the water is therefore through the medium of a metallic surface, and if one side or the other of this surface be coated with grease or scale the transmission becomes imperfect, and a feed temperature several degrees lower than that of the open heater may be expected. The closed heater is usually placed in the feed line between the pump and boiler, and for this reason may be subjected to boiler pressure, a fact that necessitates for them a physical structure of great strength, usually of steel plates in cylindrical form. The closed heater may be used as a live-steam heater and the temperature of the feed be raised to within a few degrees of that of the boiler, if so desired.

122. The Wainwright Feed-water Heater. — The Wainwright heater, an example of the closed type, is shown in Fig. 101. In arrangement it is very similar to a surface condenser. It consists of a cylindrical shell containing a series of corrugated copper tubes arranged in four nests. The water enters the feed inlet at the top, and, after passing successively through the four nests of tubes, is discharged through the feed outlet at the bottom. The exhaust steam enters at the bottom and circulates

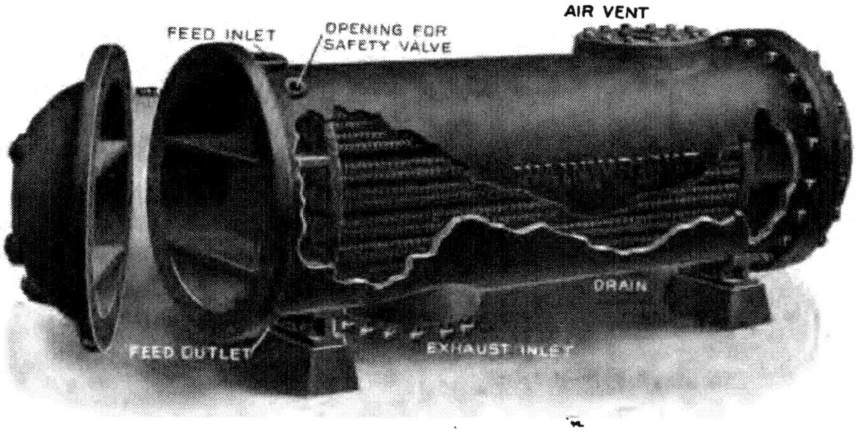

FIG. 101.

around the tubes in the space between the tube sheets, imparting its heat to the water passing through the tubes. It will be noted that the exhaust inlet is so placed that the heated water, just as it leaves the heater, receives the full benefit of the entering exhaust steam. There is a safety valve connection with the water side to afford relief from the water expansion that would take place in case the inlet and outlet water valves should become closed.

A particular feature of this heater is the corrugated copper tube. In any feed-water heater an appreciable amount of time is required to transfer the heat from the steam to the water, but it is claimed that more of this time is taken up by the heat in passing from the outside of the cylindrical body of water passing through the tube to its center or core than is used in passing through the metal of the tube, in which case the corrugations of the tube are effective in breaking up the core and bringing every part of the water into direct contact with the hot tube. The inevitable expansion of the tubes due to the changes in temperature is taken up in the corrugations, which permits the

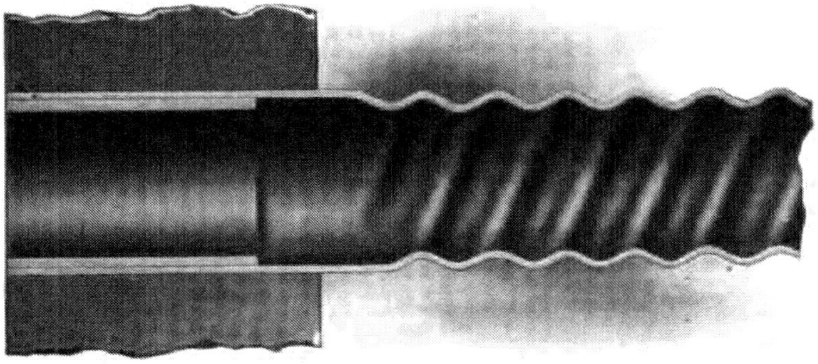

FIG. 102.

ends of the tubes to be rigidly expanded in the tube sheets, an arrangement much to be preferred to the ordinary stuffing-box method of allowing for expansion and for keeping the tubes tight. The tubes of this heater are fastened at each end in the tube plates by a brass thimble expanded in position after the manner shown in Fig. 102.

123. The Cochrane Heater. — An interior view of the Cochrane Steam-Stack Cut-Out Valve Heater and Receiver is shown in Fig. 103. It is a heater of the open type. Steam enters from the right into the oil separator, which is of the single, vertically-ribbed baffle type. The water and oil in the steam, and the oily emulsion flowing in the bottom of the exhaust pipe, are diverted into the well of the separator, where they are removed from the action of the steam current and whence they drain to the trap below. The water to be heated flows into a trough at the top and then over and through a series of perforated trays, inclined first one way and then the other, each tray catching the drips from the one above. The oil separator must be highly efficient,

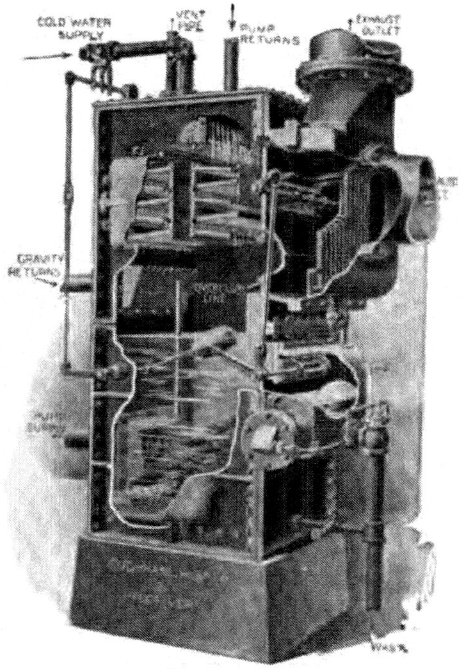

FIG. 103.

as the success of the heater depends upon it. After passing around the separator baffle, the steam flows around and in between the trays and comes in direct contact with the water, to which it imparts its heat. From the last tray the water falls into the settling chamber, which has a perforated false bottom for carrying a filter bed. The feed pump receives its supply from the space beneath the filter bed. The level of the water in the settling chamber is controlled by a copper float which acts through a system of levers on a valve in the cold water supply pipe. Should this float for any reason fail to act, any excess of water can escape from the heater through a sealed overflow, which is lower than the level of the exhaust inlet.

The body of the heater is rectangular in section, and is made of cast-iron plates reinforced by ribs and bolted together. Cast iron was selected as the material for the heater because it is not so susceptible to corrosion as is sheet iron or steel plate. The fittings are made of copper and brass.

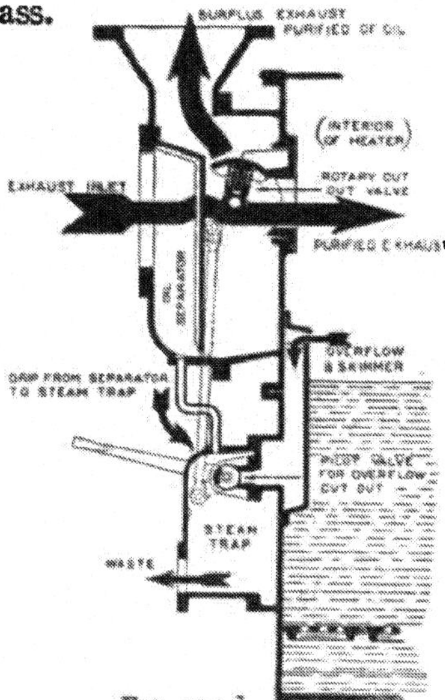

FIG. 104.

After passing around the separator baffle the exhaust steam has two paths open to it, namely, one into the heater, which will take only as much of the steam as is required to heat the water, and one through the outlet, or stack, at the top of the separator, through which the remainder of the steam must pass. The opening into the heater is controlled by a special semi-rotary cut-out valve (see Fig. 104). When this valve is open it occupies such a position that the heater has the "preference" for the steam; that is, in its open position the valve diverts a portion of the steam from the top opening and directs it into the heater, although at the same time any surplus steam can readily escape through the stack outlet. In its closed position the valve is held firmly to its seat by the pressure of the exhaust steam, as well as by the spring which holds the valve in position throughout its travel. When closed this spring-loaded valve acts as a relief for any pressure that may accumulate within the heater, as from the discharge or leakage from live-steam traps or otherwise. The trap attached to the heater receives the drips from the separator, and also takes care of the overflow from the heater. In order that the steam may not pass from the separator through the drip pipe and the trap into the heater when the cut-out valve is closed, a second semi-rotary cut-out valve is provided to close the opening between the heater overflow and the trap (see Fig. 104). Both the valves must be closed when the heater is cut out of service for cleaning or inspection, and in order to insure that both will open and close together, and to render their operation more convenient, there is a combined valve gear consisting of cranks on each valve spindle, and a rod connecting them, so arranged that the movement of the attached handle opens or closes the two valves simultaneously. On small heaters this plain crank-and-connecting-rod arrangement answers all requirements satisfactorily, but with large heaters it has been found better to arrange the gear so that the lower valve will be opened first and closed last.

The purifying action of the Cochrane heater is explained as follows:

When water is heated its power of dissolving gases decreases, and if heated up to 212° F under atmospheric pressure only, as in the Cochrane heater, all gases become entirely insoluble. If the water is heated in small particles, the expulsion of the gas takes place immediately, while if it is heated in a large body, time is required for the small bubbles to coalesce and reach the surface of the water. Inasmuch as a temperature of from 210° to 212° F is always maintained in a Cochrane heater supplied with sufficient exhaust steam to heat the water, the release of the gas is nearly complete.

The benefit derived from driving out the gases held in boiler feed water is two fold. In the first place, it has been shown by chemical experiment that the presence of air is essential to the corrosion of iron and steel by water. If the air is completely eliminated such corrosion is stopped. Air may be present in large quantities in water otherwise apparently pure, which may explain cases of corrosion by distilled water from surface condensers, heating systems, etc.

The second benefit derived from driving gases from feed water is that carbonic acid gas is expelled, including both that dissolved in the water and that combined in the bicarbonates of lime and magnesia, to which is due the condition known as "temporary hardness" and which constitute the larger part of the scale-forming matter in most natural water supplies. When the carbonic acid gas is driven off, these carbonates are converted into monocarbonates, which are insoluble, and which, if the water be permitted to stand sufficiently long, or be passed through a good filter, are precipitated and removed. It is claimed that enough of the carbonic acid gas can be driven off in the Cochrane heater to effect sufficient precipitation of carbonates to prevent trouble from them in the boiler. It will be seen from Fig. 103, that the settling chamber is of sufficient capacity for the removal of precipitated carbonates and also for the removal of sand, silt and other solid impurities carried by the water, and provision is made for the insertion of a filter of suitable material. This filter lies in the bottom of the heater and may be removed through the cleaning doors.

124. Economizers. — In order to save some of the heat that is carried away through the smoke pipe of a boiler and utilize it in heating the feed water, different forms of economizers have been devised. They consist of a series of pipes placed between the boiler and smoke pipe and then passing the feed water through them. It is difficult to keep them clear of soot and dirt, and if made too large they interfere with the draft.

The most successful of the economizers is that of Green, in which several sets of vertical tubes connected by manifolds are placed in a chamber between the boiler and smoke pipe. The feed water is forced through these tubes by the pump, entering first the set most remote from the boiler and passing finally through the set nearest the boiler, an arrangement that brings the tubes containing the hottest water in contact with the hottest gases. The success of the Green economizer is due largely to its system of scrapers, by which the tubes are kept clean. These scrapers, one on each tube, are moved up and down over the tubes by means of some suitable apparatus actuated by a small engine or motor.

125. **Pumps.** — A pump is a machine by which a fluid is made to flow against a pressure by means of the displacing action of a piston or plunger acting within a closed cylinder provided with inlet and outlet passages controlled by valves.

In the operation of a non-condensing steam engine there is but one essential pump, namely, the feed pump for supplying the boiler with water. In the surface condensing type of engine there must be, in addition to a feed pump, an air pump and a circulating pump. The office of the air pump is to remove from the condenser the air, the uncondensed vapor, and the water resulting from the condensation of the exhaust steam, while the duty of the circulating pump is to force cold water through the tubes of the condenser. In the case of the jet type of condenser the circulating pump is dispensed with, the air pump, in addition to its own functions, performs those of the circulating pump by maintaining a vacuum in the condenser, thus causing the injection water to flow into the condenser without the aid of a pump.

The steam pump used for feeding boilers is required to deliver its water into the boiler against the same pressure as that of the steam which operates it, and for this reason the steam piston is made larger than the water piston in order that the total pressure on the steam piston shall be greater than that on the water piston. This preponderance of pressure on the steam piston overcomes the resistance to the passage of the water through the valves and feed pipe and forces the entry of the water into the same boiler from which the pump gets its steam. It is the usual practice to make the steam piston of a feed pump from two to three times the area of the water piston.

The air chamber supplied to pumps is for the purpose of equalizing the pressure in the discharge pipe. The discharge of water from the pump is not regular, as its velocity changes at each reversal of the stroke, and hence there is a variation in pressure, which would cause shock to the pipes and pump were it not for the entrapped air in the top of the chamber acting as a cushion to keep the pressure uniform.

There are a number of types of feed pumps in general use, differing from each other principally in the expedients employed in operating the steam end of the pump, the operation of the water ends being practically the same in all. The difficulty with the ordinary type of steam pump is that it has no crank, like that of the steam engine, to convert its reciprocating motion into one of rotation, such as would permit the use of an eccentric to operate its valve, and it is this difficulty which has led to the employment of the ingenious valve-operating devices, each one of which marks the distinctive feature of the pump employing it. It is absolutely essential that a feed pump should be reliable in its action in order that the water in the boiler may be kept as nearly as possible at a constant level, the pump being designed when running at moderate speed, to supply feed to keep pace with the maximum rate of evaporation in the boiler. The working parts of a feed pump should be as few as possible and simple in action, and the one is to be preferred that furnishes the most positive action in admitting and exhausting steam to and from the cylinder, with the least liability of derangement of the device actuating the steam valve.

A few examples of the successful pumps in use will be given.

126. The Knowles Pump. — A longitudinal section through the steam and water ends of a Knowles single direct-acting steam pump is shown in Fig. 105. Steam is supplied to the pump through a pipe at *s*. A *chest piston* p is driven back and forth by the pressure of the steam, carrying with it the main valve v which controls the admission of steam into the cylinder to drive the piston P, and thus operate the pump. The main

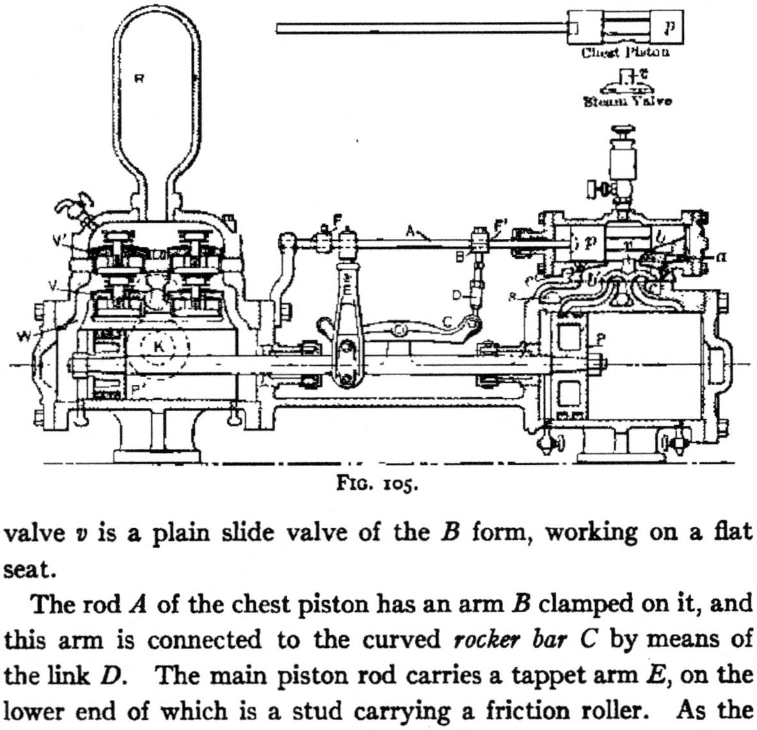

Fig. 105.

valve v is a plain slide valve of the B form, working on a flat seat.

The rod A of the chest piston has an arm B clamped on it, and this arm is connected to the curved *rocker bar* C by means of the link D. The main piston rod carries a tappet arm E, on the lower end of which is a stud carrying a friction roller. As the main piston moves back and forth the friction roller comes in contact with the ends of the curved rocker bar at the termination of each stroke, and in doing so raises the bar slightly, thus giving to the chest piston a small motion of rotation. On the under side of each end of the chest piston is a small port a (shown in part section at the right end), and four corresponding small ports b, b' and c, c' lead from the chest piston cylinder, the former to the steam chest and the latter to the exhaust. In the position shown in the figure, the main piston has reached the end of its stroke to the left and the friction roller is raising the left end of the rocker bar, the consequent movement of rotation of the chest piston bringing the port a in connection with b, thus admitting steam into the right end of the valve chest piston, driving that piston to the left, carrying the main valve with it so as to open the right end of the main cylinder to the exhaust and open the left end to steam in order that the main piston may be driven on its stroke to the right. The same motion of rotation that brought port a in connection with steam port b simultaneously brought port a' (not shown) in the left end of the chest piston in connection with exhaust port c' so as to exhaust from the left end of the chest-piston cylinder the steam that was used on the preceding stroke; but before this steam is entirely exhausted the port c' is closed by the chest piston's movement to the left, entrapping some of the steam to act as a cushion in preventing the chest piston from striking the head of its cylinder. On the arrival of the main piston at the end of its stroke to the right the friction roller will have come in contact with and raised slightly the right end of the rocker bar, thus giving to the chest piston a movement of rotation in a direction opposite to that in the first instance, and the operations just described will take place at the other end. Should the steam pressure from any cause fail to move the chest piston, the upper end of the tappet arm will come in contact with the tappets F, F' (F' is partially concealed by arm B) and move it mechanically.

It will be observed that when the movement of rotation is given to the chest piston by the outside mechanism, the opening of the port for the admission of steam to one end of the main cylinder and the opening of the port to exhaust at the other end immediately follows, thus making the action of the pump *positive.*

The water piston P' has for its rod a prolongation of the rod of the steam piston P. On the stroke of the piston from left to right a partial vacuum is formed in the left end of the water cylinder causing the suction valve V to lift and water to flow into the cylinder through K from the source of supply. On the return stroke of the piston this water is forced out of the cylinder through the port W, on top of and closing the valve V, up through the delivery valve V' into the air chamber R, and thence to the delivery pipe. A similar action takes place at the other end for the stroke of the piston from right to left and return.

127. The Blake Pump. — The illustrations of Fig. 106 represent the steam end of the Blake pump as arranged for boiler feed and pressure pumps. The valve movement, without any regard to the gear by which it is accomplished, will first be considered in the manner authorized by the makers.

The main valve, which controls the admission and exhaust of steam from the main cylinder, is carried by the auxiliary piston, and moves on the back of the movable seat. This movable seat is shown in Fig. 4 in perspective, and the passages A, B, C serve as steam ports to the main cylinder; while the lugs GG' control the admission of steam to the auxiliary cylinder and the holes HH' control the exhaust from that cylinder.

With the valves in the position shown, the course of the steam is through live-steam passage N, through the port C to the right-hand end of the main cylinder, thus forcing the piston over to the left. Now, when the piston nearly reaches the left end of the cylinder the movable seat, by a means described later, is shifted over to the left, so that the lug G covers the port E, while the lug G' moves off from the port E', thus admitting steam behind the auxiliary piston, at the left-hand side. At the same time the exhaust port K of the auxiliary cylinder is put into communication with the hole *S* which leads to the exhaust. The auxiliary piston is therefore forced over to the right, and uncovers the port *A* to live steam. Near the right-hand end of the stroke the operation is reversed. That is, the movable seat,

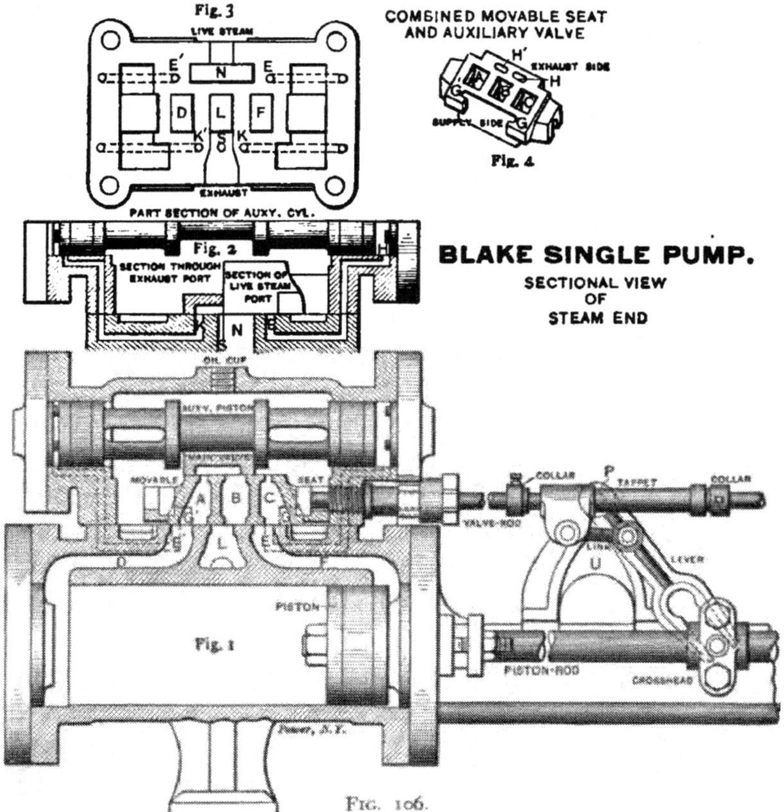

Fig. 106.

which is then at the left, is moved over to the right, assuming the position shown in Fig. 1. The lug *G* then uncovers the port *E*, while *E'* is covered by *G'*. This admits live steam to the right of the auxiliary piston. At the same time, the hole *H'* in the auxiliary valve or movable seat places *K'* and *S* in communication, thus exhausting the steam from the left of the auxiliary piston. This drives the auxiliary piston over to the left, until it assumes the position shown in Fig. 1.

The auxiliary piston is cushioned on steam, because the exhaust port is not out at the end of the auxiliary cylinder and consequently there is steam imprisoned when the piston covers the exhaust, as shown at the left in Fig. 2. The main piston is cushioned on live steam, because the valve has lead; that is, the operation of admitting steam is performed before the piston reaches the end of its stroke.

It will be seen that if means are provided to shift the movable seat from one end of its travel to the other, the rest of the operation is automatic. Fig. 1 shows the valve gear provided for this operation. The piston rod is provided with a crosshead, the latter having a pin as shown. The frame of the pump is built with an upright piece, *U*, to which is pivoted at *P* a lever whose lower end is slotted and engages with the crosshead. The valve rod, which is secured to the movable seat, is provided with two collars as shown. These collars are made of split nuts which work on a thread cut on the valve rod for a short distance on each side of their ordinary positions. Between these two collars is a tappet, which is free to slide on the valve rod. The link shown connects the tappet with the lever. When the piston rod moves, the lever rotates about *P*, carrying the tappet with it; and when the tappet strikes either collar it moves the movable seat in the direction in which the tappet is moving. By placing the collars so that the tappet strikes them before the piston reaches the end of its stroke, the movable seat will be shifted in the required manner.

No valve adjustment is required to be made inside the steam chest, and the only adjustment which can be performed is that of altering the distance between the collars, thus changing the travel of the valve. This is done by loosening the set screws in the collars, and rotating the latter until they come to the required point. Changing the distance between the collars alters the length of the stroke. This is easily seen, because the action of the tappet in striking the collars is what admits and exhausts the steam; and if the distance which the tappet has to travel is varied, the time at which the valve is actuated is varied, and the stroke varies as well.

The adjustment of these collars is very simple, and can be performed while the pump is running. In adjusting them it is desirable to make the stroke as long as possible and secure enough cushioning, for the shorter the stroke the greater the amount of the clearance, and the steam required to fill the clearance is wasted on every stroke.

If the collars on the valve rod are not set at equal distances from the center line of the lever when the latter is vertical, the movable seat will be reversed sooner on one stroke than on the other, and consequently the piston will travel further in one direction than in the other.

128. The Worthington Duplex Pump. — A duplex pump is one in which two single pumps of similar construction are placed side by side, the slide valve of each steam cylinder receiving its motion by means of a lever which is rocked by the piston rod of the adjacent cylinder; that is, the steam piston of one cylinder, instead of operating its own steam valve as in the single pump, operates the valve of the adjacent cylinder. Thus, one piston, after finishing its own stroke, waits for its steam valve to be acted upon by the rod of the other piston before it can renew its motion. This pause allows the water valve to seat quietly and obviates harshness of motion. As one or the other of the steam valves is always open, there is no dead point, and therefore the pump is always ready to start when steam is admitted.

The duplex pump, the invention of Henry R. Worthington, is shown in longitudinal section in Fig. 107. The simplicity of the

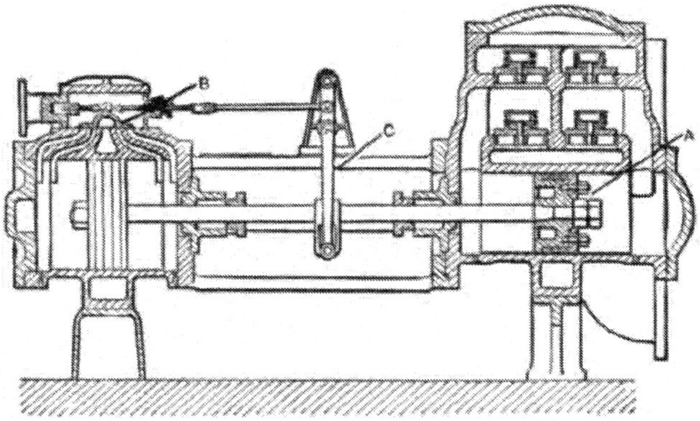

FIG. 107.

FIG. 108.

pump is apparent. The movement of the steam valve *B*, which is of the ordinary *D*-slide type, is controlled by arm *C*, which receives its vibrating motion from the piston rod of the adjacent cylinder. Since the arm *C* swings through the whole length of the stroke the parts actuating the valve are always in contact, thus avoiding the blow inseparable from the tappet system of other pumps. The suction and discharge action of the water piston *A* is the same as that of other pumps. The perspective view of the pump, Fig. 108, shows clearly how the vibrating motion given to the arm *C* by the piston rod of one pump operates the steam valve of the other pump.

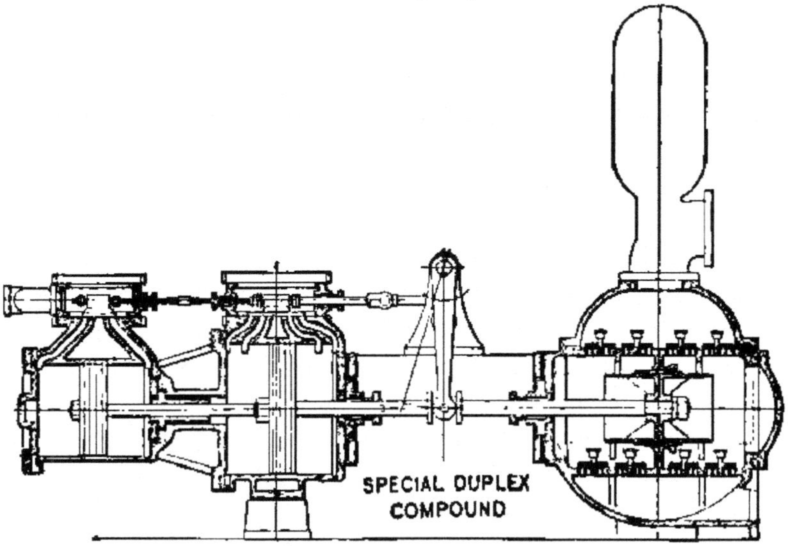

FIG. 109.

129. The Compound Pump. — In the simple direct-acting steam pump the piston is acted on by the full steam pressure during its entire stroke, there being no cut-off to utilize the expansive force of the steam. It is, therefore, a machine that is very extravagant in the consumption of steam, and to overcome this waste many of the larger pumps have their steam ends made in the form of the tandem-compound and triple-expansion engines. The volume of steam used at each stroke is equal to the volume of the high-pressure cylinder, and as this steam finally occupies the low-pressure cylinder the number of expansions is the volumetric ratio between the high-pressure and low-pressure cylinders.

A longitudinal section of a Blake & Knowles tandem-compound-duplex pump is shown in Fig. 109. It is seen that the valve mechanism is similar to that of the Worthington duplex pump of Art. 128.

130. The Centrifugal Pump. — The action of a centrifugal pump depends upon the pressure produced by the centrifugal force of a rotating mass of water. The revolving part of the pump, known as the impeller, consists of a wheel resembling a fan, on the hub of which are mounted vanes or arms, the whole being mounted on a shaft and inclosed in a chamber of variable section area, being largest in section at the point at which the water is discharged. When the wheel revolves the vanes produce a partial vacuum in the chamber, followed by an inrush of water through the suction orifice about the hub and filling the spaces between the vanes. The revolving vanes pick up this water, and, with a rapidly accelerated motion, cause it to flow outward radially between them, until, at the outer circumference, it has absorbed the power transmitted to the shaft by its acquired velocity and pressure, and is then discharged. The vanes may be straight or curved, and vary in number from six to eight.

Impellers are of two general forms, known as open and closed types. The former consists of radial vanes attached to a central hub and disk and open at the sides, the whole revolving between two fixed side plates of the pump, while in the latter the vanes form, with two circular disks, closed passages extending from the inlet opening at the hub to the outlet openings to the inclosing casing at the periphery of the impeller.

Centrifugal pumps may be driven by a belt from a steam engine, or may be driven by a direct connection to a turbine or other motor. They are very efficient when not pumping against excessive lifts, and with a free and uniform supply of water to the inner ends of all the vanes the discharge will be in the nature of a continuous and steady stream.

The advantages claimed for the centrifugal pump are:

It has an easy rotary motion, acquired without the assistance of valves or other contrivances that consume power in friction; it discharges continuously without the aid of an air chamber; it is of simple and durable construction and economical in use; and it is especially applicable where the water to be pumped is filled with foreign matter, since there are neither valves nor packing to be affected by such impurities.

In connection with the steam engine, centrifugal pumps are frequently used to circulate water through the tubes of a surface condenser, the pump being driven by a directly-connected motor or by a belt from a small engine. There is a distinct advantage, particularly in marine practice, of having the circulating pump operated by means entirely independent of the engine, inasmuch as the condenser can then be kept cool and a vacuum maintained while the engine is at rest, the vacuum enabling the engine to be started easily.

131. **The Alberger Centrifugal Pump.** — The Alberger volute centrifugal pump, so named from the volute or spiral form of its casing, is shown in longitudinal section in Fig. 110. The casing *B*, *B* contains the diffusing space of the pump, and is of gradual increasing section area, being largest at the discharge orifice. Bolted to the casing are the heads *C*, *C*, the head away from the source of power containing the suction opening *A*, which allows the greatest facility for connecting the suction pipe and for inspecting the interior of the pump. The impeller is of the closed type, and as it revolves at high rotational speed under

the action of the motor, the water enters the suction orifice *A*, is then drawn into the impeller through the openings *D*, *D*, and thence, by centrifugal force, it is thrown with pressure and high velocity into the diffusing space of the casing. The velocity energy of this water must be transformed into pressure before discharge into the feed pipe, and the accomplishment of this is

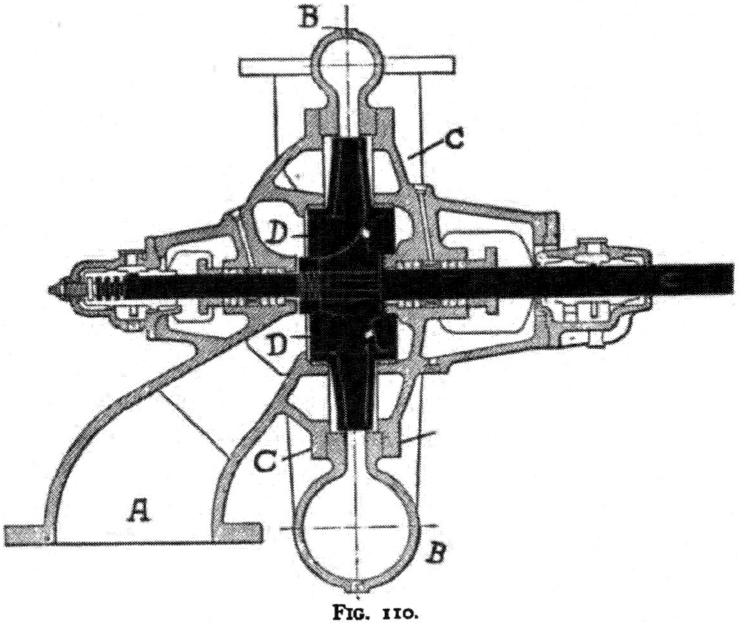

FIG. 110.

the function of the volute casing. The change from velocity to pressure is effected during the passage of the water through the diffusion ring of the casing, the velocity decreasing and transforming to pressure as the section area of the diffusing ring increases. With a proper design of the diffusion space and volute ring it is possible to transform practically the whole of the velocity into pressure.

The passage from the suction to the inlet opening *D* to the impeller is ingeniously worked in the head so that the bearing is brought close to the pump, and at the same time the passage is so proportioned that the water is directed uniformly to every inlet passage of the impeller. The discharge pipe of the pump is bolted to the flanged end of the volute casing.

132. The Centrifugal Feed Pump. — The use of motor-driven centrifugal pumps for boiler-feeding purposes is becoming extended. They require less space and less cost of maintenance than the reciprocating pump, as there are no valves to give out

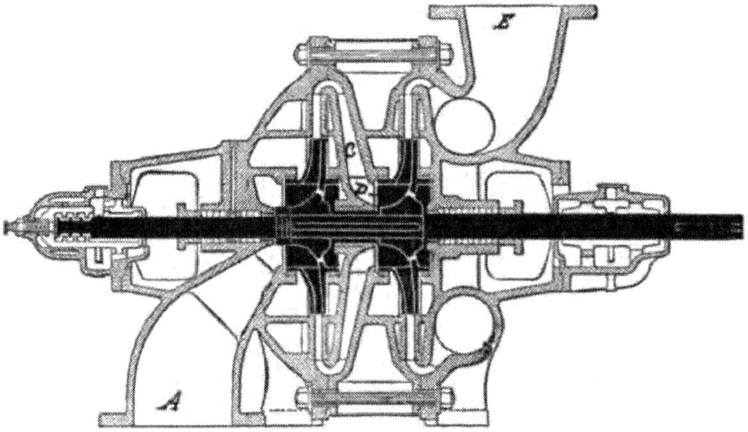

SIDE SECTIONAL VIEW

FIG. 111.

nor plungers to wear. The nearly constant speed at which they run insures a steady and even delivery, avoiding shocks in the feed piping. They may be made to deliver against the highest boiler pressures without undue impeller rotational speed by means of the multi-stage system; that is, by using two, three, or more impellers instead of one. In such an arrangement the additional impellers and diffusion rings are placed between the heads in a construction such as that shown in Figs. 111 and 112, which represent a two-stage Alberger turbine pump, so named from the resemblance of its interior construction to that of water turbines. Like the single-stage pump, the water enters the first impeller from the suction *A*. After being discharged from the first impeller, the water has nearly all its velocity transformed into pressure while passing through the diffusing space, the series of diffusing vanes, *B*, *B* . . . (see Fig. 112) assisting in the transformation by dividing the swiftly-moving current into parts before entering the diffusion ring, so that at its entrance through the passage *C* into the inlet *D* of the second impeller the velocity of the water is very low and its pressure much increased.

END SECTIONAL VIEW
FIG. 112.

After passing through the second impeller and its diffusing space the water undergoes a similar transformation and is discharged at *E* at a still further augmented pressure. In this manner the pressure of discharge to the feed pipe is increased as the number of stages of the pump is increased. These Alberger pumps are driven by a specially designed steam turbine.

133. **Injectors.** — The injector is a boiler-feeding device exclusively used on locomotives and extensively used in connection with stationary boilers. It is very efficient as a boiler feeder, as nearly all the heat of the steam used in its operation is imparted to the feed water and carried back into the boiler. The capacity of an injector decreases as the lift and temperature of the feed water increases, its greatest capacity obtaining with cold water and a lift not exceeding 8 feet.

The general principle governing the action of an injector is that of bringing a jet of steam in contact with a relatively smaller jet of water, the steam, in condensing, imparting its high velocity to the water, and in mingling with it forms a continuous jet of water of such high velocity as to overcome the pressure in the boiler.

FIG. 113.

Of the injectors at present in use the improved self-acting one made by Wm. Sellers & Co. is typical and as efficient as any. It is shown in standard form in Fig. 113, and in longitudinal section in Fig. 114. It is simple in construction and has few operating parts. Its operation is as follows:

Assume the starting lever *A* to be at its extreme throw to the right, the plug *B* on the end of the spindle *C* entering and closing the steam nozzle *D* and the valve *E* on the spindle closing the diagonal holes *F* leading into the lifting nozzle *G*. Draw the starting lever about an inch to the left. This will not entirely withdraw the plug *B* from the steam nozzle *D*, but will cause the valve *E* to uncover the holes *F* for the admission of steam from the pipe *H* into the nozzle *G*. This steam discharges into the combining tube *K*, thence through the openings *a*, *b*, *c* into the chamber *L*, lifting the overflow valve *M* and escaping through the waste pipe *N*. In passing between the nozzles *D* and *G* the steam creates a vacuum in the space *O*, causing water

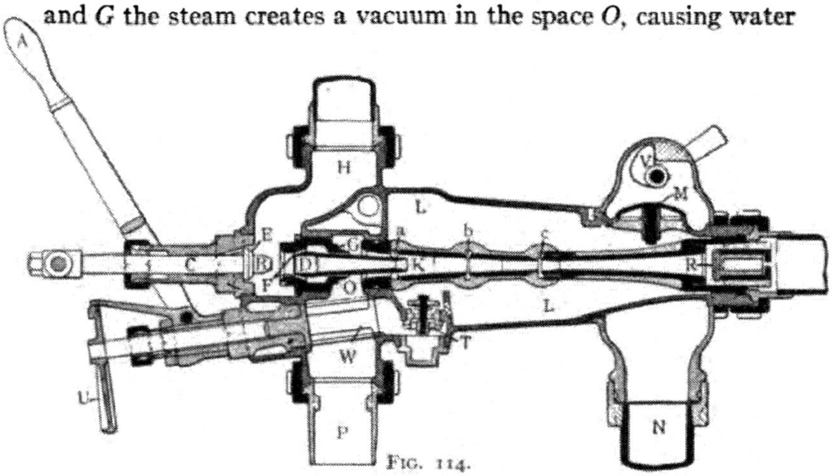

Fig. 114.

to rise from the supply through the pipe *P*. When water is lifting, the starting lever is drawn to the left, releasing the plug *B* from the steam nozzle and admitting the full supply of steam through *D* into the combining tube *K*. This steam, coming in contact with the water entering *K*, combines with it in condensing and imparts to it so much of its velocity as to cause it to enter the boiler through the check valve *R*. Should the steam or water be temporarily interrupted for any cause, the injector will automatically restart without the assistance of the operator. The openings *a*, *b* and *c* are so proportioned as to permit sufficiently free discharge of both steam jets to produce a continuous vacuum in the suction pipe *P* at all steam pressures between 20 pounds and 250 pounds. With the injector operating at high steam pressure, a vacuum is produced in the overflow chamber L, causing the valve T to open and admit an additional supply of water, which enters the combining tube through the openings a, b, c, and is forced into the boiler. By turning the plug valve W by means of the handle U, the feed supply may be regulated. Should it be desired to heat the supply of water the overflow valve may be held shut by the cam V and steam be permitted to blow down the pipe P.

134. **The Duty of a Pump.**—The method of rating pumps is to state their capacities in gallons discharged in twenty-four hours.

The usual method of stating the performance of a pump is to express in foot pounds the work it does for each 1,000,000 B.t.u. supplied. This is called the *duty* of the pump, and its value ranges from 60,000,000 to 180,000,000.

Owing to the escape of water through the valves while they are closing, and to piston leakage, a pump discharges from 0.85 to 0.95 only of its cylinder capacity per stroke, and this is known as the coefficient of slip.

If L denotes the length of stroke in feet, A the net area of piston in square feet, W the weight of one cubic foot of the water pumped, N the number of discharge strokes per hour, C the coefficient of slip, we shall have

$$\text{Water discharged per hour} = LAWNC \text{ pounds.}$$

If this water is discharged against a pressure head of H feet, which is equivalent to raising the water through a height of H feet, we shall have

$$\text{Work per hour} = LAWNCH \text{ foot pounds.}$$

Then $$\text{Duty} = \frac{1{,}000{,}000\ LAWNCH}{\text{B.t.u. supplied per hour}},$$

and

$$\text{Efficiency of pump} = \frac{\text{Useful work performed}}{\text{Total energy expended}} = \frac{\text{Duty}}{778{,}000{,}000}.$$

Example. — A single-acting pump, 80″ × 48″, makes 20 strokes per minute. Reading of pressure gauge on discharge pipe, 200 pounds; reading of vacuum gauge on suction pipe, 11 pounds below atmospheric pressure; height between gauges, 6 feet; coefficient of slip, 0.95; heat units supplied per hour, 16,593,000. Find the duty of the pump.

Solution. —

$$A = \frac{\pi \times (40)^2}{144} = 34.91 \text{ square feet.}$$

Since one pound pressure is equivalent to $\frac{33.913}{14.7} = 2.307$ feet of water head, we have

$$H = 2.307\,(200 + 11) + 6 = 492.78 \text{ feet.}$$

$$\text{Duty} = \frac{1{,}000{,}000\,LAWNCH}{\text{B.t.u. supplied per hour}},$$

$$= \frac{4 \times 34.91 \times 62.4 \times 20 \times 60 \times 0.95 \times 492.78}{16.593 \times 2},$$

$$= 147{,}300{,}000 \text{ foot pounds.}$$

PROBLEM

A double-acting pump, 8″ × 12″, makes 30 strokes per minute. Reading of pressure gauge, 150 pounds; reading of vacuum gauge, 12 pounds; height between gauges, 5 feet; coefficient of slip, 0.92; heat units supplied per hour to steam cylinder of pump, 195,200. Find the duty of the pump.

Ans. 70,000,000 foot pounds.

CHAPTER IX

THE INDICATOR AND ITS DIAGRAM. THE PLANIMETER. CLEARANCE. RATIO OF EXPANSION.

135. **The Steam-engine Indicator.** — The steam-engine indicator is an instrument devised to show primarily the steam pressure within the cylinder at every point of the stroke.

It consists essentially of a small steam cylinder containing a piston whose vertical movement, when acted upon by the pressure of the steam beneath it, is opposed by a spiral spring of known tension. The movement of the piston actuates a parallel motion consisting of a system of light levers, a point in which is constrained to move in a vertical line parallel to the motion of the piston. There is also a cylinder, or drum, to which a paper is attached and which receives a forward and backward motion of rotation on its axis, the one by means of a string attached to the crosshead or other part of the engine having a motion coincident with that of the engine piston, and the other by means of the reaction of a coiled spring attached to the base of the drum which is drawn into tension by the preceding motion of the drum.

The parallel point of the parallel motion carries a pencil which is made to reproduce the vertical motion of the indicator piston, magnified from three to five times by means of the system of levers, so it can readily be understood that when the instrument is in operation and the pencil pressed gently against the paper on the drum, the combination of the vertical motion of the pencil and the rotating motion of the drum will cause a closed diagram to be described, the area of which, it will be shown, represents the effective work done by the steam on one side of the engine piston during one revolution of the engine.

When communication is opened between one end of the engine cylinder and the lower end of the indicator cylinder, the vertical motion of the indicator piston during the forward rotation of the drum is that due to the pressure of the steam in the engine cylinder during the forward stroke of the piston; and during the backward rotation of the drum it is that due to the pressure on the same side of the piston during its return stroke. During the forward stroke of the piston the upper boundary of the diagram is described and its ordinates, measured from the base line of no pressure and to the scale of the indicator spring, show the varying pressure of the steam on the piston during the stroke; and during the return stroke of the piston the lower boundary is described, its ordinates showing the pressures opposing the piston. The length of the mean ordinate of the upper boundary of the diagram will then represent the mean pressure urging the piston forward, and the length of the mean ordinate of the lower boundary will represent the mean pressure opposing the piston on its return stroke. It is evident that the difference in length of these two ordinates, expressed in pounds to the scale of the indicator spring, is the mean effective pressure on one side of the piston during its forward and return strokes. In like manner, when communication is opened between the indicator and the other end of the engine cylinder a similar diagram will be described from which the mean effective pressure on the other side of the piston during one revolution (two successive strokes) may be determined. The mean of these two effective pressures is the mean effective pressure in the cylinder during one revolution and is a factor in the power of the engine.

The pistons of the indicators in ordinary use are exactly one-half square inch in area and are fitted in the cylinders with great nicety, requiring only water packing. The upper side of the indicator piston has communication with the atmosphere by means of a small hole in the upper part of the cylinder.

The location of the indicator and the manner of its connection with the engine cylinder is governed largely by the construction of the engine to which it is to be applied; but wherever its location or the manner of its connection, the principles governing its action are always the same. In the case of the ordinary horizontal engine the usual location is midway in a pipe connecting the two ends of the cylinder, a three-way cock, to which the indicator is attached, enabling steam from either end of the engine cylinder to be admitted to the steam cylinder of the indicator.

136. Reducing Motion. — An important feature of the indicator on which the accuracy of the instrument depends equally with the parallelism of the motions of the piston and pencil is the reducing motion, which provides for the drum cylinder a motion that corresponds to the motion of the engine piston in all parts of its stroke.

The forward and backward motion of rotation on its axis of the drum cylinder that carries the diagram must be exactly coincident with the forward and backward motion of the engine piston, and this is obtained by a suitable connection by means of a cord and pulley with the engine crosshead, the tension of the coiled spring in the drum being sufficient to keep the cord taut during the return stroke of the engine piston.

The length of the diagram desirable, as well as the height, depends largely upon the speed of the engine, slow rotational speeds permitting longer and higher cards. The ordinary length is from 3 to 4 inches, so it is obviously necessary that the motion of the crosshead be reduced in the ratio

$$\frac{\text{Length of stroke of engine}}{\text{Length of diagram desired}}.$$

This arrangement is effected by a reducing gear peculiar to the different makes of indicators, care being taken to avoid long stretches of cord, and that the cord in some part of its course be made to travel in a direction parallel to the line of motion of the crosshead of the engine, the object being to secure an equal motion of the cord at the drum and at the point of its attachment to the reducing motion.

137. The Lazy Tongs. — A form of reducing motion especially adapted to low-speed engines is a modification of the pantograph, known as the *lazy tongs*. It is shown in Fig. 115, and consists of a system of levers arranged so as to form a flexible frame. The point C pivots in a vertical hole in the crosshead,

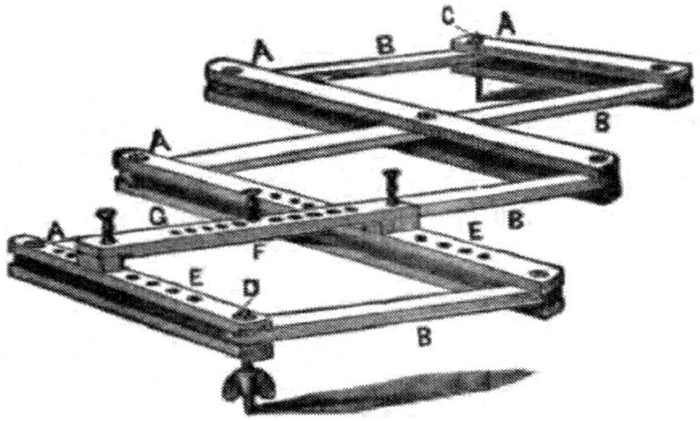

Fig. 115.

and the point D pivots in a socket in the top of a post, so placed on the floor near the crosshead guides that point D will be directly opposite point C when the crosshead is at mid-stroke. When the engine is in motion, every point in a line joining C and D has a true motion parallel to the path of the crosshead, varying in distance from nothing at D to length of stroke at C. The *hitch strip* G is arranged so that it may be shifted to similarly located holes in the bars E, E. For each position of strip G, one of its holes will be in line with the points C and D, and in this hole must be placed the *hitch pole* F, to which the indicator cord is attached. The floor post should be so placed that the cord shall lead in a line parallel to the line of motion of the cross-head; otherwise a guide pulley will be necessary.

To adjust the lazy tongs for a definite length of diagram, we have the proportion

$$\frac{DC}{DF} = \frac{\text{Length of stroke}}{\text{Length of diagram}},$$

which is true for any distance between the points D and C.

138. Indicator Springs. — The springs accompanying an indicator are numbered to correspond to the number of pounds pressure per square inch required to cause a vertical movement of one inch to the pencil. For example, a spring numbered 80 would require a steam pressure of 80 pounds per square inch to cause a vertical movement of one inch to the pencil. The tension of the spring to be used in any case depends upon the height of the diagram desired. For example, with a steam pressure of 160 pounds per square inch per gauge, an 80 spring would give a diagram 2 inches in height above the atmospheric line. In order, however, to obtain satisfactory diagrams from the modern engines of high rotational speed it has been found necessary to use springs of high tension to limit the movement of the piston and thus obviate the loss from friction that would result from a long and rapid movement. This provision for a small piston movement is all the more necessary with the high pressures now in use, in order that the pencil movement shall not be so great as to occasion vibrations that would cause undulations in the lines of the diagram.

The height of the diagram must be sufficient for practical use, and to provide for this the parallel motion is designed to give to the pencil a motion of from three to five times that of the piston, and it is a matter of importance that the designed motion shall not only insure the straight line motion of the pencil, but should also maintain the fixed ratio of movement between piston and pencil, for otherwise the vertical movement of one inch to the pencil would not correspond in pounds pressure to the number on the spring.

139. **The Thompson Indicator.** — There are a number of steam-engine indicators on the market, each claiming some distinctive merit, but of those in general use that of Thompson, manufactured by the American Steam Gauge and Valve Manufacturing Co., combines all of the newest and best features of the instrument.

An outside view of this indicator is shown in Fig. 116, *A* being the steam cylinder; *B* the drum cylinder which carries the paper on which the diagram is to be traced, the clips *C*, *C* holding the paper in place; *D* the pulley which leads the cord in any desired direction to get motion from the engine crosshead; *E* a coupling by means of which the indicator is connected to the engine; and *F* the lever operating the detent motion by means of which the motion of the drum of the paper cylinder may be stopped while that of its carriage continues. The system of small levers shown at the top of the steam cylinder comprises the parallel motion which insures a vertical motion to the pencil point *K*.

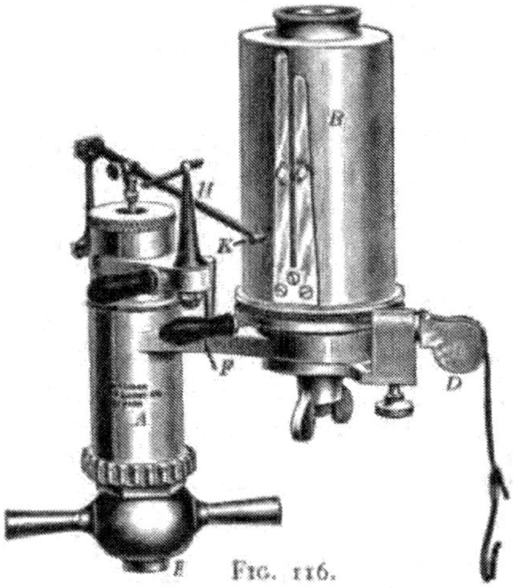

FIG. 116.

Figure 117 shows the indicator in vertical section. The steam from the engine cylinder enters at *E* and acts on the piston *P*, compressing the spiral spring on top of the piston to a degree depending on the steam pressure within the engine cylinder. The movement of the piston is transferred to the pencil point and there multiplied three times by means of the combination of the levers of the parallel motion. The paper cylinder is connected by a cord to the reducing motion, the cord pulling the cylinder around on its axis during one stroke of the engine piston, the movement of the cylinder on its axis in the opposite direction during the return stroke of the engine piston being accomplished by the reaction of the coiled spring *G* which the cord had drawn in tension during the first axial movement. The tension of the coiled spring *G* may be increased or decreased according to the speed of the engine, the idea being to give to the spring sufficient tension only to keep the cord taut.

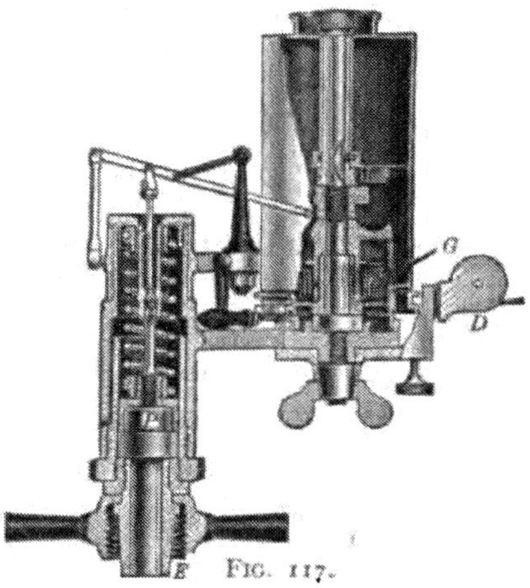

FIG. 117.

The pencil lever *H*, Fig. 116, is guided by a short connecting link pivoted at the top of the stationary post of the parallel motion, and its connection with the piston rod is made by means of a yoke and screw.

FIG. 118.

Figure 118 shows the piston, the spring, and the piston rod. The piston is grooved for water packing and is made as light as is consistent with the strength necessary to prevent expansion from pressure. The spring is screwed fast to the piston at the lower end and to the cap of the cylinder at the upper end. The piston rod is connected to the pencil lever by means of the yoke and screw as shown. The springs are made of the finest quality of steel wire and are scaled to a piston of one-half square inch area and to provide for a vacuum. The capacity of a spring to be used with a piston of one-half square inch area may be obtained by multiplying the scale number by $2\frac{1}{2}$ and subtracting 15 from the product. For example, the capacity of a 40-pound spring is 85 pounds when used with a one-half square inch area piston, but its capacity would be 170 pounds with a piston of one-quarter square inch area. Springs are made to any desired scale, and for pressures up to 600 pounds per square inch.

To avoid the necessity of disconnecting the cord in stopping the drum cylinder for each diagram it is desired to take, it is only necessary to move the lever *F*, Fig. 116, of the detent motion until the drum cylinder releases itself from its carriage, after which the lever must be returned to its original position. The motion of the drum being stopped a paper for a new diagram may readily be put on it and its motion renewed by turning the drum by means of the milled rim at the top until it reëngages itself with the carriage. The detent motion is a very important feature of the instrument, for the diagrams for the two ends of the cylinder should be taken as nearly simultaneous as possible, and in the many cases where it is necessary to take successive diagrams the delays resulting from disconnecting and reconnecting the cord are not only vexatious but impair the accuracy of results.

The reducing motion is shown in Fig. 119. The cord from the lower wheel *A* runs from the lead pulley to the crosshead of the engine. The wheel *B* is geared to wheel *A* in the ratio of 3 to 1. The cord of wheel *B* leads to the paper cylinder of the indicator, its winding on the disk *C* projecting from a bushing in wheel *B* being in a direction opposite to that of the winding of the cord on wheel *A*. The wheel *B* makes but one revolution while the wheel *A* makes three, thus reducing the velocity of the back-and-forth movement of the paper cylinder. Disks are made of different diameters for the bushing of wheel *B* in order to get the length of diagram desired. Within the wheel *B* is a coiled spring which is drawn in tension during one stroke, its reaction keeping the cords taut during the return stroke.

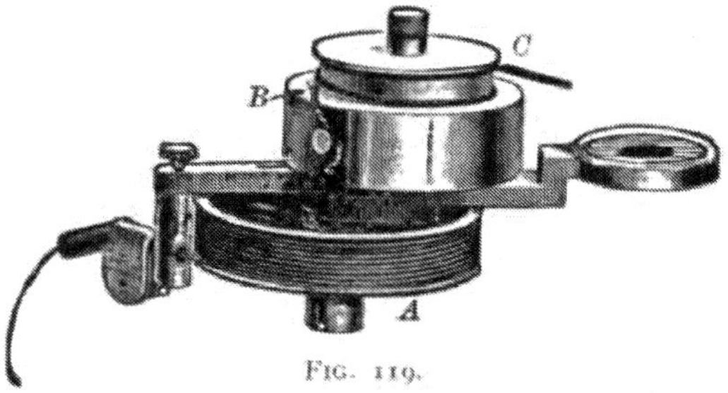

FIG. 119.

To avoid the injurious effects of high temperatures upon the springs, the Thompson indicator is made in the form shown in Fig. 120, the spring being outside the steam cylinder and out of contact with the steam; in all other respects this form of the indicator is the same as the inside spring type.

The outside spring type of indicator is specially adapted to gas engine practice where the temperature runs very high, and to steam engines using superheated steam.

140. Information from the Indicator Diagram. — The proper interpretation of indicator diagrams reveals so many facts that are essentially necessary to secure an economical and efficient performance of a steam engine that a careful study leading to a complete understanding of them is a matter of the greatest importance.

The principal value of the indicator diagram is that it furnishes the means of ascertaining the mean effective pressure exerted on the engine piston throughout a revolution, thus furnishing the factor which permits the power of the engine to be determined; but an inspection of the diagram reveals information concerning the engine's design and performance, of which the following are the most important:

1. Whether the valves are properly set; whether the admission of steam is early or late; whether the initial pressure is

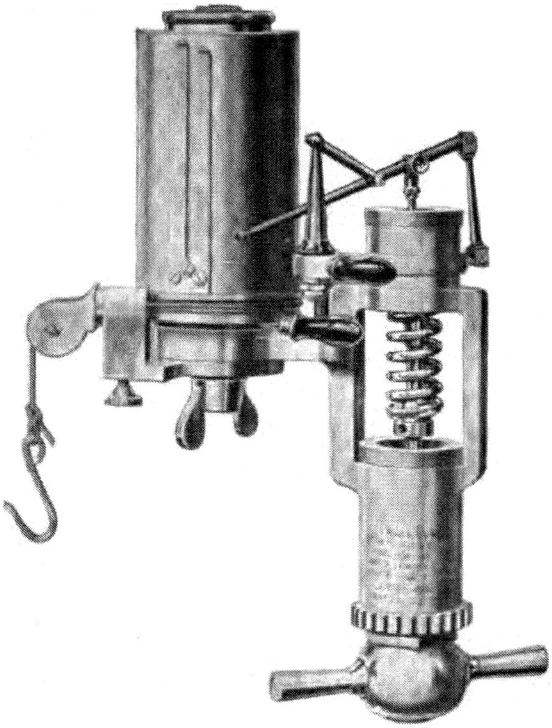

FIG. 120.

unduly lower than the boiler pressure, and the degree to which it is maintained up to the point of cut-off.

2. The point of the stroke at which the admission of steam to the cylinder is cut off, and whether the cut-off is sharp or gradual.

3. The point of the stroke at which release takes place, and the pressure of the steam at that instant.

4. The amount of back pressure opposed to the exhaust, the point of the stroke at which the exhaust is closed, and the amount of compression at the end of the stroke.

5. Whether the steam ports are of adequate size, and whether the valve or the piston leaks.

6. The approximate amount of steam consumed in a given time, and a number of vital features concerning the balance of the engine.

141. Taking the Indicator Diagram. — When about to take a diagram the indicator should be warmed and the pipes blown through by alternately connecting the indicator with the two ends of the engine cylinder by means of the three-way cock. Then, the length of the drum cord being adjusted so that the drum does not bring up against the stops, it is connected with the reducing motion, and if the pencil (lead or metallic) be sharpened to a finely rounded point, everything is in readiness to take the diagram.

Closing the three-way cock to both ends of the engine cylinder, a small hole in the cock then admits the pressure of the atmosphere under the indicator piston, thus placing the piston in a state of equilibrium with respect to the atmosphere. The *atmospheric line AA'*, Fig. 121, must then be traced by applying the pencil gently to the paper on the moving drum.

If the cock now be turned so as to admit steam from one end of the engine cylinder to the indicator cylinder, and the pencil be pressed against the paper during one revolution of the engine, a diagram approaching *CDFGHI*, Fig. 121, will be traced. The lines of this diagram and certain of its points have distinctive names in the language of the indicator.

Immediately upon the admission of steam into the indicator cylinder the *admission line CD* is traced, its height above the atmospheric line, measured to the scale of the indicator spring, showing the initial gauge pressure of the steam admitted to the engine cylinder.

The engine piston starting on its stroke, the ***steam line DF*** is traced during the time steam is being admitted into the engine cylinder.

At the point of cut-off, *F*, the valve closes, preventing any further admission of steam into the cylinder. The exact point of cut-off, when effected by a slide valve, is difficult to locate, owing to the fall of steam pressure due to the gradual closing of the port by the valve, shown by the curving of the diagram about *F*.

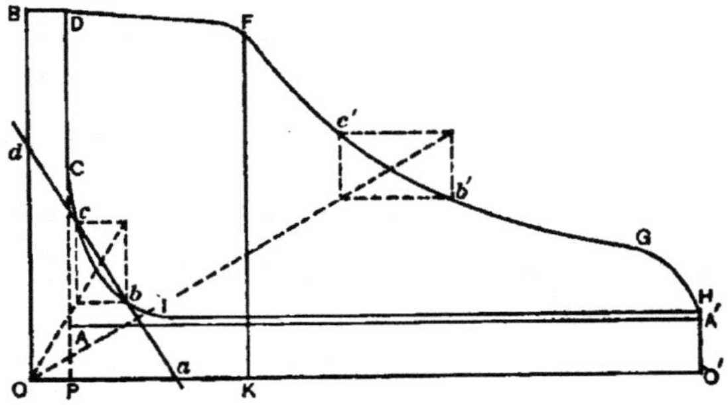

Fig. 121.

The *expansion curve FG* represents the fall in pressure of the steam confined in the cylinder after cut-off in forcing the piston to the end of the stroke.

At *G*, *the point of release*, the valve opens to the exhaust, releasing the steam from the cylinder. The higher the rotational speed of the engine the earlier the steam must be released to enable its pressure to fall to that of the back pressure before the piston commences its return stroke.

The *exhaust line GH* is traced in the interval between release and the end of the stroke, the pressure falling rapidly to that of the back pressure opposed to the exhaust.

In order that the exhaust steam may flow from the cylinder of a condensing engine to the condenser, or into the atmosphere from the cylinder of a non-condensing engine, the actual back pressure must be greater than the pressure in the condenser in the one case or greater than the atmospheric pressure in the other, and this excess of pressure depends largely upon the freedom of passage for the exhaust steam from the cylinder to the condenser or to the atmosphere. The release of the steam at from 88 per cent to 90 per cent of the stroke assists materially in the freedom of the exhaust; this is necessary in a condensing engine to insure a nearly complete vacuum when the piston starts on its return stroke, and with a non-condensing engine it enables the exhaust steam to begin its flow into the atmosphere before the return stroke commences.

The *back-pressure line HI* shows the pressure opposed to the piston on its return stroke. In non-condensing engines this line is slightly above the atmospheric line, as in Fig. 121, and in condensing engines it is below the atmospheric line a distance corresponding to the vacuum obtained; but in either case it is back pressure. Vacuum is expressed in inches of mercury, and since one cubic inch of mercury weighs 0.491 pound, the inches of vacuum multiplied by 0.491 will give the pressure equivalent to the vacuum.

At *I*, the *point of exhaust closure*, the valve closes the port to the exhaust and the compression of the steam entrapped in the cylinder begins.

The *compression curve IC* represents the rise in pressure of the entrapped steam due to its compression into the clearance space by the piston. The advancing piston compresses the steam, its pressure rising to some point *C* where the valve opens to lead, the pressure rising suddenly to *D* and a new stroke commences.

For the study of the diagram, and for computations involving pressures, it is necessary to locate the *vacuum line OO′*, or line of no pressure, from which *all pressures must be measured to make them absolute*. The vacuum line is parallel to the atmospheric line and at a distance below it equal to the pressure of the atmosphere measured to the scale of the indicator spring. The average value of the atmospheric pressure is 14.7 pounds.

Of equal importance to the vacuum line in computations involving the indicator diagram is the *clearance line OB*. It is perpendicular to the atmospheric line, and at a distance from the end of the diagram equal to the same percentage of the length of the diagram that the volume of the clearance space of the cylinder bears to the volume displaced by the piston.

The diagrams from the two ends of the cylinder should be taken simultaneously if two indicators are used, or one immediately after the other if only one be used.

Diagrams taken from engines of proper design and adjustment do not differ very materially from the theoretical diagram, but it requires careful study and discriminating judgment to make proper use of the information presented by them, a fact that may be appreciated when it is considered that the only absolute information a diagram gives is the varying pressure of the steam in the cylinder.

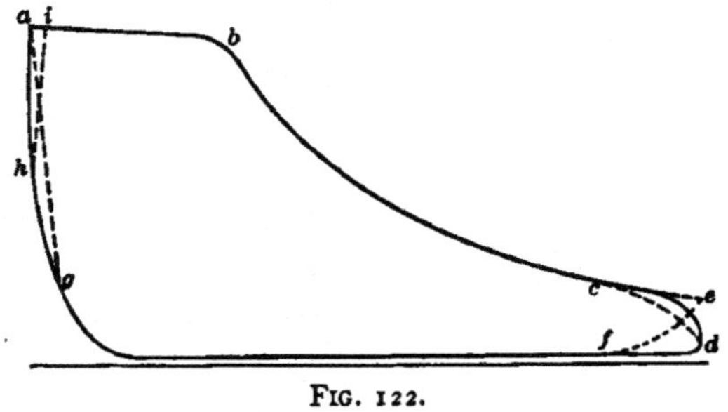

FIG. 122.

The full-line diagram of Fig. 122 would indicate a very satisfactory performance. The gradual fall in pressure in the steam line from *a* to *b* indicates *wire-drawing*, the technical name given the reduction in pressure due to friction in the passages. Improper design of the ports may cause this loss to be excessive.

The dotted lines illustrate some possible defects of an engine which would readily be detected by the indicator. The line *cd* would show that the release was too early, and the line *ef* that it was too late; the inclination of the admission line to the left, as at *ga*, would show the lead to be too great, and its inclination to the right, as at *hi*, would show insufficient lead.

Should a diagram be looped as in Fig. 123, the area *adc* represents negative work, and in obtaining the mean pressure from such a diagram the lengths of the ordinates included in the loop must be subtracted from the total length of those within the

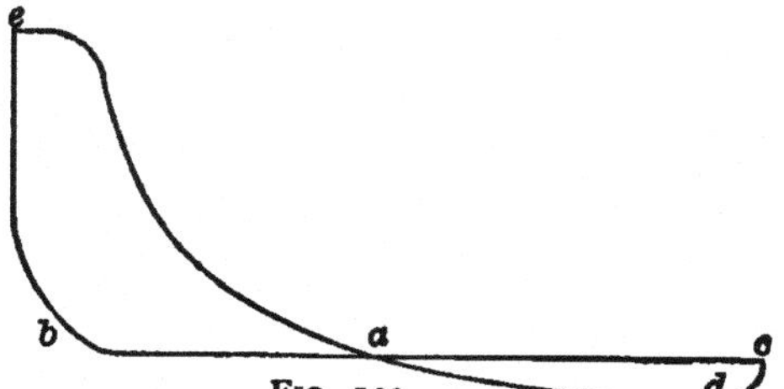

FIG. 123.

area *eba*. A loop like that in Fig. 123 would be occasioned by excessive expansion. At the point *a*, where the expansion curve crosses the back-pressure line, it is evident that the pressures on both sides of the piston are equal and a cut-off which would occasion an expansion so excessive as to reduce the steam pressure to a point below the back pressure opposed to the piston would be manifestly too early. The theoretical limit of expansion is such that the terminal pressure should be just equal to the back pressure, but practical considerations make it exceed this, varying from 24 to 28 pounds absolute in non-condensing engines and from 10 to 15 pounds in condensing engines. In actual practice a loop in a diagram would very likely indicate that the engine was underloaded.

It has been shown that mechanical work is produced by a force working through a distance. In the case of any gas working within a cylinder against a piston, the force will be the mean value of the pressure of the gas multiplied by the area of the piston, and the distance will be the stroke of the piston. In order that the work may be expressed in foot pounds the force must be expressed in pounds and the distance in feet. It is seen then that the area of an indicator diagram is the measure of the work performed on one side of the piston during one revolution; for this area is the product of the length of the mean ordinate of the diagram and the length of the diagram, the first factor expressing the mean effective pressure on the piston in pounds per square inch and the second factor expressing the length of the stroke in feet.

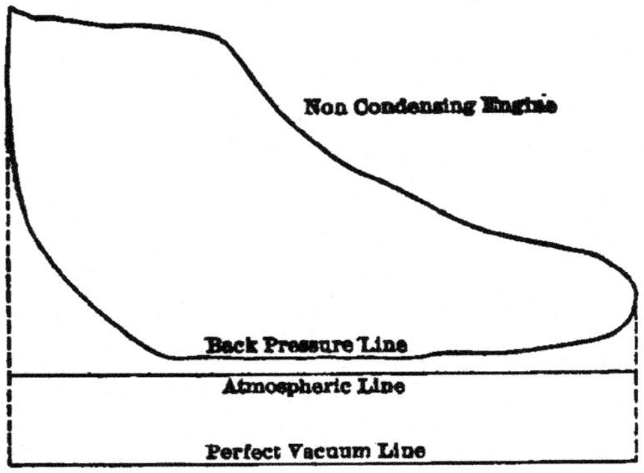

FIG. 124.

Figures 124 and 125 are diagrams from a non-condensing and a condensing engine respectively.

142. Mean Effective Pressure from the Diagram. — There are two methods of obtaining the mean effective pressure from the indicator diagram: (a) By the measurement of ordinates. (b) By the use of the planimeter.

143. M.E.P. by the Method of Ordinates. — To obtain the mean effective pressure from the indicator diagram by the method of ordinates, erect perpendiculars to the atmospheric line touching the extreme ends of the diagram. Divide the space between these perpendiculars into ten equal parts and at the middle points between these divisions erect ordinates to the diagram perpendicular to the atmospheric line. The first and last of these ordinates will be $\frac{1}{20}$ of the length of the diagram from the

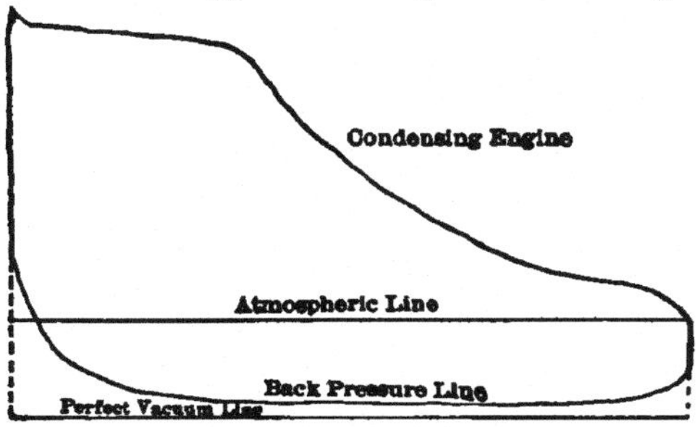

FIG. 125.

ends, and the common interval between the ordinates will be $\frac{1}{10}$ of the length of the diagram. One-tenth of the sum of the lengths of the ordinates will be the length of the mean ordinate, and the length of the mean ordinate multiplied by the scale of the indicator spring gives the mean effective pressure on the piston throughout the stroke in pounds per square inch.

The diagrams of Fig. 126 were taken from a high-speed engine of the Harrisburg type. The sum of the lengths of the ordinates

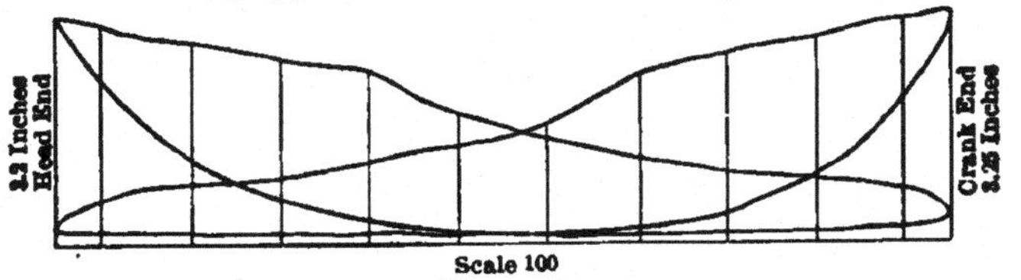

FIG. 126.

of the diagrams from the two ends of the cylinder is 3.2 inches and 3.25 inches, and the scale of the indicator spring is 100 pounds to the inch. Then, for one revolution

$$\text{M.E.P.} = \frac{100\,(3.20 + 3.25)}{20} = 32.25 \text{ pounds.}$$

144. M.E.P. by the Planimeter. — The planimeter is an instrument designed primarily to measure the areas of plane figures. Its application to finding the area of an indicator diagram, from which the length of the mean ordinate is readily obtained, enables the mean effective pressure to be found more quickly and accurately than by the method of ordinates.

The instrument most commonly used is some form of the polar planimeter of Amsler. The one illustrated in Fig. 127 is manufactured by Keuffel & Esser. It consists essentially of the pole arm *AB* of fixed length and the tracer arm *AC* whose length is

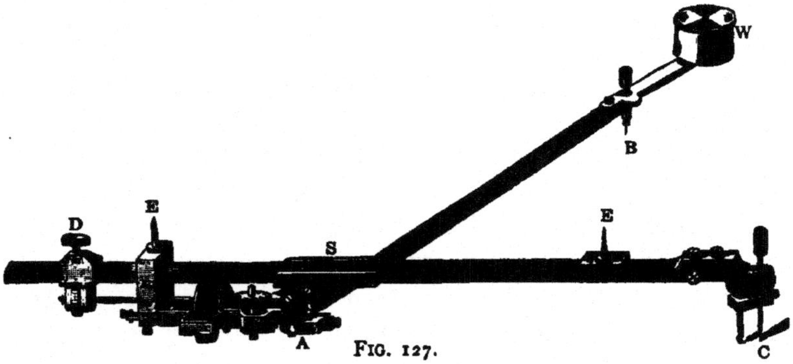

Fig. 127.

adjustable. At the pole *B* is a needle point by which the instrument is attached to the paper, the other end of the pole arm being pivoted to the slide *S*. The tracer arm, whose length may be varied by moving it in the slide, has at its outer end the tracing point *C* which is moved over the perimeter of the figure to be measured. The carriage of the instrument, of which the slide *S* is a part, contains a graduated record wheel and vernier whose reading gives the area of the figure measured in units depending upon the scale of the figure and the adjustment of the arm *AC* in the slide *S*. The graduated wheel is flanged at one side and on this flange the whole system rides, the only other moving part of the instrument touching the paper being the tracing point C. In addition to the wheel and vernier the carriage contains a graduated disk which records the number of revolutions of the wheel.

As usually made the flange of the record wheel is 0.796 inch in diameter, so that its circumference is $0.796\pi = 2.5$ inches. This circumference is divided into fortieths of an inch, making $2.5 \times 40 = 100$ divisions. The divisions are not actually made on the circumference of the flange but on the face of the wheel itself, the length of each division being the intercepted part of the circumference of the wheel made by the radii drawn from the extremities of one of the fortieth inch divisions on the circumference of the flange. Then the reading of the scale on the wheel very properly gives the distance in fortieths of an inch passed over by the flange.

It is evident that for one revolution of the wheel any one point on the circumference of the flange will travel a distance of 2.5 inches. In a mathematical discussion of the theory of the planimeter it is shown that the area registered by the instrument is equal to the distance passed over by a point in the circumference of the flange multiplied by the arm whose length is the distance between the pivot end of the arm AB and the tracing point C. As the instrument is designed the arm is the distance between the points E, E when the measurement is made. If this distance is 4 inches the area measured for one revolution of the wheel will be $2.5 \times 4 = 10$ square inches.

145. Operation of the Planimeter. — Bring the 10-square-inch mark on the tracer arm AC in coincidence with the fixed mark on the slide S and clamp the arm in that position by means of the milled screw D. The distance between the points E, E will then be 4 inches. Anchor the pole B to the paper by means of the needle point and weight W, being careful to select the point of anchorage so that the flange of the record wheel will not run off the paper during the tracing. Place the tracing point

C at some chosen starting point on the perimeter of the figure to be measured and then bring the zeros of the scale and vernier together. Move the tracing point in a clockwise direction over the perimeter of the figure until the starting point is reached. Suppose then that the scale reading is 52.5, showing that the distance passed over in a positive direction was $\frac{52.5}{40}$ inches. The area would then be $\frac{52.5}{40} \times 4 = 5.25$ square inches. The divisions on the scale are divided into groups of ten and numbered consecutively from 0 to 9 in order to facilitate the reading of results. Thus, the reading of the result in the example showed that 5.25 of the groups had been passed over, hence an area of 5.25 square inches.

If the 15-square-inch mark on the tracer arm be brought into coincidence with the mark on the slide the distance between the points E, E will be found to be 6 inches, and if, in this adjustment, the tracing point be run over the figure of the above example the reading of the record wheel will be $\frac{35}{40}$ inch, and the area then will be $\frac{35}{40} \times 6 = 5.25$ square inches, as before. In this instance the area cannot be read directly from the scale by means of the group divisions, since that applies only when the tracer arm is adjusted in the slide at the 10-square-inch mark. Engraved also on the tracer arm of the Keuffel & Esser instrument is a 12.5-square-inch mark and a 16-square-inch mark. If the former is adjusted to the mark on the slide the distance between the points E, E is 5 inches, and with the latter in adjustment the distance is 6.4 inches. Areas may be measured by any one of the adjustments as shown, but for small areas the 10-square-inch adjustment is most convenient. Should the area exceed 10 square inches the record wheel would make more than one revolution. In such case the revolutions would be recorded on the graduated flat disk, and to the reading of the record wheel should be added 10 square inches for each revolution so recorded.

In the measurement of large areas from figures drawn to scale care must be taken in the interpretation of results. The instrument is designed on the basis of a 2.5-inch movement of the record wheel and a 4-inch arm, making 10 square inches for one revolution of the wheel. When the tracer arm is adjusted to the 10-square-inch mark, the result, as we have seen, is given in square inches, if the figure measured be drawn to full scale. If the figure be drawn to any other scale, say, to the scale of 1″ = 10′, then 1 square inch of the figure represents 100 square feet, and the reading of the record wheel must be multiplied by 100 to get the area in square feet, the vernier unit then being 1 square foot. If the figure of the example given above had been drawn to a scale of 1″ = 10′ the area would have been $5.25 \times 100 = 525$ square feet.

If the tracer arm be adjusted in the slide at the 15-square-inch mark, the distance between the points *E*, *E* being then 6 inches, the reading would have to be converted to the basis of the 4-inch arm. We found in the measurement of the figure of our example that for the 15-square-inch adjustment the reading of the record wheel was $\frac{35}{40}$ inch. If the figure had been drawn to a scale, say, of 1″ = 20′, then 1 square inch of the figure would represent 400 square feet for the 10-square-inch adjustment, and $400 \times \frac{6}{4} = 600$ square feet for the 15-square-inch adjustment, and the area would be $\frac{35}{40} \times 6 \times 600 = 3150$ square feet, the vernier unit then being 6 square feet. In like manner the readings for other adjustments of the tracer arm and for other scales may be interpreted.

In this explanation of the operation of the planimeter it is understood that the pole *B* is outside the figure measured. If the figure be too large to permit this it may be divided into parts and the measurement of each part taken separately.

The use of this instrument in connection with indicator diagrams greatly facilitates the finding of the mean effective pressure. If the tracer arm be so adjusted that the distance between the points *E*, *E* is just equal to the length of the diagram, and if the zeros then be brought together and the tracing point moved over the outline of the diagram, the reading of the record wheel will be the length of the mean ordinate of the diagram in fortieths of an inch, and the product of this reading and the scale of the indicator spring will be the mean effective pressure in pounds per square inch on the piston throughout the stroke.

This will be understood from the fact that the area of the diagram is equal to the product of its length and the length of its mean ordinate; and since, by the theory of the instrument, the area measured is the product of the movement of the record wheel and the distance between the points *E*, *E*, it follows that when the distance *EE* is equal to the length of the diagram the reading of the record wheel must necessarily be the length of the mean ordinate in fortieths of an inch. For example, suppose the reading of the record wheel to be $\frac{32.8}{40}$, and the scale of the indicator spring to be 50 pounds to the inch. The mean effective pressure would then be $\frac{32.8}{40} \times 50 = 41$ pounds.

The smaller instrument illustrated in Fig. 128 has no sliding arm and is limited in its radius of action. The record wheel, vernier, and recording disk are the same as in the larger instrument. Its design is based on the 10-square-inch adjustment of the larger instrument; that is, the areas are given in square inches when the figure measured is drawn to full scale. Should the figure be drawn to any other scale, say 1″ = 10′, then 1 square inch of the area would represent 100 square feet, and the reading of the record wheel multiplied by 100 would be the area in square feet.

Despite the fact that this small instrument cannot be adjusted to the length of indicator diagrams, it is none the less handy and convenient in finding mean pressures. In its operation, the pole *B* is anchored as in the larger instrument, the zeros of the record wheel brought together, and the tracing point *C* moved in a clockwise direction over the outline of the diagram.

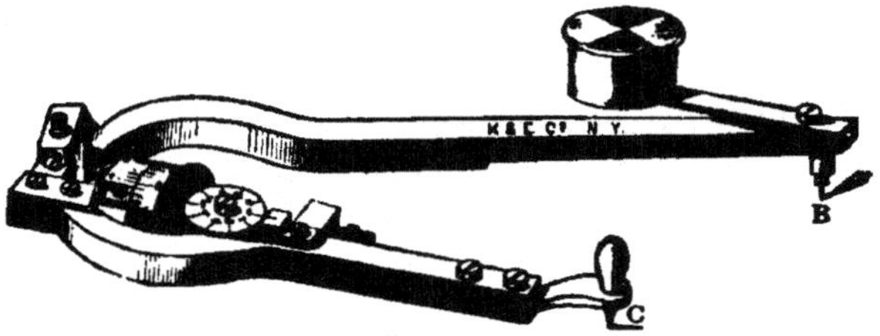

FIG. 128.

The reading of the record wheel gives the area in square inches, and this area, divided by the length of the diagram, gives the length of the mean ordinate. The product of the length of the mean ordinate and the scale of the indicator spring gives the mean effective pressure. For example, if the reading were 2.85 square inches, the length of the diagram 3.8 inches, and the scale of the spring 40 pounds to the inch, the mean effective pressure would be $\frac{2.85}{3.8} \times 40 = 30$ pounds.

146. The Power of the Engine. — Having found from the indicator diagram the mean effective pressure in pounds per square inch acting on the engine piston throughout one revolution, the product of this pressure and the area of the piston in square inches will be the total pressure acting on the piston in pounds. If this total pressure be multiplied by the distance in feet moved through by the piston in one minute the product will be an expression in foot pounds of the work performed by the engine in a minute, and this product, divided by 33,000, will be the horse-power of the engine. The mean effective pressure having been found from the indicator diagram, the power thus obtained is called *indicated horse-power*, usually denoted by the initials I.H.P., and is equal to the useful work delivered by the engine and the work expended in overcoming the friction of the engine itself.

The factors in the determination of the indicated horse-power of an engine are:

p_e = M.E.P. = mean effective pressure on the piston in pounds per square inch,
L = length of stroke of the piston in feet,
A = area of piston in square inches,
N = number of revolutions of the engine per minute.

The product p_eA is the effective pressure on the piston in pounds, and since there are two strokes of the piston for each revolution of the engine, the product $2\,LN$ is the speed of the piston in feet per minute, so that

$$\text{I.H.P.} = \frac{2\,p_eLAN}{33{,}000}.$$

Only the net area of the piston must be used in calculating the horse-power of an engine. The steam on one side of the piston acts only on an area equal to that of the piston diminished by the cross-sectional area of the piston rod, so if a denotes the area of the cross section of the piston rod, we shall have

$$\text{Net area of piston} = \frac{A + (A - a)}{2} = A - \frac{a}{2}.$$

It should be noted that p_e and N are the only variables in the horse-power formula, so if the constant $C = \dfrac{2\,LN}{33{,}000}$ for any engine be known, we shall have for its horse-power

$$\text{I.H.P.} = Cp_eN.$$

In the stage-expansion engine the I.H.P. for each cylinder must be obtained and their sum taken as the I.H.P. of the engine.

147. Clearance. — The volume of all the space between the piston when at the end of its stroke and the valve face is known as the *clearance* of the engine. Clearance is expressed in terms of percentage of the volume proper of the cylinder, that is, of the volume displaced by the piston in one stroke. The amount of clearance varies in the different types of engines. In engines of slow speed and long stroke the variation is from 2 per cent to 4 per cent; in engines of high rotational speed and short stroke it may be as much as 8 per cent; and in marine engines a clearance of 15 per cent is not uncommon.

Engine builders usually measure the clearance very carefully but in the absence of any infonnation on the subject an approximation of the clearance of an engine may be determined graphically from its indicator diagram. Thus, in Fig. 121, select two points, *b* and *c*, in the compression curve as far apart as possible, and through them draw an indefinite line intersecting the vacuum line at *a*. From *c* lay off $cd = ba$, and through *d* draw *OB* perpendicular to *OO'*. Then *OB* will be the clearance line, and *BD* will represent the volume of the clearance to the same scale that *AA'* represents the volume of the cylinder. This construction, as well as the two that follow, assumes the curves of compression and expansion to be rectangular hyperbolas, the asymptotes of which are the vacuum and clearance lines. The method just given depends upon the property of the hyperbola that, if any secant line be drawn, the portions of the secant intercepted between the curve and its asymptotes are equal.

The two other constructions are: (a) The line *bc* joining the two points selected on the compression curve is a diagonal of a rectangle constructed on the curve, two adjacent sides of which are parallel respectively to the vacuum and clearance lines. If this rectangle be constructed and the other diagonal drawn and prolonged, its intersection *O* with the vacuum line will be the center of the curve, or the point of intersection of the asymptotes;

hence OB perpendicular to OO' is the clearance line. (b) If two points, b' and c', be taken on the expansion curve the construction is similar. Though these constructions are mathematically correct when the curves are hyperbolas, they give results by no means reliable when applied to curves of actual diagrams. When applied to an actual diagram all three methods should be adopted and the mean of the results taken.

The clearance space at each end of the cylinder must be filled with steam for each revolution of the engine, and this steam must come from the boiler, or from the steam left in the cylinder by the exhaust closure, or from both. Since the piston does not traverse the clearance space the clearance steam performs no initial work; that is, it does no work during the period of admission, but after cut-off its effect is to raise the pressure during expansion and thus increase the area of the expansion part of the diagram. If there were neither expansion nor compression the clearance steam would perform no work at all and would be a total loss in the exhaust. On the other hand, if the expansion curve were carried down to the back pressure and the compression curve carried up to the initial pressure there would be absolutely no loss from clearance. Such conditions are never realized in practice, therefore there is always a loss from clearance, and this loss is greater as the clearance is proportionally large.

One effect of cushioning is that it reduces the loss from waste of steam in the clearance space, but its most important effect is that it provides for smooth running of the engine by preventing shocks at the end of the stroke. It is especially desirable that the diagram of an engine of high rotational speed have its compression curve well rounded.

Clearance in an engine occasions a loss when the consumption of steam per unit of power is considered, but there are practical considerations which make its existence highly desirable, if not necessary. The clearance space between the piston and the cylinder head, when the piston is at the end of its stroke, gives space for the variable amount of water which is always present in a cylinder and doubtless prevents serious accidents which might otherwise occur.

148. Ratio of Expansion. — The ratio of expansion of the steam used in an engine is the quotient derived from dividing the final volume of steam found in the cylinder by the initial volume admitted. By initial volume is meant the volume of steam admitted to the cylinder up to the point of cut-off, plus the clearance volume, and by final volume is meant the volume of the cylinder, plus the clearance volume. Denoting the ratio of expansion by r, the initial volume by v_1, and the final volume by v_2, we shall have

$$r = \frac{v_2}{v_1}.$$

Clearance has a marked effect on the ratio of expansion and should never be neglected. Referring to Fig. 121, the ratio of expansion if clearance were neglected would be $\frac{PO'}{DF}$. The volume of steam that expands after cut-off at F is $DF + BD$, BD denoting the clearance, and the final volume found in the cylinder is $PO' + PO$. The true ratio of expansion is therefore $\frac{PO' + PO}{DF + BD}$.

Since the cross-section area of the cylinder is uniform, the volume displaced by the piston at any point is directly proportional to the fractional part of the stroke completed at the point, so that the volumes may be represented by their corresponding fractions of stroke. In like manner, the clearance volume, when divided by the cross-section area of the cylinder, will be expressed as a fractional part of the stroke. Then, if we denote the full stroke of the piston by unity, it may also represent the volume displaced by the piston in one stroke, in which case the fraction of the stroke denoting the cut-off will represent the volume displaced up to the point of cut-off.

If we denote the fraction of the stroke completed up to the point of cut-off by a, and if c be the fractional part of the stroke denoting the clearance, we shall have $a + c$ for the initial volume and $1 + c$ for the final volume. Then, for the ratio of expansion, we shall have

$$r = \frac{v_2}{v_1} = \frac{1 + c}{a + c}.$$

The question of expansion in the compound, triple, or any other form of stage-expansion engine is not different from that in the single-cylinder engine, so far as the total ratio of expansion is concerned. As in the simple engine, the total ratio of expansion is the ratio of the final volume of steam found in the last cylinder to the initial volume admitted to the first cylinder. That is, $r = \frac{v_2}{v_1}$, in which v_2 is the volume of the L.P. cylinder and its clearance and v_1 is the initial volume, or volume up to cut-off, including clearance, of the H.P. cylinder.

If, in any stage-expansion engine, we denote the volumetric ratio of the L.P. cylinder to the H.P. cylinder by ϕ, we shall have

$$\phi = \frac{\text{Volume of L.P. cylinder + its clearance volume}}{\text{Volume of H.P. cylinder + its clearance volume}}.$$

If r_1 denotes the ratio of expansion in the H.P. cylinder, c the clearance of the H.P. cylinder, and a the fractional part of the stroke at which cut-off takes place in the H.P. cylinder, then, as in the single-cylinder engine, we shall have

$$r_1 = \frac{1 + c}{a + c}.$$

It is evident that the volume of steam resulting from this expansion in the H.P. cylinder, that is, the volume of steam occupying the H.P. cylinder and its clearance at the end of the stroke, must eventually occupy the whole of the L.P. cylinder and its clearance, a volume ϕ times as great, it follows that for the total ratio of expansion in the stage-expansion engine, we shall have

$$r = \frac{\phi(1 + c)}{a + c}.$$

Neither the volume of the receiver nor the cut-off in the L.P. cylinder has anything to do with the question of the total ratio of expansion in stage-expansion engines. The effect of the receiver is to make the initial pressure lower in the L.P. cylinder than it otherwise would be if the exhaust from the H.P. to the L.P. cylinder were direct, and this reduction in pressure is due to the *drop* occasioned by the unrestricted expansion of the steam when it enters the receiver space. The receiver only plays the part of a large clearance space.

The low-pressure cut-off will increase the receiver pressure and therefore the power of the L.P. cylinder, as has been shown, and this increase in receiver pressure increases the back pressure on the piston of the next preceding cylinder in the expansion and therefore decreases the power of that cylinder. So it is seen that the function of the L.P. cut-off is to equalize the power between the cylinders and has nothing to do with the total ratio of expansion. Whether the steam is or is not cut off in the L.P. cylinder, the same weight of steam must find its way into that cylinder at each stroke, and if, by means of the cut-off, a less space be provided for the reception of the steam, the pressure will increase accordingly. The question of expansion in stage-expansion engines may be understood better with the aid of an example.

Example. — A non-condensing two-stage-expansion engine has a high-pressure cylinder 9 inches in diameter, a low-pressure cylinder 17 inches in diameter, and a stroke of 14 inches. The clearance of the H.P. cylinder is 5 per cent and that of the L.P. cylinder 3 per cent. The initial absolute steam pressure is 150 pounds, and the cut-off in the H.P. cylinder is at 0.5 stroke. It is required to find the total ratio of expansion.

Solution. — The cross-section areas of the H.P. and L.P. cylinders are 63.62 square inches and 226.98 square inches respectively.

The H.P. clearance of 5 per cent is equivalent to the addition of $14 \times 0.05 = 0.7$ inch to the stroke. Neglecting compression and cylinder condensation, we shall have, since the cut-off is at 0.5 stroke,

$$\frac{63.62 \times 7.7}{1728} = 0.2835 \text{ cubic feet}$$

of steam supplied to the H.P. cylinder per stroke, which is the initial volume v_1.

The L.P. clearance of 3 per cent is equivalent to the addition of $14 \times 0.03 = 0.42$ inch to the stroke. The initial volume v_1 must eventually occupy the L.P. cylinder and its clearance at one end, so that we shall have for the final volume

$$v_2 = \frac{226.98 \times 14.42}{1728} = 1.894 \text{ cubic feet.}$$

Therefore $$r = \frac{v_2}{v_1} = \frac{1.894}{0.2835} = 6.681.$$

To apply the formula $r = \frac{\phi(1 + c)}{a + c}$, we have, by denoting the L.P. clearance by k,

$$\phi = \frac{\text{Volume L.P. cylinder + its clearance volume}}{\text{Volume H.P. cylinder + its clearance volume}}$$

$$= \frac{\text{Area L.P. cylinder } (1 + k)}{\text{Area H.P. cylinder } (1 + c)} = \frac{(17)^2 \times 1.03}{(9)^2 \times 1.05} = 3.5.$$

Then $$r = \frac{3.5\,(1 + 0.05)}{0.5 + 0.05} = 6.681.$$

The specific volume of steam at 150 pounds pressure is 3.012 cubic feet, so that in the above example the weight of steam used per stroke was $\frac{0.2835}{3.012} = 0.0945$ pound. At the commencement of the return stroke of the H.P. piston the delivery of this steam into the receiver will begin and all of it will have been delivered when the return stroke is completed. Its withdrawal from the receiver will begin with the commencement of the stroke of the L.P. piston and must all be drawn out by the time the L.P. cut-off takes place. If it were not all drawn out the pressure in the receiver would increase as the engine ran until no exhaust from the H.P. cylinder could take place; if more were drawn out a point would soon be reached at which a vacuum would exist in the receiver. Both these suppositions being impossible, it follows that all the steam admitted to the H.P. cylinder eventually finds its way to the L.P. cylinder.

We have found that 0.0945 pound of steam is used per stroke and that the ratio of expansion is 6.681. If these figures are correct, the calculated volume of steam found in the L.P. cylinder and clearance at the end of the stroke should be equal to the combined volume of the L.P. cylinder and its clearance.

For the terminal pressure, p_2, in the L.P. cylinder, we have

$$p_2 = \frac{p_1}{r} = \frac{150}{6.681} = 22.45 \text{ pounds.}$$

The specific volume of steam at pressure of 22.45 pounds is 18.03 cubic feet, and since there is 0.0945 of a pound it will occupy $18.03 \times 0.0945 = 1.704$ cubic feet. Allowing for abbreviated decimals, this corresponds to the calculated volume of the L.P. cylinder and clearance, as it should.

In the above calculations the effect of compression has been neglected, but the only way this could affect the question would be to reduce slightly the quantity of steam withdrawn from the boiler at each stroke, which may be regarded as virtually increasing very slightly the ratio of expansion, because a less weight of fresh steam would be used each stroke.

The effect of cylinder condensation has also been neglected, but this also would occasion a virtual augmentation of the ratio of expansion, because a smaller weight of steam than that delivered to the H.P. cylinder would be found in the L.P. cylinder at the end of its stroke.

PROBLEMS

1. The ratio of expansion is 3.2727, and the clearance 8 per cent. Find the point of cut-off. *Ans.* 0.25.

2. The cylinder ratio of a compound engine is 4, the cut-off in the H.P. cylinder is at 0.4 stroke, and the clearance of each cylinder is 6 per cent. Find the total ratio of expansion. *Ans.* 9.22.

3. The diameter of the H.P. cylinder of a triple-expansion engine is 28 inches and its clearance 15 per cent; the diameter of the L.P cylinder is 87 inches and its clearance 12 per cent. If the cut-off in the H.P. cylinder is at 0.5 stroke, find the total ratio of expansion. *Ans.* 16.6.

4. Diameter of cylinder, 9 inches; stroke, 10 inches; cut-off, 0.28 stroke; clearance, 6.5 per cent. Initial pressure of steam, 95 pounds per square inch absolute, the specific volume of which is 4.65 cubic feet. Find the ratio of expansion and the terminal pressure, assuming the steam used to be saturated. Prove the accuracy of the results by showing that the calculated volume of the steam in the cylinder and clearance at the end of the stroke is equal to the combined volume of the cylinder and clearance. The specific volume of steam at the terminal pressure is 14.04 cubic feet. *Ans.* $r = 3.087$; $p_2 = 28.68$.

CHAPTER X

EXPANSION OF GASES. MEAN PRESSURE OF AN EXPANDING GAS

149. Adiabatic Expansion. — If a portion of gas expands before a weighted piston in a cylinder, both piston and cylinder being made of materials that are absolute non-conductors of heat, and if the gas neither receives nor parts with heat during the expansion, except that due to the external work performed in moving the piston, the expansion is called *adiabatic* (without transference).

During adiabatic expansion the temperature will fall as the volume increases and the pressure decreases, because the external work can be done only at the expense of the heat energy of the gas.

If, instead of allowing the gas to expand, it were compressed by adding weights to the piston, the temperature would rise in consequence of the work done on the gas by the piston, and in this instance the process would be called adiabatic compression.

The equation of the curve of adiabatic expansion and compression is of the form $PV^{\kappa} = C$, in which κ is the ratio of the specific heat of the gas at constant pressure to the specific heat at constant volume. For air, $\kappa = 1.406$ and for dry steam, $\kappa = 1.135$.

150. Isothermal Expansion. — If a portion of gas expands, or is compressed, before a piston in a cylinder, receiving during the expansion the exact amount of the heat expended in the external work of moving the piston in the cylinder, or rejects during the compression the heat equivalent to the mechanical energy spent upon it during the process, the temperature will remain constant, but the pressure will fall owing to the increase in volume, according to Boyle's law. Expansion of this character is called *isothermal* (equal temperature), the expansion curve being a rectangular hyperbola, the equation of which is $PV = C$, a constant.

General Equation for the Expansion of Gases. — In practice neither the expansion nor compression of gases is either exactly adiabatic or exactly isothermal, the conditions of the actual expansion probably being intermediate between the two. It is usual and sufficient to express the expansion of gases by a general equation of the form $pv^n = C$, in which n varies in value from 1 to 1.406. The equations $PV^K = C$ and $PV = C$ for adiabatic and isothermal expansions are but special cases of $pv^n = C$, n equaling K in the first case and unity in the second case. In these equations it should be remembered that P denotes the pressure in pounds per square foot and p the pressure in pounds per square inch; both V and v denote volumes in cubic feet.

151. Mean Pressure of a Working Gas. — With the understanding of work and of the indicator diagram obtained from Chapters I and IX, it may now be shown how the mean pressure of any gas working within a cylinder may be determined, the gas expanding according to any law.

In Fig. 129 let

p_1 = absolute initial pressure in pounds per square inch,
p_2 = absolute terminal pressure in pounds per square inch,
p_3 = absolute back pressure in pounds per square inch,
p_m = mean absolute pressure in pounds per square inch,
p_e = M.E.P. = mean effective pressure = $p_m - p_3$,
v_1 = initial volume of gas admitted to the cylinder,
v_2 = final volume of the gas in the cylinder,

$r = \frac{v_2}{v_1}$ = ratio of expansion of the gas.

The area of the diagram represents the gross work performed by the volume of gas v_1 of initial pressure p_1 in driving a piston in a cylinder through a stroke AE against a back pressure p_3,

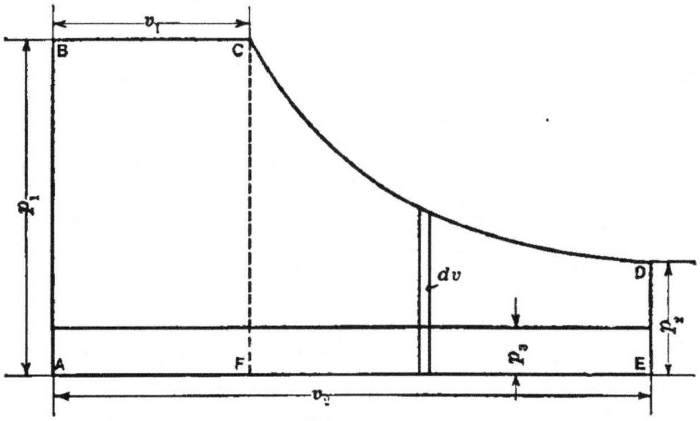

FIG. 129.

the volume v_1 of pressure p_1 expanding to the volume v_2 of pressure p_2.

If p_m denotes the mean absolute pressure of the gas, then

$$\text{Gross work} = p_m v_2,$$

whence $$p_m = \frac{\text{Gross work}}{v_2} = \frac{\text{Area of diagram}}{rv_1}.$$

The area of the rectangle $ABCF$ is p_1v_1, and if the equation of the expansion curve is $pv^n = C$, then the expansion area $FCDE$ is

$$\int_{v_1}^{v_2} p\, dv. \quad \text{But } pv^n = p_1v_1^n, \text{ whence } p = \frac{p_1v_1^n}{v^n}.$$

Substituting this value of p we get

$$\text{Expansion area } FCDE = \int_{v_1}^{v_2} \frac{p_1v_1^n\, dv}{v^n}.$$

$$\text{Whole area of diagram} = p_1v_1 + p_1v_1^n \int_{v_1}^{v_2} \frac{dv}{v^n} = p_1v_1 + p_1v_1^n \int_{v_1}^{rv_1} \frac{dv}{v^n}$$

$$= p_1v_1 + p_1v_1^n \int_{v_1}^{rv_1} v^{-n}\, dv.$$

By integration we get

$$\text{Whole area} = p_1v_1 + p_1v_1^n\left[\frac{1}{-n+1}v^{-n+1}\right]_{v_1}^{rv_1}$$

$$= p_1v_1 + p_1v_1^n\left[\frac{(rv_1)^{1-n} - v_1^{1-n}}{1-n}\right] \qquad (3)$$

$$= \frac{p_1v_1 - p_1v_1n + p_1v_1(r^{1-n} - 1)}{1-n}$$

$$= \frac{p_1v_1(r^{1-n} - n)}{1-n},$$

whence $$p_m = \frac{p_1v_1(r^{1-n} - n)}{rv_1(1-n)} = \frac{p_1(r^{1-n} - n)}{r(1-n)}, \qquad (4)$$

and $$p_e = \text{M.E.P.} = p_m - p_3.$$

A convenient expression for the expansion area, or for the work done during expansion, is deducible from equation (3), thus,

$$\text{Expansion area } FCDE = p_1v_1^n\left[\frac{(rv_1)^{1-n} - v_1^{1-n}}{1-n}\right]$$

$$= \frac{p_1v_1^n(v_1^{1-n} - v_2^{1-n})}{n-1} = \frac{p_1v_1 - p_1v_1^n v_2^{1-n}}{n-1}$$

$$= \frac{p_1v_1 - p_2v_2^n v_2^{1-n}}{n-1} = \frac{p_1v_1 - p_2v_2}{n-1}.$$

Equation (4) is the general expression for the value of the mean pressure, whatever the value of n.

If steam be the gas to be considered, there are three cases:

(1) **Adiabatic Expansion of Steam.** — The consideration of the adiabatic expansion of steam is of use only in theoretical investigations, since non-conducting cylinders do not exist in practice. The law governing the adiabatic expansion of steam is expressed approximately by the equation $pv^{\frac{10}{9}} = C$.

To find the expression for the mean pressure of steam expanding adiabatically we substitute $\frac{10}{9}$ for n in equation (4), thus

$$p_m = \frac{p_1(r^{1-\frac{10}{9}} - \frac{10}{9})}{r(1 - \frac{10}{9})} = \frac{p_1(r^{-\frac{1}{9}} - \frac{10}{9})}{-\frac{r}{9}} = \frac{p_1(10 - 9r^{-\frac{1}{9}})}{r}.$$

(2) **Saturated Steam Expansion.** — When saturated steam expands in a cylinder in the performance of work, receiving from a jacket or other external source sufficient heat to prevent liquefaction, the pressure at all points of the stroke is that due to the volume and temperature of saturated steam. According to Rankine's experiments the law governing such expansion of steam is expressed by the equation $pv^{\frac{17}{16}} = C$.

To find the expression for the mean pressure for saturated steam expanding in a cylinder we let $n = \frac{17}{16}$ in equation (4) thus

$$p_m = \frac{p_1 \left(r^{1-\frac{17}{16}} - \frac{17}{16}\right)}{r\left(1 - \frac{17}{16}\right)} = \frac{p_1 \left(r^{-\frac{1}{16}} - \frac{17}{16}\right)}{-\frac{r}{16}} = \frac{p_1 \left(17 - 16\, r^{-\frac{1}{16}}\right)}{r}.$$

(3) **Isothermal Expansion of Steam.** — When steam expands isothermally in a cylinder, receiving from an external source an amount of heat equivalent to the external work performed, the temperature remains constant. The expansion under such conditions is expressed by the equation $pv = C$, which is the equation of the rectangular hyperbola referred to its asymptotes as axes.

The complex thermal actions of steam within the cylinder of the steam engine can be conjectured only, but since the temperature falls with the increase in volume it is certain that the expansion is not isothermal. Experience has shown, however, that the curve of steam expansion approximates the form of $pv = C$, and no great error can result from so considering it. For this reason, and the additional one of the ease of construction of the hyperbola, the curve of expansion of the indicator diagram is assumed to be hyperbolic, an assumption that greatly facilitates calculations concerning the steam engine.

In the equation $pv = C$ we have the general equation $pv^n = C$ in which $n = 1$. If this value of n be substituted directly in equation (4) in an effort to find an expression for p_m, the result is $\frac{0}{0}$, or indeterminate; but if we evaluate by differentiation before substituting the value for n, we get

$$p_m = \frac{p_1(r^{1-n} - n)}{r(1-n)}\Bigg]_{n=1} = \frac{d\,[p_1(r^{1-n} - n)]}{d\,[r(1-n)]}\Bigg]_{n=1}$$

$$= \frac{p_1(-\log_e r \cdot r^{1-n}\,dn - dn)}{-r\,dn}\Bigg]_{n=1} = \frac{p_1(1 + \log_e r \cdot r^{1-n})}{r}\Bigg]_{n=1}.$$

Substituting now the value of 1 for n, we get

$$p_m = \frac{p_1(1 + \log_e r)}{r}.$$

Or if, in the expression

Whole area of diagram $= p_1v_1 + p_1v_1{}^n \int_{v_1}^{rv_1} \frac{dv}{v^n}$, we let $n = 1$,

we get

$$\text{Whole area} = p_1v_1 + p_1v_1 \int_{v_1}^{rv_1} \frac{dv}{v} = p_1v_1 + p_1v_1 [\log_e v]_{v_1}^{rv_1}$$

$$= p_1v_1 + p_1v_1(\log_e rv_1 - \log_e v_1) = p_1v_1 + p_1v_1 \log_e \frac{rv_1}{v_1}$$

$$= p_1v_1 + p_1v_1 \log_e \frac{v_2}{v_1} = p_1v_1(1 + \log_e r),$$

then $$p_m = \frac{p_1v_1(1 + \log_e r)}{rv_1} = \frac{p_1(1 + \log_e r)}{r}. \qquad (5)$$

With hyperbolic expansion we have $pv = p_1v_1 = p_2v_2$, from which we have

$$\frac{p_1}{p_2} = \frac{v_2}{v_1} = r.$$

If $\frac{p_1}{p_2}$ be substituted for r in equation (5), we get

$$p_m = p_2(1 + \log_e r),$$

which is frequently a very convenient expression for the mean pressure.

152. Construction of the Equilateral or Rectangular Hyperbola. — Draw the rectangular asymptotes OX and OY, Fig. 130, and regard them as the lines of *no pressure* and *no volume* respectively. Let P be a known point of the curve. Through P draw PS and PT parallel to and produced to meet OX and OY

respectively. From O, the intersection of the asymptotes, draw a secant line, cutting PS and PT at A and B respectively. Complete the rectangle $APBP'$.

From the similar triangles OBC and OAD we have

$$\frac{BC}{OC} = \frac{AD}{OD}, \text{ or } \frac{PD}{OC} = \frac{P'C}{OD}, \text{ or } \frac{p}{v'} = \frac{p'}{v},$$

whence $pv = p'v' = C$, the equation of a rectangular hyperbola referred to its asymptotes; hence P' is a point of the curve.

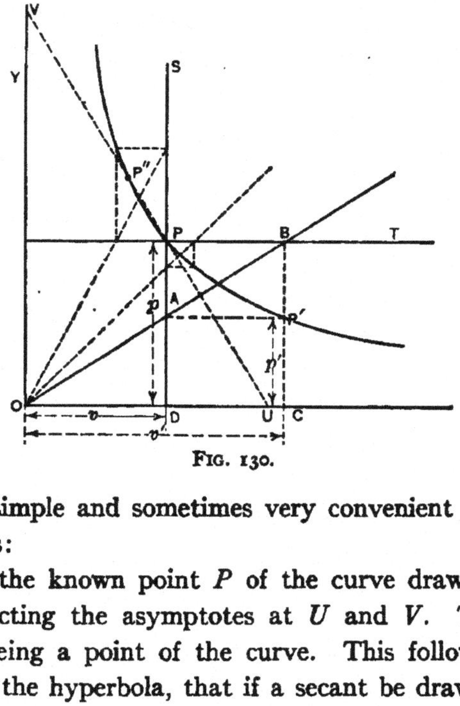

FIG. 130.

Another simple and sometimes very convenient construction is as follows:

Through the known point P of the curve draw any secant line, intersecting the asymptotes at U and V. Then $PU = VP''$, P'' being a point of the curve. This follows from the property of the hyperbola, that if a secant be drawn the parts of it intercepted between the curve and asymptotes are equal.

153. Comparison of the Isothermal, Saturated, and Adiabatic Curves. — Suppose steam of a volume v_1 and pressure $p_1 = 140$ pounds absolute to expand in a cylinder under the limitations imposed successively for isothermal, saturated and adiabatic expansions, the final volume and pressure to become v_2 and p_2 respectively.

Draw the lines of no pressure *OX* and of no volume *OY* of Fig. 131, and on *OY* take *OA* three and one-half inches in length

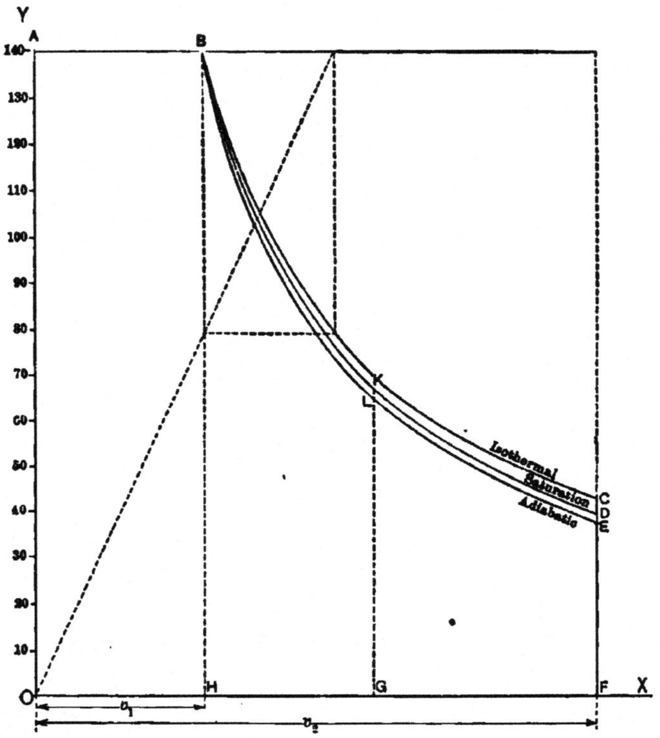

FIG. 131.

to represent 140 pounds to the scale of 40 pounds to the inch. Take $AB = OH = v_1$ and $OF = v_2$. Of course *B* will be a point of each of the curves of expansion.

First construct the rectangular hyperbola *BC* by the method already explained. It will be the curve of isothermal expansion.

To plot the saturation curve $p_1v_1^{\frac{17}{16}} = C$, we proceed as follows:

We have $p_1 = \frac{C}{v_1^{\frac{17}{16}}}$, whence $\log p_1 = \log C - \frac{17}{16} \log v_1 = \log C - 1.0625 \log v_1$. If the initial volume v_1 be denoted by unity we shall have $C = p_1 v_1^{\frac{17}{16}} = p_1 = 140$. Having now a value for C we shall have, for any pressure p and volume v,

$$p = \frac{C}{v^{\frac{17}{16}}} = \frac{p_1}{v^{\frac{17}{16}}}, \text{ whence } \log p = \log p_1 - 1.0625 \log v.$$

Thus, if the volume v_1 be denoted by unity the volume v_2 must be denoted by the ratio $\frac{OF}{OH} = 3.3333$ by measurement. Then, for the terminal pressure p_2 of the saturation curve we shall have

$$\log p_2 = \log 140 - 1.0625 \log 3.3333,$$

whence $$p_2 = 38.956 \text{ pounds} = \frac{38.956}{40} \text{ inch} = FD.$$

The logarithmic computation for p_2 is as follows:

	3.3333 log 0.52288	llog 9.71840 − 10
	1.0625	log 0.02633
	(1.0625 log 3.3333) log 0.55556	llog 9.74473 − 10
	140 log 2.14613	
$p_2 = 38.956$	log 1.59057	

For the pressure p when the volume is OG we must represent the volume OG by the ratio $\frac{OG}{OH} = 2$ by measurement.

Then $$\log p = \log 140 - 1.0625 \log 2,$$

whence $$p = 67.037 = \frac{67.037}{40} \text{ inches} = GK.$$

Any number of points may thus be determined and the saturation curve BD be drawn.

To construct the adiabatic curve $p_1 v_1^{\frac{10}{9}} = C$ we proceed thus:

We have $p_1 = \frac{C}{v_1^{\frac{10}{9}}}$, whence $\log p_1 = \log C - \frac{10}{9} \log v_1 = \log C - 1.1111 \log v_1$. If we denote the initial volume v_1 by unity we get $C = p_1v_1^{\frac{10}{9}} = p_1 = 140$. Having now a value for C we have, for any pressure p and volume v,

$$p = \frac{C}{v^{\frac{10}{9}}} = \frac{p_1}{v^{\frac{10}{9}}}, \text{ whence } \log p = \log p_1 - 1.1111 \log v.$$

If the volume v_1 be denoted by unity the volume v_2 must be denoted by the ratio $\frac{OF}{OH} = 3.3333$ by measurement. Then, for the terminal pressure p_2 of the adiabatic curve we have

$$\log p_2 = \log 140 - 1.1111 \log 3.3333,$$

whence $\quad p_2 = 36.743 \text{ pounds} = \frac{36.743}{40} \text{ inch} = FE.$

For the pressure p when the volume is OG we represent the volume OG by the ratio $\frac{OG}{OH} = 2$ by measurement.

Then $\quad \log p = \log 140 - 1.1111 \log 2,$

whence $\quad p = 64.81 \text{ pounds} = \frac{64.81}{40} \text{ inches} = GL.$

Any number of points may thus be determined and the adiabatic curve BE be drawn.

Figure 131 affords a means of comparing the three expansion curves. They represent the expansions, under the respective conditions, of an initial volume AB of steam of absolute pressure $OA = p_1$ to a final volume OF and a final, or terminal, pressure $p_2 = FC$, FD, and FE respectively for isothermal, saturated, and adiabatic expansion. The three curves always will be relatively as shown.

As an example of computing the mean pressure, suppose a portion of steam of an absolute pressure of 180 pounds to expand to 6 times its volume under the conditions imposed for saturated steam. To find the mean pressure, we have

$$p_m = \frac{p_1(17 - 16r^{-\frac{1}{16}})}{r} = \frac{180\left(17 - \frac{16}{(6)^{\frac{1}{16}}}\right)}{6} = 30\left(17 - \frac{16}{(6)^{\frac{1}{16}}}\right)$$

$$= 30(17 - 14.31) = 80.7 \text{ pounds.}$$

Let $x = (6)^{\frac{1}{16}}$.

Then $\log x = \frac{1}{16}\log 6$, and $\text{llog}\, x = \text{llog}\, 6 - \log 16$

	6 log 0.77815	llog 9.89107 − 10
		16 log 1.20412
	$(6)^{\frac{1}{16}}$ log 0.04864	llog 8.68695 − 10
	16 log 1.20412	
$\frac{16}{(6)^{\frac{1}{16}}}$ =	14.31 log 1.15548	

PROBLEMS

1. Steam of absolute initial pressure of 88 pounds per square inch expands in a cylinder under the conditions, respectively, of isothermal, saturated, and adiabatic expansions. The ratio of expansion being 4, it is required to find the terminal pressure in each case.

Ans. 18.86 pounds; 20.177 pounds; 22 pounds.

2. A portion of steam of absolute initial pressure of 160 pounds per square inch expands to four times its initial volume under the conditions successively of isothermal, saturated, and adiabatic expansion. Find the mean pressure for each case.

Ans. 95.452 pounds; 93.16 pounds; 91.4 pounds.

CHAPTER XI

COMPOUND OR STAGE-EXPANSION ENGINES

154. General Description. — A stage-expansion engine is one so designed that the steam, after entrance into one cylinder and there partially expanded in the performance of work, is exhausted into a second, third, and not infrequently into a fourth larger cylinder for further expansion and work before being finally exhausted into a condenser.

The term *stage-expansion* is used to avoid the confusion which sometimes arises from the use of *compound*. All stage-expansion engines are compound, but by common consent the term *compound* is restricted to engines in which the expansion takes place in only two cylinders, or in two stages, while the terms *triple-expansion* and *quadruple-expansion* are applied to engines in which the expansion takes place in three and four cylinders respectively. The first and last cylinders in the order of the expansion are called, respectively, the high-pressure and the low-pressure cylinders, any cylinder or cylinders intervening being known as intermediate cylinders.

There has been much discussion concerning the relative merits of the simple and stage-expansion types of engines, but the results of comparative tests clearly indicate an economical advantage of from 10 per cent to 20 per cent in favor of the compound system.

Perhaps the most convincing evidence of the advantage in the use of the stage-expansion engine is, that it has almost entirely displaced the simple engine for marine purposes, and has come largely into use in mill engines, locomotives, and in stationary plants where large powers are developed.

It has been shown that a given weight of steam in working from a higher to a lower pressure is capable of doing a definite amount of work, and while the number of cylinders through which the expansion takes place can make no increase in this theoretical limit of work, it does make a very material difference in the manner of its performance.

A prerequisite to the use of the stage-expansion system, and the one to which its economy is due, is a high pressure of steam; and since the terminal pressure desired is the same, whatever the system, the higher the pressure the greater the rate of expansion. The attainment of a high rate of expansion in a single cylinder would be accompanied with such a wide range of temperature as to occasion excessive initial condensation and final re-evaporation, and the consequent wide variation in pressure would lead to very objectionable irregularity of rotational effort on the crank-pin and excessive strains on the framing. With a stage-expansion engine this high rate of expansion would be distributed through two or more cylinders, thus greatly reducing the range of temperature and pressure in each, resulting in a reduction of condensation and the production of a more uniform turning moment.

There are two classes of stage-expansion engines, known as the *continuous-expansion* and the *receiver* types. The essential difference is that with the continuous-expansion type the cranks must be placed with each other or directly opposite each other on the shaft while with the receiver engine the cranks may be placed at any angle with each other.

The continuous-expansion engine is one of the two-stage type in which the steam enters the low-pressure cylinder as fast as it is exhausted from the high-pressure cylinder, a familiar example of which is that of the tandem engine where the two pistons are on one rod.

The receiver engine derives its name from the fact that originally it was designed to have an intermediate reservoir to receive the steam as it was exhausted from the high-pressure cylinder, and in which it remained until the valve opened to admit it into the low-pressure cylinder. The size of this receiver equalled that of the high-pressure cylinder, and even larger. This separate reservoir was found to be unnecessary, inasmuch as the low-pressure steam chest, the exhaust pipe from one cylinder to the other, and the portion of the high-pressure cylinder yet filled with steam when the low-pressure valve opened, were found to furnish ample receiver space.

A type of engine known as the *cross compound* has the high-pressure and low-pressure cylinders placed side by side and the cranks set at right angles on the shaft.

The common form of the compound, or two-stage expansion, engine with cranks at right angles is shown in Fig. 132. When this form of engine is designed for such large powers as require

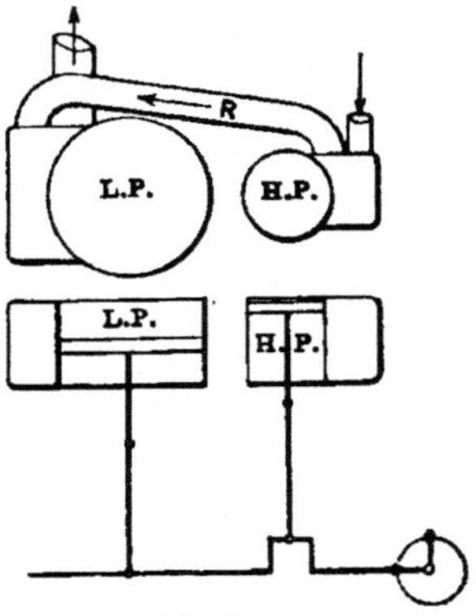

FIG. 132.

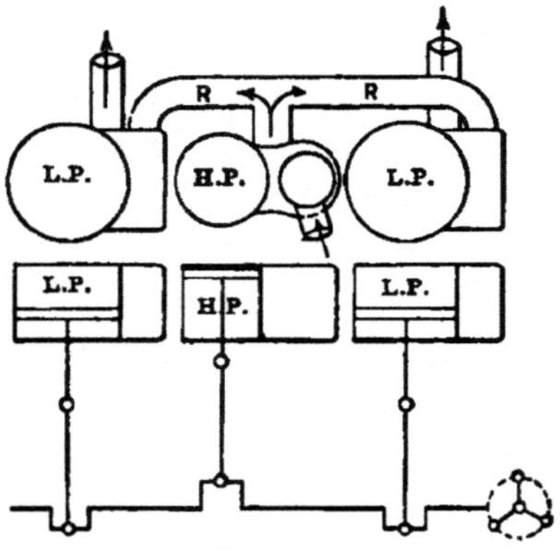

FIG. 133.

a low-pressure cylinder greater than 90 inches in diameter, the three-cylinder compound engine with two low-pressure cylinders, as shown in Fig. 133, would be adopted. In this case the excessively large low-pressure cylinder is replaced by two cylinders whose combined volume equals that of the large cylinder they replace, and notwithstanding the increased space occupied and the greater number of parts required with no increase in the power developed, this arrangement is very advantageous, as the pieces are lighter and more easily made and the increase of the number of cranks from two to three enables them to be placed at angles of 120° on the shaft, thus insuring reduced straining and a steadier motion on the engine. With this engine the work may be equally divided between the three cylinders by a proper adjustment of the receiver pressure. The receiver pressure may be controlled by means of the cut-off in the L.P. cylinder. For example, shortening the L.P. cut-off will increase the receiver pressure, and by thus increasing the back pressure on the H.P. piston will decrease the power developed in the H.P. cylinder; the increased receiver pressure will increase the power developed in the L.P. cylinder. Making the L.P. cut-off later will occasion opposite results.

The two-stage compound engine is advantageously employed where the pressure varies from 100 pounds to 150 pounds absolute. With pressures varying from 160 pounds to 280 pounds, the triple-expansion type should be adopted, and it is this type which is very largely used in the merchant marine and in vessels of war. For pressures varying from 275 pounds to 300 pounds, the quadruple-expansion type is used. This type has not as yet come into general use and its employment is confined to small vessels where lightness of the machinery admits of high rotational speeds.

The common arrangement of the three-cylinder triple-expansion engine is shown in Fig. 134. The first receiver, R, leads from the exhaust of the high-pressure cylinder to the valve chest of the intermediate cylinder, and the second receiver, R', connects the intermediate exhaust with the low-pressure valve chest. The cranks are placed at angles of 120° with each other and their sequence on the shaft is a matter of dispute, but the order of high, low, intermediate, seems to obtain the greatest favor from the fact that such arrangement secures a more uniform receiver pressure.

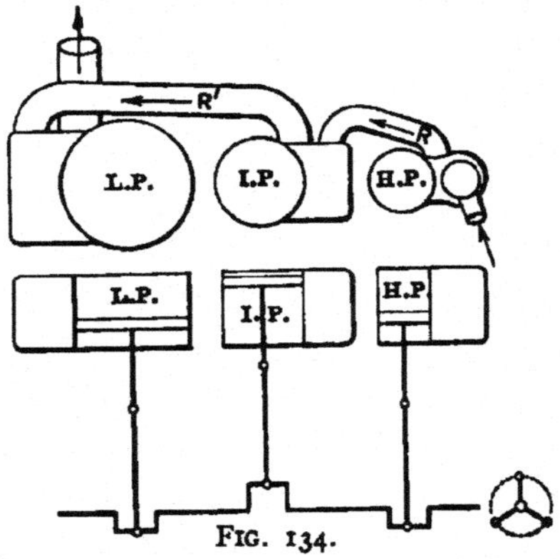

FIG. 134.

It is inadvisable to make cylinders larger than 90 inches in diameter in the interests of weight and space, so for the large powers required for modern marine practice it is usually the practice to have one H.P. cylinder, one I.P. cylinder, and two L.P. cylinders for marine triple-expansion engines, the cranks being 90° apart, the H.P. crank leading, followed by the I.P., forward L.P., and after L.P. The power in such cases is distributed as follows: One-third each to the H.P. and I.P. cylinders and one-sixth to each L.P. cylinder.

The most serious inherent defect of the stage-expansion engine is the loss occasioned by *drop*. By drop is meant the fall in pressure between the terminal pressure in one cylinder and the initial pressure in the next succeeding cylinder in the expansion. This is occasioned by friction in the exhaust passages and pipes, and by the unrestricted expansion in the receiver. While it is impossible to avoid entirely the loss from drop, measures should be taken to prevent its being excessive.

The mean pressure obtained in the stage-expansion engine with a given rate of expansion is less than would be obtained were the expansion all to take place in one cylinder, and this difference is due to drop. The steam, when its expansion is completed, occupies the low-pressure cylinder only and were there no loss from drop the size of this cylinder for a given power would be the same as that of a single-cylinder engine working with the same pressure and the same ratio of expansion. The size of the low-pressure cylinder therefore must be made somewhat larger than would suffice for the single cylinder of a simple engine of the same power, and from this it is seen that the high and intermediate cylinders add nothing to the power of the arrangement and that the low-pressure cylinder is the measure of the power.

155. Cylinder Ratios of Stage-expansion Engines. — There is a lack of agreement among authorities as to the proper volumetric ratios between the cylinders of stage-expansion engines. In selecting these ratios it is the aim to equalize the power between the cylinders, but this equalization may be effected within practical limits by means of the adjustment of the points of cut-off and quite independently of the cylinder ratios, though an unduly low cylinder ratio with equality of powers may cause pronounced inequalities in the initial stresses on the pistons. The important consideration of making the weight of the machinery and the space it occupies as small as possible in vessels of war, together with the fact that engines of such vessels are seldom worked at their maximum power, make it desirable that their cylinder ratios be made smaller than those of engines of the mercantile marine where the conditions are somewhat reversed.

The proposition has been advanced that, with given initial and terminal pressures, percentages of clearance, point of cut-off in the H.P. cylinder and the pressure at that point, the ratio of the net piston areas of the L.P. cylinder to the H.P. cylinder follows as a consequence. The reasoning employed in establishing this proposition is logical and interesting, but as the operation is long and tedious, and as it involves several assumptions which can only be justified by a reference to the results of successful practice, the question very naturally arises — why not at once select for any particular case the cylinder ratio employed in successful practice? The construction of the theoretical diagram and an investigation of the initial stresses on the pistons would then reveal the fact if the selected ratio were improper.

Successful practice has shown that in the two-stage expansion engine, with the initial pressure ranging from 90 to 150 pounds, the volumetric ratio

$$\frac{\text{Volume L.P. cylinder} + \text{its clearance volume}}{\text{Volume H.P. cylinder} + \text{its clearance volume}}$$

varies from 3.25 to 3.75 for stationary engines; from 3.5 to 4.5 for engines of the mercantile marine; and from 3.5 to 4 for engines of war vessels. For triple-expansion engines of the mercantile marine, with pressures ranging from 160 pounds to 275 pounds, the ratio varies from 8 to 10; and for engines of war vessels, the pressure varying from 175 pounds to 300 pounds, the ratio varies from 7 to 10.

Good practice makes the intermediate cylinder of triple-expansion engines from 2.5 to 2.8 times the H.P. cylinder, the volumetric ratio $\frac{\text{L.P.}}{\text{H.P.}}$ varying from 7 to 10. The Navy Bureau of Steam Engineering makes the cross-section area of the I.P. cylinder equal to $\frac{A \times K \times \sqrt{R}}{1 + c}$, in which A is the cross-section area of the H.P. cylinder, K the volume, including clearance, of the H.P. cylinder up to cut-off expressed in percentage of the stroke volume of the H.P. cylinder, R the total ratio of expansion, and c the clearance of the I.P. cylinder.

According to recent practice the cylinder ratio of marine quadruple-expansion engines using steam of 270 pounds pressure is 1 : 2.5 : 5 : 10; and for 300 pounds pressure it is 1 : 3 : 6 : 12.

156. Mechanical Advantages of Stage-expansion Engines. — To illustrate the mechanical advantage of the stage-expansion engine in producing a reduction in the initial stress on the piston and a decrease in the ratio of initial to mean stress over that produced by the simple-expansion engine, a comparison will be made between a simple-expansion engine with two cylinders, each having an area $\frac{A}{2}$, and a two-stage compound engine with a cylinder ratio of 4, both engines being condensing and developing the same power. The area of the L.P. piston of the compound engine will then be A and that of the H.P. piston $\frac{A}{4}$. Each engine has the same number of working parts, but a separate expansion valve would be required on each of the cylinders of the simple engine to effect a cut-off early enough to produce a ratio of expansion of 4. In each case the absolute initial pressure will be taken as 95 pounds, and the back pressure in the condenser as 3 pounds. Clearance is neglected for simplicity.

1. The simple engine with two cylinders:

$$\text{Mean pressure in each cylinder} = \frac{95\,(1 + 1.3865)}{4} = 56.68 \text{ pounds.}$$

$$\text{M.E.P. in each cylinder} = 56.68 - 3 = 53.68 \text{ pounds.}$$

$$\text{Effective initial load on each piston} = (95 - 3)\frac{A}{2} = 46\,A \text{ pounds.}$$

$$\text{Effective mean load on both pistons} = 53.68\,A \text{ pounds.}$$

$$\text{Efficiency} = \frac{\text{Mean effective load expected}}{\text{Mean effective load obtained}} = \frac{56.68\,A}{56.68\,A} = 1.$$

2. The two-cylinder compound engine:

Suppose the H.P. cut-off to be at 0.5 stroke, making a ratio of expansion of 2 in the H.P. cylinder, and a total ratio of expansion of $2 \times 4 = 8$.

$$\text{Mean pressure in H.P. cylinder} = \frac{95\,(1 + 0.693)}{2} = 80.47 \text{ pounds.}$$

Assuming the pressure in the receiver to be maintained at 22 pounds, we must have, for equal power in the cylinders, an equality between the pressure multiplied by the volume of steam given to the H.P. cylinder and the pressure multiplied by the volume given to the L.P. cylinder. Hence

$$95 \times 0.5 = 22 \times 4 \times x,$$

whence $$x = 0.54.$$

That is, the steam must be cut off at 0.54 of the stroke in the L.P. cylinder in order to maintain the receiver pressure at 22 pounds.

$$\text{Ratio of expansion in L.P. cylinder} = \frac{1}{0.54} = 1.85.$$

$$\text{Mean pressure in L.P. cylinder} = \frac{22\,(1 + 0.6152)}{1.85} = 19.21 \text{ pounds.}$$

M.E.P. in L.P. cylinder $= 19.21 - 3 = 16.21$ pounds.

M.E.P. in H.P. cylinder $= 80.47 - 22 = 58.47$ pounds.

$$\text{Effective initial load on H.P. piston} = (95 - 22)\frac{A}{4} = 18.25\,A.$$

$$\text{Effective initial load on L.P. piston} = (22 - 3)\,A = 19\,A.$$

$$\text{Mean effective load on both pistons} = 58.47 \times \frac{A}{4} + 16.21\,A = 30.83\,A.$$

The mean pressure due to the total ratio of expansion is

$$\frac{95\,(1 + 2.079)}{8} = 36.56 \text{ pounds.}$$

$$\text{Mean effective load to be expected} = (36.56 - 3)\,A = 33.56\,A.$$

$$\text{Efficiency} = \frac{30.83\,A}{33.56\,A} = 0.9186.$$

In comparing the results, it is found that the initial load on each of the pistons of the simple engine is 46 A pounds, while that on the H.P. piston of the compound engine is 18.25 A pounds and on the L.P. piston 19 A pounds, showing that the initial loads on the pistons of the compound engine are practically equal, and that their sum is less than the initial load on either piston of the simple engine

The ratio of the mean load to the initial load on each piston of the simple engine is $\dfrac{53.68\frac{A}{2}}{46A} = 0.583$. The like ratios for the pistons of the compound engine are as follows:

$$\text{On H.P. piston} = \frac{58.47 \times \frac{A}{4}}{18.25A} = 0.8.$$

$$\text{On L.P. piston} = \frac{16.21A}{19.8A} = 0.818.$$

In consequence of the reduction in the initial and mean loads, the various parts of the compound engine whose dimensions depend on these factors may be made lighter than those of the simple engine, and the friction on the guides and bearings will be less. The ratios of the mean to the initial loads being much nearer unity with the compound engine, the turning moment is much more uniform and the engine will therefore run more steadily. The question of uniform load and steadiness in running is one of great importance in marine practice, but with stationary engines where the use of a flywheel is permissible its importance is lessened.

The mean effective loads on the pistons of the compound engine show the work to be very evenly divided between the cylinders, but the total work falls short of that of the simple engine by $1 - 0.9186 = 0.0814$, or 8.14 per cent. This loss may be attributed very largely to the drop between the terminal pressure in the H.P. cylinder and the initial pressure in the L.P. cylinder, amounting in this case to $\frac{95}{2} - 22 = 25.5$ pounds.

A consideration of the range of temperatures in the cylinders shows a very marked advantage in favor of the compound engine. The temperature of steam of 95 pounds pressure is 324° and that of steam of 3 pounds pressure is 141.5°. The range of temperature is then 324° − 141.5° = 182.5° in each cylinder of the simple engine. The temperature of steam of 22 pounds pressure is 233°, therefore $324^\circ - 233^\circ = 91^\circ$ is the range of temperature in the H.P. cylinder of the compound engine, and $233^\circ - 141.5^\circ = 91.5^\circ$ is the range of temperature in the L.P. cylinder. It is thus seen that the range of temperature in each cylinder of the compound engine is practically the same, and is but one-half that in either cylinder of the simple engine. The advantage of such a distribution of temperature is shown in Art. 170, page 299.

CHAPTER XII

BOILER EFFICIENCY. CALORIMETRY

157. Boiler Efficiency. — The performance of a boiler is usually expressed in terms of the weight of steam it produces at a stated temperature per pound of coal it consumes. As the steam is assumed to be generated under atmospheric pressure the stated temperature is 212° F, so that the result is expressed in pounds of water evaporated from and at 212° per pound of coal.

In estimating the efficiency of a boiler from the basis of coal consumption, all heat due to the coal thrown into the furnace is charged to the boiler, while it is credited only with the heat transferred through the heating surface and used in the generation of steam. The ratio of these two quantities expresses the efficiency of the boiler. If the two quantities were equal the efficiency would be perfect and would be expressed by unity, but owing to various causes the actual efficiency varies from 0.65 to 0.75. For example, if the potential energy of a pound of coal be taken as 15,000 thermal units its evaporative power from and at 212° would be $\frac{15000}{966} = 15.53$ pounds of water; but in practice no greater evaporative power than $15.53 \times 0.7 = 10.87$ pounds could be expected. If the boiler received its feed water at a temperature of 140° and generated the steam at 200 pounds absolute pressure, the actual evaporation would be $\frac{10.87}{1.207} = 9$ pounds, the factor of evaporation being 1.207 (see page 32, for factor of evaporation).

There are a number of sources to which the loss in boiler efficiency may be attributed. In the first place the losses in the furnace under the most favorable conditions vary from 2.5 to 4 per cent. The coal may be such that portions of it fall through the grate unconsumed, and frequently the finer portions, when only partly consumed, are carried from the furnace by the draft and lodged in the tubes and uptakes or carried off through the smoke pipe; particularly so is this the case with forced draft. The supply of air at times may be insufficient for complete combustion, resulting in the production of carbonic oxide of a thermal value of only 4500 thermal units per pound, or less than one-third that due perfect combustion. Imperfect combustion produces smoke containing carbon only partly consumed which is of course lost through the smoke pipe.

All the heat which is liberated in the furnace gases is not by any means transmitted through the heating surface to the water in the boiler. A small part is radiated to the surrounding atmosphere, and another and much larger part escapes through the smoke pipe. The heat contained in the gases escaping through the smoke pipe is the most serious loss of all and is inevitable, as its prevention could only be accomplished by reducing the temperature of the chimney gases to that of the outside air. This of course cannot be done, since the temperature of the gases must be higher than that of the water in the boiler to prevent a transference of heat in the wrong direction. The temperature of the escaping gases will always be considerably higher than that of the water in the boiler, for in no instance would sufficient heating surface be given to a boiler to reduce their temperature to that of the steam. Aside from this, the natural draft of the boiler, which is occasioned by the difference in weights between the columns of air within and without the smoke pipe, depends directly upon the temperature of the chimney gases and this practical consideration fixes the temperature between the limits of 500° and 600°. The utilization of some of the heat of the chimney gases in heating the boiler feed water by the use of economizers reduces the amount of fuel required for the production of steam for a given power, but if the practice be carried too far a blower will be necessary to produce the draft.

Under the most favorable conditions the losses above considered aggregate little less than 25 per cent, and under conditions of insufficient air for complete combustion and incompetent firing the aggregate may reach as much as 35 per cent.

158. Calorimetry. — In estimating the evaporative efficiency of boilers the condition of the steam with respect to its dryness should be ascertained in order to avoid erroneous results. In practice the steam produced by the average boiler contains more or less moisture held in suspension, and it is evident that the total heat required to produce such steam is less than would be required to produce dry saturated steam to the extent of the latent heat necessary to convert the suspended moisture into steam. For example, suppose a calorimeter test of the steam of 380° temperature produced by a boiler from a feed-water temperature of 180° disclosed the fact that the steam contained 5 per cent of suspended moisture instead of being dry. The heat necessary to produce a pound of this steam is $0.95\, H_l + H_s - H_f = 0.95\,[1091.7 - 0.7\,(382 - 32)] + (382 - 32) - (180 - 32) = 1006.36$ thermal units, while that necessary to produce a pound of dry steam is $1091.7 + 0.305\,(382 - 32) - (180 - 32) = 1050.45$ thermal units. If, in this instance, the 5 per cent of moisture were neglected and the steam assumed to be dry, the evaporative efficiency of the boiler would be exaggerated in the ratio $\frac{1050.45}{1006.36} = 1.0438$, or to the extent of 4.38 per cent.

In the above example the sensible heat, H_s, or the heat in a pound of the steam at the boiling point is taken as $t - 32$, in which t is the temperature of the steam. This value is less than that given in the table of the properties of saturated steam, the difference arising from the fact that in the value $(t - 32)$ the assumption is made that one degree rise in temperature per pound of water is the equivalent of one thermal unit. This is not exactly true, for the water absorbs a small quantity of heat due to the work performed in slightly expanding it and the table value includes this quantity. The difference is not great and in the absence of the table the sensible heat may be taken as $t - 32$. The value of the latent heat given by the formula $H_l = 1091.7 - 0.7\,(t - 32)$ is an approximation, giving values greater than those of the table for low temperatures and smaller values for high temperatures. In the examples that follow the table values of the latent and sensible heats will be taken, though in the absence of the table they may be calculated from the formulas.

We may modify the formula by finding the total heat required for the evaporation of one pound of water so as to meet the cases where the condition of the steam with respect to its dryness is to be taken into account. Thus, if x denotes the ***dryness fraction*** of the steam, we shall have

$$H_w = xH_l + H_s - H_f.$$

If the steam is saturated, or dry, $x = 1$, and the formula becomes $H_w = H_l + H_s - H_f = H_t - H_f = 1091.7 + 0.305\,(t - 32) - (t_f - 32)$, the formula previously given for H_w.

The dryness fraction of steam may be determined by a calorimeter test. There are three forms of calorimeters in use for this purpose, known as the barrel, the throttling, and the separating calorimeters.

159. **Barrel Calorimeter.** — The barrel calorimeter is very simple in construction, but owing to its inaccuracy its use is not advisable when other forms of the instrument are available.

It consists of a barrel of hard wood about three-fourths filled with water and placed on accurate platform scales. A half-inch pipe leads from the main steam pipe of the boiler, the part of the pipe within the steam pipe being closed at the end and perforated with holes about one-eighth inch in diameter. The pipe should be covered with some non-conducting substance and should be thoroughly heated before commencing the test. The steam is admitted to the barrel through a hose attached to the half-inch pipe. This hose reaches nearly to the bottom of the barrel, is closed at the end and perforated for about nine inches of its length from the end. During the entrance of the steam the hose should be moved about so as to promote the mixture of the condensed steam and the condensing water. When the scales show that the desired weight of steam (about one-fifteenth of the original weight of the water in the barrel) has entered the barrel the resulting temperature of the water is taken by a thermometer deeply immersed. The pressure of the steam is taken from the gauge, and the corresponding temperature and latent heat is taken from the table. The temperature of the water in the barrel before and after the mixture should be taken with great care and with a thermometer graduated to at least one-quarter of a degree, and it is not advisable to have the temperature of the mixture much over 110° in order that radiation may be avoided.

Let W = original weight of water in barrel,
W_1 = weight of steam and suspended moisture blown in,
w = weight of dry steam supplied,
t = original temperature of water in barrel,
t_1 = temperature of steam,
t_2 = temperature of the water in barrel after the addition of the steam.

The heat lost by the steam will be the latent heat of the dry steam supplied plus the heat given up in cooling the liquefied dry steam and the water (moisture) carried in by the steam from the temperature t_1 to the temperature t_2. The former quantity is wH_l, and the latter quantity is $W_1\,(t_1 - t_2)$.

The total heat lost by the steam will then be

$$wH_l + W_1\,(t_1 - t_2).$$

The heat gained by the water will be $W\,(t_2 - t)$.

Since the heat lost by the steam must equal that gained by the water in the barrel, we shall have

$$wH_l + W_1\,(t_1 - t_2) = W\,(t_2 - t),$$

whence $$w = \frac{W\,(t_2 - t) - W_1\,(t_1 - t_2)}{H_l} = \text{weight of dry steam}$$

supplied. Calling y the fraction of moisture in the steam, we shall have $y = \dfrac{W_1 - w}{W_1}$; and if x denotes the dryness fraction of the steam, then

$$x = 1 - y = 1 - \frac{W_1 - w}{W_1} = \frac{w}{W_1}.$$

Example. — The barrel of a calorimeter contains 175 pounds of water at a temperature of 60°. After 10 pounds of the boiler steam at an absolute pressure of 170 pounds had been blown into the barrel the temperature of the mixture was found to be 122°. Find the dryness fraction of the steam.

Solution. — From the steam table the temperature of steam at 170 pounds pressure is found to be 368.5° and the latent heat 854.7 units. Then

$$w = \frac{W\,(t_2 - t) - W_1\,(t_1 - t_2)}{H_l} = \frac{175\,(122 - 60) - 10\,(368.5 - 122)}{854.7} = 9.81$$

pounds of dry steam supplied. Then

$$\text{Dryness fraction} = x = \frac{w}{W_1} = \frac{9.81}{10} = 0.981.$$

160. Throttling Calorimeter. — The operation of the throttling calorimeter depends upon the principle that the increase in volume of steam, unaccompanied by the performance of work, occasions a liberation of heat which may be utilized in the evaporation of any moisture in the steam and in superheating the steam by raising its temperature above that due its pressure.

The form of this calorimeter designed by Prof. R. C. Carpenter and manufactured by the Shaeffer & Budenberg Co. is shown in Fig. 135. As described by the designer its form and action is substantially as follows:

The sample steam from the main steam pipe passes through a globe valve and throttling orifice into the interior of the calorimeter and there surrounds a thermometer cup before passing out of the calorimeter through the bottom opening. In many instances a gauge or thermometer cup is inserted between the

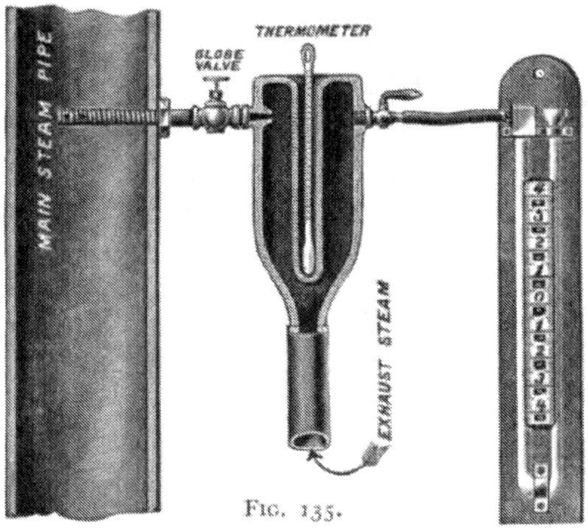

FIG. 135.

valve and throttling orifice in order to determine the pressure of the sample steam as accurately as possible. The sample steam entering the calorimeter through the orifice has its pressure greatly reduced by throttling and its pressure is measured by a low-reading manometer, connected as shown. Where the exit from the calorimeter is comparatively free, that is, not obstructed by long pipes, bends, or valves, the so-called back pressure in the calorimeter should not be more than 0.75 inch mercury, and a manometer is therefore the best instrument with which to determine it. The throttling orifice has a diameter of about 0.08 inch, and the calorimeter itself is about 3 inches in diameter and 6 inches long. For the purpose of reducing radiation the body of the calorimeter is nickel-plated, but it is nevertheless advisable to cover the instrument thoroughly with some non-conducting material. The sample pipe leading to the calorimeter should also be thoroughly covered. If the instruments for determining the pressure of the sample be inserted between the valve and the throttling orifice, it is essential that the valve be opened wide during a trial.

In the theory of the instrument it is assumed that in the operation of the throttling calorimeter there is exactly as much heat in one pound of steam ahead of the throttling orifice as there is in the same weight of steam after throttling; that is, no heat is lost in radiation and no work is done in the expansion process in the orifice. There is, however, less total heat in saturated steam at the lower pressure than there is in the higher, and the excess heat thus liberated in the expansion through the orifice will first evaporate any moisture that the sample contains and will next superheat the dry steam, if any heat is left after the drying process is completed. Hence, provided enough heat has been liberated, the steam in the calorimeter will be superheated. As a matter of fact, this must be the case or the instrument cannot be used, for if the thermometer shows the saturation temperature of the steam in the calorimeter, there is no guarantee that even the original moisture in the sample has been evaporated.

If x denotes the dryness fraction of the sample steam, then $xH_l + H_w$ expresses the total heat of one pound; and if t denotes the temperature of steam due to the pressure in the calorimeter as registered by the manometer, and t_1 the temperature of the superheated steam in the calorimeter as registered by the thermometer of the instrument, then $C (t_1 - t)$ is the measure of the superheat per pound of steam in the calorimeter, C being the specific heat of the superheated steam in the calorimeter. The total heat of the steam in the calorimeter is then $H_t + C(t_1 - t)$ per pound, and we shall have

$$xH_l + H_s = H_t + C(t_1 - t),$$

whence

$$x = \frac{H_t + C(t_1 - t) - H_s}{H_l}.$$

A value of x greater than 1 shows that the sample of steam was originally superheated. If $t_1 = t$ in the equation for x, then there is no superheat in the calorimeter and the result is doubtful, as there is no certainty that all the moisture in the sample has been evaporated. If the difference $t_1 - t$ is less than 10° the results given by the throttling calorimeter are uncertain.

Example. — Pressure of sample steam is 160 pounds absolute; pressure in calorimeter, as registered by manometer, is 2.75 inches mercury; barometer pressure is 29.89 inches mercury; and the temperature in the calorimeter is 286°. Find the dryness fraction of the sample steam.

Notes. — 1. To convert inches of mercury into pounds, divide by 2.04.

2. The specific heat of superheated steam at or near atmospheric pressure may be taken as 0.48 without very material error:

Solution. — The absolute pressure in the calorimeter is $\frac{29.98 + 2.75}{2.04} = 16$ pounds. The saturation temperature at 16 pounds pressure is found from the table to be 216° and the total heat 1152 B.t.u. The degree of superheat is therefore $286° - 216° = 70°$ and the specific heat 0.48. The latent heat of steam at 160 pounds absolute pressure is found from the table to be 858.8 B.t.u. and the sensible heat 335.6 B.t.u. Then

$$x = \frac{1152 + 0.48\,(286 - 216) - 335.6}{858.8} \quad 0.9889 \text{ or } 98.89 \text{ per cent.}$$

161. Separating Calorimeter. — The separating calorimeter is an instrument devised to determine the amount of moisture in steam by a mechanical process dependent upon the density of water being greater than that of steam.

As devised by Prof. R. C. Carpenter, and manufactured by Shaeffer & Budenberg, the separating calorimeter is shown in Fig. 136. It consists essentially of two vessels, one being interior to the other, the space 4 between the two serving as a steam jacket. The inner vessel 2 has a water gauge-glass 10 and a graduated scale 12 by means of which the weight of water separated from the steam sample, and contained in the inner vessel, is measured. The outer vessel 1 has a flow gauge 9 by means of which the weight of the sample steam in its dry state is measured after having been divested of its water in the inner vessel. The sample of steam, the quality of which is to be determined, enters through the pipe 6 into a meshed cup 14 in the upper part of the interior vessel. Here the course of the steam and water of the sample is changed through an angle of nearly 180°, which causes the water to be thrown outward through the meshes of the cup by the force of inertia and enters the chamber 3 of the interior vessel, the steam of the sample in its dry state entering the top 7 of the chamber 4 of the outside vessel. The fins of the meshes of the cup project upward in the inside, so that any intercepted water will drip into the chamber 3. The dry steam is discharged from the chamber 4 through a small orifice 8 of known area. This orifice is much smaller in section than that of any of the passages through the calorimeter, so that the steam in chamber 4 suffers no sensible reduction in pressure in passing through the calorim-

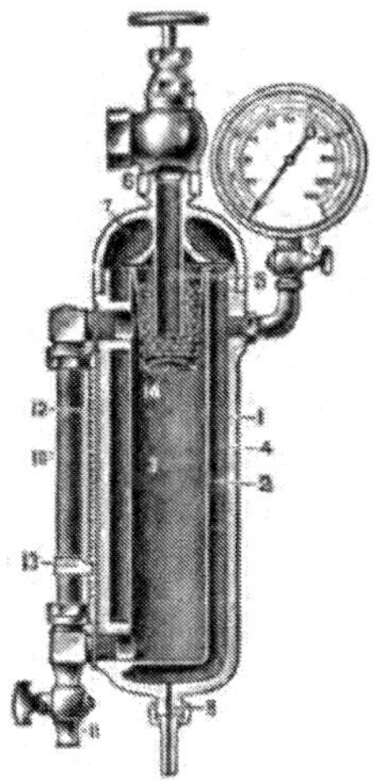

FIG. 136.

eter. Thus, the pressure in the two chambers being practically the same, the temperatures are the same, so that there can be no loss from radiation from the interior chamber except that which takes place from the exposed surface of the gauge-glass which is so slight as to be negligible.

The pressure in the outer chamber 4 and also the flow of steam in a given time is shown by suitably engraved scales on the gauge 9. The scale for showing the flow of steam, the outer one on the gauge, is graduated by trial and indicates the weight of steam discharged in ten minutes of time.

In operating the separating calorimeter it should be thoroughly wrapped with hair felt to prevent radiation, and connected to a felt-wrapped pipe leading to the main steam pipe so as to get the fairest sample of the steam. Permit the steam to blow through the instrument until it is thoroughly heated before making observations.

Take the initial and final readings on the scale 12 at the beginning and ending of a period of ten minutes, and note the average position of the hand on the dial of gauge 9 during this time. The pressure should be kept constant as nearly as possible during the period of discharge, in which case the position of the hand will remain constant.

To compute the dryness fraction or quality of the sample we proceed as follows:

Let y denote the percentage of moisture in the sample of steam, w the weight of water in pounds that have been separated from the sample (found from scale 12), and W the weight of dry steam in the sample (average reading from the outer scale of gauge 9).

Then $$y = \frac{w}{W + w},$$

and $$x = 1 - y = 1 - \frac{w}{W + w} = \frac{W}{W + w},$$

in which x is the dryness fraction of the steam.

Example. — The reading of scale 12 at the beginning of a test was 0.005 pound and at the end of the test it was 0.2 pound. The average reading of the outer scale of gauge 9 during the test was 6.6 pounds. It is required to find the dryness fraction or quality of the steam.

Solution. — $w = 0.2 - 0.005 = 0.195$ pound, and $W = 6.6$ pounds. Then

$$x = \frac{W}{W + w} = \frac{6.6}{6.6 + 0.195} = 0.9713, \text{ or } 97.13 \text{ per cent.}$$

It should be observed that the separating calorimeter is adaptable to all degrees of moisture in steam, while the use of the throttling calorimeter is limited to cases in which the steam contains comparatively small amounts of moisture.

162. Actual Boiler Efficiency. — The real efficiency of a boiler is the product of two efficiencies — that of heating surface and that of combustion. If we denote the heating surface by H.S., we shall have

Efficiency of H.S.

$$= \frac{\text{Heat absorbed by water per pound of fuel}}{\text{Heat available per pound of fuel}}.$$

Efficiency of combustion

$$= \frac{\text{Heat available per pound of fuel}}{\text{Heat contained in pound of fuel}}.$$

Then boiler efficiency

$$= \text{Efficiency of H.S.} \times \text{Efficiency of combustion}$$

$$= \frac{\text{Heat absorbed by water per pound of fuel}}{\text{Heat contained in pound of fuel}}.$$

Example. — A boiler using coal containing 92 per cent of C, 3 per cent H, 4 per cent O, and 0.9 per cent S evaporates 9 pounds of water per pound of coal into steam of 352° temperature from feed water of the temperature of 140°. It is estimated that 4400 units are lost by radiation and through the smoke pipe.

It is required to find: (a) Efficiency of the heating surface; (b) efficiency of the combustion; (c) efficiency of the boiler.

Solution. —

$$h = 14{,}500\left[0.92 + 4.28\left(0.03 - \frac{0.04}{8}\right)\right] + 4050 \times 0.009 = 14{,}928 \text{ B.t.u.}$$

contained in a pound of the coal.

$$H_w = 1091.7 + 0.305\,(352 - 32) - (140 - 32) = 1081.3 \text{ B.t.u.}$$

required to evaporate one pound of water.

$$1081.3 \times 9 = 9731.7 \text{ B.t.u.}$$

absorbed by the water per pound of coal.

$$14{,}928 - 4400 = 10{,}528 \text{ B.t.u.}$$

available for generating steam per pound of coal.

$$\text{Efficiency of heating surface} = \frac{9731.7}{10{,}528} = 0.9243.$$

$$\text{Efficiency of combustion} = \frac{10{,}528}{14{,}928} = 0.7053.$$

$$\text{Efficiency of boiler} = 0.9243 \times 0.7053 = 0.672.$$

We might have proceeded as follows:

$$\frac{\text{Heat available per pound of coal}}{H_w} = \frac{10{,}528}{1081.3} = 9.74 \text{ pounds}$$

of water which should have been evaporated from the available heat per pound of coal.

$$\text{Efficiency of heating surface} = \frac{9}{9.74} = 0.924.$$

$$\frac{\text{Heat units in pound of coal}}{H_w} = \frac{14{,}928}{1081.3} = 13.81 \text{ pounds}$$

of water that would be evaporated per pound of coal if all the heat in the coal were utilized.

$$\text{Efficiency of combustion} = \frac{9.74}{13.81} = 0.7053.$$

PROBLEMS

1. The efficiency of a boiler is 0.7; steam pressure, 135 pounds per gauge; temperature of the feed water, 120°; thermal value of the coal used, 14,200 units per pound. The temperature of steam at 135 pounds pressure is 358.3°. Find the evaporation of saturated steam per pound of coal.

Ans. 8.99 pounds.

2. The efficiency of a boiler is 0.75; steam pressure, 150 pounds per gauge; temperature of feed, 130°; dryness fraction of the steam, 0.95; thermal value of the coal, 14,300 units per pound. The temperature of steam at 150 pounds pressure is 365.9°. Find the number of pounds of steam of the given quality evaporated per pound of coal.

Ans. 10.49 pounds.

3. A boiler under certain conditions evaporates 9.25 pounds of water per pound of coal into steam at 364° from feed water at 70°, the dryness fraction of the steam being 0.9. Under different circumstances 8.75 pounds of water are evaporated per pound of coal into steam at 380° from feed water at 102°, the fraction of moisture in the steam being 0.05. Which of the two cases is the more efficient?

Ans. First case is 4.29 per cent more efficient.

4. Steam containing 4 per cent of moisture is generated at 165 pounds gauge pressure from feed water at 152°. The evaporation per pound of coal is 8.5 pounds. In another instance 8 pounds of water per pound of coal are evaporated into steam at 125 pounds pressure from feed water at 132°, the dryness fraction being 0.97. The temperature of steam at 165 pounds pressure is 373°, and at 125 pounds pressure is 352.9°. Find the equivalent evaporations from and at 212°, and under the supposition that coal in the first instance cost $4.10 per ton and in the second $3.40 a ton, show which case is the more economical.

Ans. Equivalent evaporation first case, 9.13 pounds.
Equivalent evaporation second case, 8.13 pounds.
First case 7.4 per cent more expensive.

5. The barrel of a calorimeter contains 185 pounds of water at a temperature of 60°. Eight pounds of steam at a pressure of 150 pounds absolute are blown into the barrel, after which the temperature of the mixture is found to be 107°. The temperature of steam at 150 pounds is 358.5°. Find the dryness fraction of the steam. *Ans.* 0.931.

6. A test made with a throttling calorimeter gave the following data: Pressure of sample steam, 90 pounds per gauge; manometer reading, 0.6 inch of mercury; barometer reading, 29.6 inches; temperature of steam in calorimeter, 260°. Find the per cent of moisture in the sample steam.

Ans. 1.56 per cent.

7. A boiler is to evaporate 45 cubic feet of water per hour into steam of 300° temperature from feed water of 100°. Average rate of transmission per square foot of heating surface per hour is 3000 units. Efficiency of boiler 70 per cent, and thermal value of one pound of the coal used is 15,000 units. Required: (a) Heating surface; (b) the evaporative power of the coal from and at 212°.

Ans. (a) 1036.55 square feet; (b) 10.87 pounds.

8. A boiler using coal containing 88 per cent C, 5 per cent H, and 4 per cent O, evaporates 9.5 pounds of water per pound of coal into steam of 366° from feed water of 152°. 4550 units are wasted per pound of coal used. Find: (a) Efficiency of heating surface; (b) efficiency of combustion; (c) efficiency of the boiler.

Ans. (a) 92.75 per cent; (b) 70.73 per cent; (c) 65.60 per cent.

9. An engine and boiler test shows a heating surface efficiency of 96 per cent. There are 11,400 units available for generating steam, and 3100 units wasted, per pound of coal. The temperature of the steam was 342°, and of the feed water 122°. Find: (a) Efficiency of combustion; (b) efficiency of boiler; (c) the number of pounds of water evaporated per pound of coal.

Ans. (a) 78.62 per cent; (b) 75.48 per cent; (c) 9.98 pounds.

CHAPTER XIII

ENGINE EFFICIENCY. THE CARNOT CYCLE

163. The Perfect Heat Engine of Carnot. — The work developed by an engine is not all usefully employed, as there are always causes which occasion a waste of work, notably that expended in overcoming the friction of the engine itself. The ratio of the useful work performed by an engine to the total energy expended is the efficiency of the engine.

Before any further consideration of the question of engine efficiency, a brief reference will be made to what is known as Carnot's cycle.

Any series of operations in physics by which a substance is brought back to its initial state is a cycle. Thus, if a gas in a working process passes through a series of heat changes and finally returns to its initial condition, the changes constitute a cycle.

The ideal heat engine of Carnot is imaginary in conception, consisting of a cylinder and piston of perfectly non-conducting material, except the bottom head of the cylinder, which is a conductor. Imagine three bodies, A, B, and C, which can be applied at will to the bottom head of the cylinder. The body A, which is the source of heat, is constantly kept at a temperature T_1; the body B is a non-conductor of heat; and the body C is a receiver of heat, or cooler, kept constantly at a temperature T_2, lower than T_1. Suppose the space between the piston and the bottom of the cylinder to contain as the working substance a perfect gas of temperature T_1.

The cycle consists of four operations, as follows:

1. Apply the hot body A to the bottom head of the cylinder, allowing the piston to move, the gas expanding and doing work, the pressure falling but the gas receiving sufficient heat from A to maintain its temperature constant at T_1. When this isothermal expansion, represented by ab in Fig. 137, has proceeded sufficiently the first operation is completed.

2. Remove A and apply the non-conductor B, the piston continuing to move owing to the work done upon it to the extent only of the internal energy of the expanding gas, the temperature of the gas falling to T_2. This completes the second operation, represented by bc in Fig. 137, and since no heat has been allowed to enter or leave the gas, the expansion has been adiabatic.

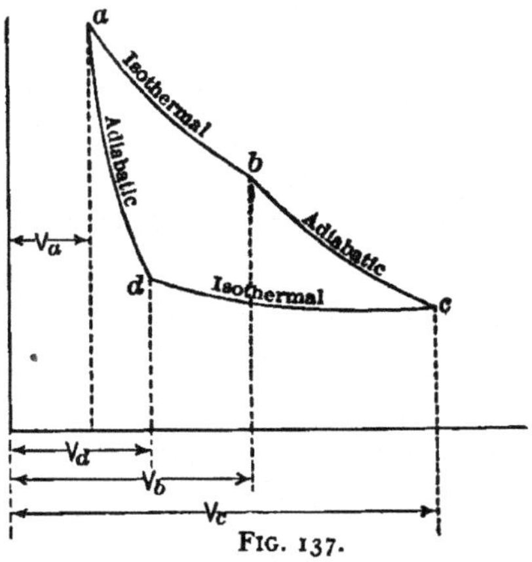

FIG. 137.

3. Remove B and apply C and force the piston back, compressing the gas. The slightest increase in the temperature of the gas due to the compression causes sufficient heat to pass from the gas to the cooler C to maintain the temperature of the gas at T_2.

When the compression has proceeded to the point d of Fig. 137, the third operation, cd, is completed, and since the temperature of the gas was maintained at T_2 the compression has been isothermal.

4. Remove C and apply B and continue the compression, the temperature and pressure of the gas rising, and if the point d has been properly chosen the gas will have been compressed to its original pressure and will have its original temperature T_1 restored when the piston returns to its original position; that is,

the point d must be so chosen that an adiabatic curve through d will pass through a, the compression having been adiabatic. This last operation completes the cycle.

In the performance of the cycle just described the source of heat A, the non-conductor B, and the cooler C were applied in the order A, B, C, B. If now the order of their application be changed to B, C, B, A, it will readily be seen that the diagram of Fig. 137 will be traced in the opposite direction, the reversals of work being accompanied by exact reversals of heat transfers at temperatures T_1 and T_2. This constitutes a ***reversible*** or ***perfect heat engine***. That is, the perfect heat engine, if made to trace its diagram in the reverse direction, will take from the receiver the same quantity of heat it rejected to it when running direct, and will restore to the source of heat the same quantity it received when running direct. Refrigerating machinery operates on the principle of the reversed cycle.

The points b and d may be definitely fixed by a consideration of the facts that the isothermal expansion from a to b must be stopped at such times that will enable T_1 to fall to T_2 during the adiabatic expansion from b to c, and that the isothermal compression from c must be stopped at some point d such that the time remaining for the adiabatic compression of the fourth operation is sufficient to enable the temperature of the gas to rise from T_2 at d to T_1 at a. We have seen that the product of the pressure and volume of a portion of gas is proportional to the absolute temperature of the gas; that is, $PV = RT$, in which R is a constant depending upon the density of the gas. We know, too, that during adiabatic expansion the equation of the curve is PV^K = constant. By a combination of these two equations the points b and d may be located. We have

$$P_1V_1^K = P_2V_2^K, \qquad (6)$$

and

$$\frac{P_1V_1}{P_2V_2} = \frac{RT_1}{RT_2} = \frac{T_1}{T_2}. \qquad (7)$$

From equation (6) we have $\frac{P_1}{P_2} = \frac{V_2^K}{V_1^K}$; substitute in (7) and

$$\frac{T_1}{T_2} = \frac{V_2^K V_1}{V_1^K V_2} = \frac{V_2^K V_2^{-1}}{V_1^K V_1^{-1}} = \left(\frac{V_2}{V_1}\right)^{K-1} = r^{K-1}. \quad \therefore \quad r = \left(\frac{T_1}{T_2}\right)^{\frac{1}{K-1}}.$$

Then, for the adiabatic expansion during the second operation, we have from Fig. 137,

$$r = \frac{V_c}{V_b} = \left(\frac{T_1}{T_2}\right)^{\frac{1}{K-1}},$$

which locates the point *b*.

In like manner, during the fourth operation, we have

$$r = \frac{V_d}{V_a} = \left(\frac{T_1}{T_2}\right)^{\frac{1}{K-1}},$$

which locates the point *d*.

From what has just preceded, we have

$$\frac{V_c}{V_b} = \frac{V_d}{V_a}, \quad \text{whence} \quad \frac{V_b}{V_a} = \frac{V_c}{V_d};$$

that is, the ratios of the isothermal and adiabatic expansions in the Carnot cycle are equal respectively to the ratios of the isothermal and adiabatic compressions.

164. Efficiency of the Carnot Cycle. — During the first operation of the Carnot cycle the engine received a quantity of heat at the temperature T_1 of the source, and during the isothermal expansion of the operation the work done by the gas is, by Art. 151,

$$\int_{V_1}^{V_2} P\,dV = P_1V_1 \int_{V_1}^{V_2} \frac{dV}{V} = P_1V_1 \log_e \frac{V_2}{V_1} = P_1V_1 \log_e r_1,$$

in which r_1 is the ratio of isothermal expansion.

During the second operation the engine neither received nor rejected any heat during the adiabatic expansion.

During the third operation the engine performed work on the gas, rejecting a quantity of heat at the temperature T_2, and during this isothermal compression the work done on the gas was $P_2V_2 \log_e r_2$, in which r_2 is the ratio of isothermal compression.

During the fourth operation the engine neither received nor rejected any heat during the adiabatic compression.

Then $P_1V_1 \log_e r_1 = RT_1 \log_e r_1$ is the external work done by the gas during the cycle, and $P_2V_2 \log_e r_2 = RT_2 \log_e r_2$ is the external work done on the gas; and since the ratios of isothermal expansion and isothermal compression are equal, we shall have $(RT_1 - RT_2) \log_e r$ as the expression for the net external work performed by the gas during the cycle,

and $$\frac{(RT_1 - RT_2) \log_e r}{RT_1 \log_e r} = \frac{T_1 - T_2}{T_1}$$

is the fraction of the whole heat given to the engine that was converted into work and is therefore the efficiency of the ideal engine.

In general terms, if a weight W of a gas whose specific heat is c and absolute temperature T_1 be supplied to any form of heat engine and exhausted at the absolute temperature T_2, then WT_1c expresses the thermal units of heat supplied to the engine and WT_2c the thermal units rejected, so that we shall have

$$\text{Ideal efficiency} = \frac{\text{Useful work performed}}{\text{Total energy rejected}}$$

$$= \frac{WT_1c - WT_2c}{WT_1c} = \frac{T_1 - T_2}{T_1}.$$

The engine of Carnot is of course imaginary, as it is supposed to be made of perfect conductors and perfect non-conductors, which do not exist, but the results of the perfect heat engine are important from the fact that they enable us to point out the imperfections of existing engines and to assign to each its share of the responsibility for the waste of heat. The Carnot cycle of the perfect heat engine shows that no engine can be devised that will show a greater efficiency than $\frac{T_1 - T_2}{T_1}$, in which T_1

is the absolute temperature of the source of heat and T_2 the absolute temperature of the source of cold. This efficiency would have its maximum value of unity when $T_2 = 0$, or when the source of cold has the absolute zero of temperature, a result unattainable. The best that can be done is to make the efficiency of the engine approach unity by making $T_1 - T_2$ as nearly as possible equal to T_1, which is practically accomplished by making T_1 as large and T_2 as small as possible. Unfortunately, the practical range of temperature in the cylinder of a steam engine is very limited and the efficiency therefore very low. The maximum temperature which could possibly be available is that of the products of combustion of the fuel, but if there were practical means of utilizing the temperature of the furnace gases some material other than steel would have to be used in the construction of engines. We are therefore limited for the maximum temperature to that of the steam in the boiler, and for the minimum to a condenser temperature of about 126° F, corresponding to an absolute pressure of 2 pounds.

The conditions of the four operations of the Carnot cycle are far from being obtained in practice with the modern steam engine, as under the most favorable circumstances not more than from 50 to 60 per cent of the ideal efficiency is obtained in actual practice. With the present limits of temperature used in the steam engine the ideal efficiency varies from 18 to 30 per cent, so that the actual efficiency varies from 9 to 18 per cent under favorable conditions.

The cycle of operations of the steam engine approach those of the Carnot cycle only remotely. Heat is taken in at the upper limit, that of the boiler temperature; expansion follows, during which work is done, and the falling temperature approaches, but does not reach, the lower limit at which condensation takes place, during which process heat is given out. At this point the operation of adiabatic compression is wanting, the working substance in the condition of condensed water being returned to the boiler by mechanical means. The heat the feed water contains improves the efficiency of the cycle, but as such heat is supplied at a temperature below the upper limit the performance is far from approaching the ideal cycle.

165. Losses Affecting Engine Efficiency. — The principal causes which reduce the efficiency of the steam engine below that of the ideal engine may be enumerated as follows:

1. Steam, the working substance, is far from being a perfect gas, and its employment is primarily disadvantageous. It is impossible to compress the exhaust steam to its initial state, necessitating a reheating in the boiler.

2. Steam is not rejected to the condenser, the source of cold, at the condenser temperature and pressure, but suffers a fall in both particulars in consequence of incomplete expansion.

3. Owing to inevitable leakages, the equivalence in water to the steam supplied to the cylinder is not returned to the boiler, necessitating an addition of feed water at much lower temperature than that of the water in the boiler.

4. The exchange of heat between the steam and the metal of the cylinder, occasioning initial condensation and subsequent reëvaporation within the cylinder.

5. Free and unresisted expansion of the steam, resulting in the performance of no external work, as instanced in the drop in pressure from cylinder to cylinder in stage-expansion engines.

166. Cylinder Condensation and Reëvaporation. — During the period of admission the steam comes in contact with the metal of the cylinder, which has just been chilled by exhaust connection with the condenser, causing initial condensation, the latent heat of which raises the temperature of the cylinder walls. As the stroke proceeds the reduction in pressure due to the expansion lowers the temperature of the steam, with the result that abstraction of heat from the cylinder takes place, causing a partial reëvaporation of the water and the consequent raising of the expansion curve. This reëvaporation never makes up for the loss due to condensation, and there is always a quantity of water rejected at release, which, owing to the low pressure during exhaust, is partially reëvaporated and occasions back pressure.

In the processes of condensation and reëvaporation within the cylinder it is important to note that the temperature during condensation is higher than it is during reëvaporation, and as a consequence the metal of the cylinder receives more heat per pound of steam condensed than it gives up per pound reevaporated. It follows that if the abstraction from the cylinder walls of this excess of heat could be prevented, there would be no initial condensation; and it was largely this idea that suggested the use of the steam jacket for cylinders, by which the cylinder walls may be prevented from losing heat externally, and may even be made to take up heat. With such conditions, all condensation incident to the work done during expansion would be followed by immediate reëvaporation.

We have assumed thus far that the steam furnished the engine is dry, but should it contain moisture, which is not infrequently the case, the reëvaporation would no longer be confined to the water resulting from condensation during expansion, but would be extended to the moisture contained initially in the steam. Hence the importance of furnishing the engine with dry steam, a condition best secured by means of superheating.

It should be observed that the slide valve really promotes initial condensation by requiring the steam to enter the cylinder through a passage which, immediately before, was chilled by the escape through it to the condenser of the exhaust steam of the preceding stroke. This evil is remedied in engines of the four-valve type, two for steam and two for exhaust.

167. Methods of Reducing Cylinder Condensation. — There are four methods of reducing the losses from cylinder condensation, viz.: (1) Jacketing the cylinder. (2) Superheating the steam. (3) Compounding, or allowing the steam to expand in two or more cylinders, thus reducing the range of temperature in each. (4) Quick running.

168. Steam Jacketing. — The jacket of a cylinder is usually formed by the insertion of a cylindrical liner within the cylinder proper, leaving an annular space between. The liner is flanged at one end, by which it is secured to the cylinder by means of countersunk screws. At the other end is a stuffing-box to allow for expansion and to make a steam-tight joint with the cylinder.

Steam of boiler pressure is circulated through the jacket, the object being to keep the liner, in which the piston works, at a temperature as nearly as possible the same as that of the entering steam, and thus reduce initial condensation. The transfer of heat from the steam in the jacket to the liner is accompanied by a liquefaction of steam in the jacket, but the water from this liquefaction is drained and returned to the boiler without material waste. The virtue of the jacket lies in the fact that the latent heat of its liquefied steam is given to the steam within the liner, with the result that this working steam is kept comparatively dry, and being a poor conductor and radiator when in that condition, the metal of the liner is unable to give heat to or receive it from the working steam. The greatest benefit derived from jacketing is with slow-speed simple engines, its value being less with stage-expansion engines, and almost valueless with engines of high rotational speed.

The theoretical consideration of the action of steam in a jacketed cylinder may be shown by this example:

A simple expansive engine with jacketed cylinder uses steam of absolute initial pressure of 105 pounds per square inch, and exhausts against an absolute back pressure of 17 pounds. The clearance of the cylinder is 4 per cent, and steam is cut off at one-quarter stroke. The weight of a cubic foot of steam at 105 pounds pressure is 0.2365 pound, the latent heat 885 units, and the total heat 1187 units.

It is required to find: (a) The weight of steam used in the cylinder per I.H.P. per hour. (b) The weight of steam liquefied in the jacket per I.H.P. per hour, and its equivalent from a feed-water temperature of 132°. (c) The efficiency of the engine.

Solution. —

$$r = \frac{1 + c}{a + c} = \frac{1 + 0.04}{0.25 + 0.04} = 3.586,$$

$$p_m = \frac{p_1(1 + \log_e r)}{r} = \frac{105(1 + 1.277)}{3.586} = 66.67 \text{ pounds.}$$

Using a mean pressure factor of 0.9, we have

$$\text{M.E.P.} = (66.67 - 17)\,0.9 = 44.7 \text{ pounds.}$$

If we denote the volume of the cylinder in cubic feet by V, we shall have

$$\text{Steam used per stroke} = 0.29\,V \times 0.2365 \text{ pound.}$$

$$\text{Work per stroke} = PV = 144 \times 44.7 \times V \text{ foot pounds.}$$

$$\text{Work per pound of steam} = \frac{144 \times 44.7 \times V}{0.29 V \times 0.2365} = 93{,}850 \text{ foot pounds.}$$

$$\text{Weight of cylinder steam per I.H.P. per hour} = \frac{33{,}000 \times 60}{93{,}850} = 21.097 \text{ pounds.}$$

$$p_2 = \frac{p_1}{r} = \frac{105}{3.586} = 29.28 \text{ pounds.}$$

Under the assumption that the jacket steam supplies the heat necessary to maintain the working steam in a state of saturation throughout the stroke, the measure of this heat is the latent heat of the steam liquefied in the jacket. If x denotes the weight of steam liquefied in the jacket per I.H.P. per hour, then the heat expended in the jacket is x times the latent heat of steam corresponding to the pressure p_1 of the jacket steam.

The total heat of steam at pressure of 29.28 pounds is 1163 units. Denoting it by $H_2{}^t$ and that at pressure of 105 pounds by $H_1{}^t$, we shall have $(H_1{}^t - H_2{}^t)\ 21.097 + xH_1{}^l$ as the expression of the total heat expended in the performance of effective work per I.H.P. per hour, in which $H_1{}^l$ denotes the latent heat of steam at p_1 pressure. This expenditure of heat is balanced by the effective work done, that is, by $\frac{33{,}000 \times 60}{778} = 2545$ thermal units necessary to develop one I.H.P. per hour.

Then, we shall have

$$2545 = xH_1{}^l + (H_1{}^t - H_2{}^t)\ 21.097 = 885\,x + (1187 - 1163)\ 21.097,$$

whence $$x = 2.3 \text{ pounds.}$$

Supposing the liquefied jacket steam to be returned to the boiler at the temperature of the water in the boiler, there would be an expenditure for each pound of jacket steam of only the latent heat of the steam, or of 885 units, while each pound of cylinder steam requires an expenditure of H_w units, or of $1187 - (132 - 32) = 1087$ units. The weight of the liquefied jacket steam reduced to an equivalent weight of cylinder steam is $\frac{2.3 \times 885}{1087} = 1.873$ pounds. The total weight of steam per I.H.P. per hour is then $21.097 + 1.873 = 22.97$ pounds.

The efficiency of the engine

$$= \frac{\text{Useful work performed}}{\text{Total heat expended}}$$

$$= \frac{2545}{1087 \times 21.097 + 2.3 \times 885} = 0.102, \text{ or } 10.2 \text{ per cent.}$$

169. Superheated Steam. — Steam in the condition in which it is ordinarily produced and used is said to be *saturated*, the name implying that it is a saturated vapor having the maximum density — and hence the smallest volume per pound — consistent with its pressure or with its temperature; it is steam at the point of condensation, when any reduction in the temperature will cause liquefaction. If saturated steam be heated so that its temperature rises above that at the saturation point, the pressure remaining constant, it is said to be *superheated;* or, steam is superheated when, at any given boiler pressure, its temperature is higher than that of the water from which it was evaporated. There can, of course, be no moisture in superheated steam, and its behavior approaches much nearer that of a perfect gas than does that of saturated steam, and since its specific heat at constant pressure is 0.48, it requires a little less than half a thermal unit to raise the temperature of a pound of it to the extent of one degree.

It should be understood that the addition of heat to saturated steam will raise the temperature and leave the pressure unchanged only when the steam is allowed to expand as the heat is added. The practical method of superheating steam is to cause it, while on its way from boiler to cylinder, to pass through coils which are surrounded by hot gases from the furnace. The steam absorbs heat from the gases with a consequent rise in temperature, the pressure remaining constant because of the fact that the steam is used almost as fast as it is generated and that the displacement of the piston in the cylinder causes a virtual extension of the volume of the boiler, thus providing for expansion.

The total heat of a pound of superheated steam is equal to its total heat in the saturated condition plus the quantity $0.48\ (t_s - t)$ required to raise the temperature t of the steam in the saturated condition to the temperature t_s of the superheated state.

In consequence of the difficulties attending the construction of superheaters, and of the inability of hemp packing to withstand high temperatures, the early attempts to use superheated steam in reciprocating engines were abandoned and attention directed to the economies incident to the use of high steam pressures, high ratios of expansion, high piston speeds, and steam jacketing; but as it is not likely that the pressures and piston speeds now in common use will soon be exceeded, it is believed that the hope of further economy in the steam engine lies largely in the use of superheated steam.

The introduction of metallic packing and of improved methods of lubrication, together with the satisfactory construction of superheaters, have removed the objections first urged against the use of superheated steam, and the fear originally entertained that the high temperatures due to superheat would burn lubricants and cause valves and pistons to grip has been abandoned.

The actual behavior of steam in an engine cylinder is too complex to speak of with certainty in all its phases, but no doubt exists that condensation before and after cut-off, and reëvaporation during exhaust do take place, and that both of these operations are wasteful and only partially prevented by the use of jackets. Any usage, therefore, that prevents or lessens these operations is conducive to economy, and it is to this accomplishment that the use of superheated steam in the steam engine is directed.

As saturated steam is a vapor at the point of condensation, it is seen that the effect of superheating is to convert it into a gas which is capable of parting with all of its superheat without liquefying, and it has been demonstrated that such a gas loses less of its heat to the metal of a cylinder which incloses it than do liquefiable gases under similar circumstances. It is to these properties that the economies from the use of superheated steam in the steam engine are due. It is obvious that the ideal efficiency $\frac{T_1 - T_2}{T_1}$ is increased by the use of superheated steam.

It must not be understood that the use of superheated steam increases the power developed by an engine, but rather that it increases the economy with which that power is developed, a result directly attributable to the lessening of cylinder condensation and the consequent decrease in steam consumption per unit of power. Steam consumption depends largely, of course, on the design of the engine, but in so far as its reduction can be attributed to superheated steam it has been shown in practice that for every 7 degrees of superheat there may be expected a reduction of 1 per cent in steam consumption.

Very careful trials were made in 1906 of two identical ships, the *Garonne* and the *Rance*, both fitted with similar triple-expansion engines, the *Garonne* using saturated steam and the *Rance* superheated steam. These trials showed a reduction in coal consumption of 20 per cent in favor of the *Rance*, her consumption per I.H.P. being only 0.9 pound.

The economic advantages of superheated steam apply as well to the steam turbine as to the reciprocating engine, as experience has shown that with the turbine a saving of from 10 to 12 per cent in fuel consumption may be expected from a superheat of from 108 to 126 degrees, a saving of about 1 per cent in every 11 degrees of superheat.

The temperature of the chimney gases is ordinarily high enough to provide a superheat of from 75 to 180 degrees, a degree which may safely be permitted in average practice, so that any necessity for superheaters with independent furnaces does not appear likely to arise in the near future.

170. Compounding. — The introduction of higher steam pressures, with increased rates of expansion, produced such wide ranges of temperature in a single cylinder that the remedies for cylinder condensation already considered failed in their application. A division of this wide range of temperature was then resorted to by allowing the steam to expand successively through two, three, and sometimes four cylinders, thus introducing the double, the triple and the quadruple expansion engines.

For example, the range of temperature in a triple-expansion engine using steam of initial pressure of 250 pounds would be from 401° to 153°, the latter temperature corresponding to a back pressure in the condenser of 4 pounds. A ratio of expansion of 14 would not be unlikely under such conditions, and if such expansion were attempted in a single cylinder the fall in temperature would be $401 - 153 = 248°$, a range which would cause excessive condensation. With the triple-expansion engine this range of temperature could be divided equally between the three cylinders and the initial condensation much reduced.

171. **Quick Running.** — It is obvious that, with engines making a large number of revolutions per minute, there is insufficient time to transfer to and from the cylinder walls the heat involved in the processes of cylinder condensation and reëvaporation.

172. **Condensation and Production of Vacuum.** — Much of the efficiency of the modern steam engine is due to the application of the principle of condensation, by which the steam, after being used in the cylinder, is exhausted into a condenser and there converted into water and then returned to the boiler for reëvaporation. In the type known as the *jet condenser* the condensation is effected by means of the exhaust steam coming in contact with a jet of water within the condenser, the mixture of condensing water and the water resulting from the condensation of the steam being then removed by a pump known as the *air pump*. In the *surface-condenser* type the exhaust steam on entering the condenser comes in contact with a series of metallic tubes through which cold water is kept circulating by means of a *circulating pump*, the water resulting from the condensation of the steam being removed by the air pump. The result in the use of either type of condenser is the almost instant

condensation of the steam and the production of a vacuum, the perfection of which depends upon the pressure within the condenser. If the temperature were 32°, the corresponding pressure would be 0.0886 pound per square inch and the vacuum nearly perfect. No such vacuum is ever attained in a condenser, nor would it be desirable, inasmuch as there must always be an excess of pressure in the condenser over that in the air-pump cylinder in order that the pump valves may open. With either type of condenser more or less air, due to leakage and from being liberated from the water, is always present and if allowed to accumulate would destroy the vacuum. For this reason the air pump is necessary, its function being to remove from the condenser the air and uncondensed vapor, as well as the water.

The water circulating through the tubes of a surface condenser is usually allowed to go to waste, but in case the plant is situated at a place inconvenient to the water supply the circulating water may be discharged into cooling towers and there robbed of its heat so that it can be reused.

With the non-condensing engine the steam, after doing its work in the cylinder, is exhausted into the air at a pressure usually of little less than 4 pounds above atmospheric pressure, thereby occasioning an absolute back pressure on the piston of nearly 19 pounds. With the condensing engine the moving piston is opposed only by the pressure in the condenser due to the imperfection of the vacuum, amounting usually to about 3 pounds absolute, so that the increased efficiency of the condensing engine is at once apparent. This may be shown by a comparison of the indicator diagrams of two identical engines working with the same ratio of expansion and with the same initial pressure of 100 pounds absolute. The ratio of expansion being identical in the two cases, the point of cut-off, and therefore the weight of steam used per stroke, will be the same for each engine.

The non-condensing engine will produce the diagram *ABCD*, Fig. 138, the back-pressure line *CD* being about 4 pounds above atmospheric pressure, while diagram *ABCEF* will be that of the condensing engine, its back-pressure line *EF* being but 3 pounds above the perfect vacuum line *OO'*. The shaded area represents the increased work done by the same weight of steam, due to the use of the condenser.

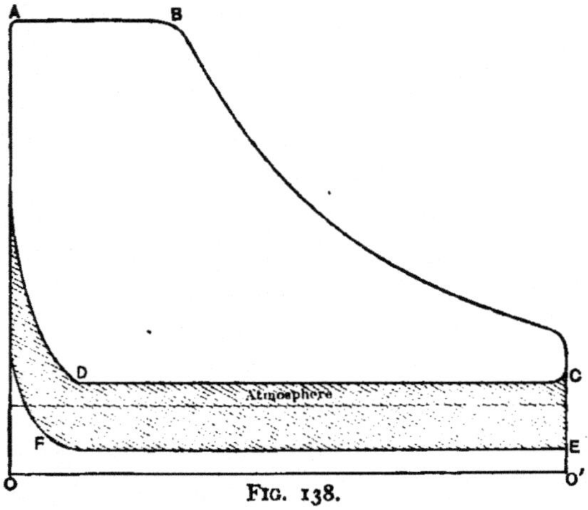

FIG. 138.

A comparison of the ideal efficiencies of the two engines above considered conclusively shows the superior economy of the condensing engine.

The temperature of steam at 100 pounds absolute pressure is 328°, at 19 pounds 225°, and at 3 pounds 142°. The ideal efficiency $\frac{T_1 - T_2}{T_1} = \frac{t_1 - t_2}{t_1 + 461}$, for the non-condensing engine will be $\frac{328 - 225}{789} = 0.13$, or 13 per cent, and that for the condensing engine will be $\frac{328 - 142}{789} = 0.2357$, or 23.57 per cent.

173. Heat Rejected into Condenser. — As stated in Art. 172, steam is commonly exhausted into the condenser at an absolute pressure of 3 pounds with a corresponding temperature of 142°, and assuming its temperature after condensation to be 100°, it follows that each pound of exhaust steam entering the condenser yields 42 thermal units in consequence of its reduction in temperature. The latent heat of a pound of steam at the temperature 142° is 1012 units, so that this additional amount is liberated because of the condensation, making a total of 1054 units given up to the condenser by each pound of exhaust steam, all of which heat must be absorbed by the condensing water. This quantity of heat is known as the heat rejected to the condenser per pound of steam used. Neglecting the loss from radiation, the difference between the total heat of the initial steam and the heat rejected to the condenser is the amount converted into work per pound of steam, and this quantity multiplied by the number of pounds of steam used per minute gives the heat units converted into work per minute, the mechanical equivalence of which is obtained by multiplying by 778, and the product divided by 33,000 will be the I.H.P. of the engine.

174. Weight of Condensing Water Required. — Using the figures of the preceding article, and assuming the temperature of the condensing water to be 60°, there would be, in the case of the jet condenser, an absorption of $100 - 60 = 40$ units by each pound of condensing water, so that to absorb the 1054 units given up by each pound of steam condensed would require $\frac{1054}{40} = 26.35$ pounds of condensing water.

With the jet condenser the condensing water and the condensed steam form a mixture at one temperature, but with the surface condenser the condensing water passing through the tubes must always be at a considerably lower temperature than that of the exhaust steam, and in consequence a greater quantity of condensing water is required for the surface condenser. Suppose in our example that the condensing water after passing through the tubes was discharged at a temperature of 85°, showing an absorption by each pound of 85 − 60 = 25 units of heat. Then for each pound of steam condensed there would be required $\frac{1054}{25} = 42.16$ pounds of condensing water.

175. Expressions for Engine and Boiler Efficiencies. — There are a number of ways of expressing the efficiencies of engines and boilers, and as such expressions are the means of estimating and comparing the economies of engine and boiler systems there should be no confusion in their application. The most commonly used expressions follow:

176. Thermal Efficiency of Engine. — The number of thermal units expended per hour per I.H.P. is a standard of engine efficiency. The development of one horse-power per minute is equivalent to the expenditure of $\frac{33000}{778} = 42.42$ B.t.u., so that we have

Thermal efficiency of engine

$$= \frac{42.42 \times 60}{\text{B.t.u. supplied engine per I.H.P. per hour}}.$$

As an example, take the case of a non-condensing engine working with steam of 95 pounds initial pressure, the temperature of which is 324°. The engine consumes 30 pounds of steam per I.H.P. per hour and the feed-water temperature is 160°. We shall then have

Thermal efficiency of engine

$$= \frac{42.42 \times 60}{30\,[1091.7 + 0.305\,(324 - 32) - (160 - 32)]} = 0.08.$$

If the example be that of a triple-expansion marine engine using steam of 401° temperature, corresponding to an initial absolute pressure of 250 pounds; and if the engine consumes 14 pounds of steam per I.H.P. per hour, and the feed-water temperature be 160°, we shall have

Thermal efficiency of engine

$$= \frac{42.42 \times 60}{14\,[1091.7 + 0.305\,(401 - 32) - (160 - 32)]} = 0.168.$$

It is seen from this that 8 per cent and 16.8 per cent are average thermal efficiencies of the simple non-condensing and the triple-expansion marine types of engines respectively.

177. Number of Pounds of Steam per I.H.P. per Hour. — This standard of engine efficiency is easily understood and very commonly used. This standard varies from 12.5 to 20 pounds for the different types of stage-expansion engines, and from 28 to 34 pounds for simple non-condensing engines.

178. Relative Engine Efficiency. — If the thermal efficiency of an engine be compared with the ideal efficiency of the same range of temperature, the result is the standard of relative engine efficiency.

Suppose in the case of the non-condensing engine considered above under the head of thermal efficiency, the back pressure was 18 pounds, the corresponding temperature being 222°. Then

$$\text{Ideal efficiency} = \frac{t_1 - t_2}{t_1 + 461} = \frac{324 - 222}{785} = 0.13,$$

whence Relative efficiency $= \frac{0.08}{0.13} = 0.616$, or 61.6 per cent of the ideal efficiency.

In the case of the triple-expansion marine engine, suppose a condenser pressure of 3 pounds with the corresponding temperature of 142°. Then

$$\text{Ideal efficiency} = \frac{401 - 142}{862} = 0.3,$$

whence Relative efficiency $= \frac{0.168}{0.3} = 0.56$, or 56 per cent of the ideal efficiency.

179. Mechanical Efficiency of Engine. — Brake horse-power (B.H.P.) is the effective horse-power of an engine; that is, it is the power developed by the engine independently of the power absorbed in friction in driving the engine itself. The I.H.P. includes this frictional work, hence

$$\text{Mechanical efficiency of engine} = \frac{\text{B.H.P.}}{\text{I.H.P.}}.$$

180. Boiler Efficiency. — A standard of boiler efficiency in common use is that of the number of pounds of water it evaporates per pound of coal. Assuming a pound of good coal to contain 15,000 B.t.u., we should expect from it an evaporation of $\frac{15000}{970}$ = 15.46 pounds of water from and at 212°; but we have seen in the chapter on fuels and combustion that not more than 70 per cent of the potential energy of the coal becomes available for generating steam, so that an evaporation of only 15.46 × 0.7 = 10.82 pounds may be expected.

181. Pounds of Coal per I.H.P. per Hour. — This is a very common standard of economy of an engine and boiler considered as one machine. Its value may be averaged as from 3 to 4 for simple non-condensing engines, from 2.22 to 2.5 for simple condensing engines, and from 1 to 2 for stage-expansion engines.

182. Thermal Efficiency of Engine and Boiler. — If an engine develops one I.H.P. per hour with an expenditure of 2.25 pounds of coal, the thermal value of which is 15,000 B.t.u. per pound, it is obvious that the

Thermal efficiency of the engine and boiler

$$= \frac{42.42 \times 60}{15{,}000 \times 2.4} = 0.0707, \text{ or } 7.07 \text{ per cent.}$$

183. Mechanical Efficiency of the System. — It has been shown that the percentage of the potential energy of a given weight of coal converted into work is the thermal efficiency of the engine and boiler, so that if this combined efficiency be multiplied by the mechanical efficiency of the engine the result will be the percentage of the energy of the coal that is applied to the shaft, and therefore the mechanical efficiency of the engine and boiler considered as one apparatus.

184. Imperfection of the Steam Engine. — To illustrate the imperfection of the steam engine we shall consider its efficiency by means of an example from the conditions of actual practice.

Example. — A high-speed non-condensing engine works with an absolute initial pressure of 95 pounds. The feed water has a temperature of 180°, and it is found by measurement that 29 pounds of steam are used per I.H.P. per hour. A boiler test shows that 9.5 pounds of water are evaporated per pound of coal, the thermal value of which is 15,000 B.t.u. per pound. The I.H.P. of the engine is 50 and the B.H.P. 43.5. It is required to find the percentage of the heat energy of the fuel that is applied to the shaft.

Solution. — The total heat of steam at 95 pounds pressure is found from the table to be 1185.4 units; therefore $H_w = 1185.4 - (180 - 32) = 1037.4$ B.t.u., so that one pound of the coal should evaporate $\frac{15,000}{1037.4} = 14.46$ pounds of water.

$$\text{Efficiency of boiler} = \frac{9.5}{14.46} = 0.657.$$

The number of B.t.u. required from the fuel per I.H.P. per minute is $\frac{1037.4 \times 29}{60} = 501.41.$

Hence

$$\text{Thermal efficiency of engine} = \frac{42.42}{501.41} = 0.0846.$$

$$\text{Mechanical efficiency of engine} = \frac{43.5}{50} = 0.87.$$

The efficiency of the boiler is 65.7 per cent, and as only 8.46 per cent of the heat supplied to the engine is converted into work, we have $0.657 \times 0.0846 = 0.0556$, or 5.56 per cent of the heat of the fuel converted into work; and finally we have but $0.0556 \times 0.87 = 0.0484$ or 4.84 per cent of the heat energy of the fuel applied to the shaft.

PROBLEMS

1. An engine develops one I.H.P. per hour on an expenditure of 1.8 pounds of coal, the thermal value of which is 15,000 B.t.u. per pound. What is the thermal efficiency of the system? *Ans.* 9.4 per cent.

2. An engine using 29 pounds of steam per minute exhausts into a surface condenser at a pressure of 3 pounds, the initial pressure being 100 pounds and the terminal pressure 9 pounds, all pressures being absolute. The total heats of steam at the initial and terminal pressures are 1186 B.t.u. and 1141 B.t.u. respectively, and the latent heat of steam at the pressure of exhaust is 1012 B.t.u. The temperature of the exhaust steam is 142° and of the condensed water 100°. The condensing water enters at a temperature of 62° and is discharged at 88°. It is required to find the I.H.P. of the engine and the number of pounds of condensing water required per pound of steam condensed. *Ans.* 30.76 I.H.P.; 40.5 pounds.

3. A non-condensing engine working with an initial absolute steam pressure of 100 pounds, the temperature of which is 328°, uses 26 pounds of steam per I.H.P. per hour. The boiler evaporates 9.5 pounds of water per pound of coal, the temperature of the feed being 202°. The thermal value of the coal used is 14,400 units per pound, and it is found that 6 per cent of the heat energy of the fuel is applied to the shaft. Find: (a) Efficiency of the boiler. (b) Thermal efficiency of the engine. (c) Mechanical efficiency of the engine.

Ans. (a) 66.76 per cent. (b) 9.67 per cent. (c) 93.00 per cent.

4. What would be the coal consumption per I.H.P. per hour in the ideal engine working between the temperatures of 369° and 120°, assuming the thermal value of the coal to be equal to that of pure carbon, and the efficiency of the boiler to be perfect? *Ans.* 0.585 pound.

5. How many pounds of water must be evaporated per hour per I.H.P. with the ideal engine, the consumption of fuel being 0.585 pound of carbon per I.H.P. per hour, the temperature of the steam 369°, and that of the feed water 120°? *Ans.* 7¾ pounds.

6. The efficiency of an engine is 14 per cent, and of the boiler 70 per cent. The coal used has a thermal value of 14,300 units per pound. Find the number of pounds of coal required per I.H.P. per hour.

Ans. 1.816 pounds.

7. A 9" by 10" engine uses steam at 80 pounds gauge pressure. Cut-off, 0.25 stroke; revolution, 375 per minute; clearance, 6 per cent; I.H.P. developed, 46. The specific volume of the initial steam is 4.66 cubic feet. The dryness fraction of the steam is 0.97, the thermal value of the fuel 14,000 units, the temperature of the feed water 182°, and the efficiency of

the boiler 70 per cent. Find: (a) The pounds of steam per I.H.P. per hour. (b) The pounds of water evaporated per pound of coal. (c) The pounds of coal per I.H.P. per hour.

Ans. (a) 23.97 pounds. (b) 9.76 pounds. (c) 2.46 pounds.

8. Diameter of cylinder, 14 inches; stroke, 13 inches; cut-off, 0.25 of the stroke; clearance, 5 per cent; initial absolute steam pressure, 95 pounds; back pressure, 17 pounds; revolutions per minute, 300. The specific volume of steam at 95 pounds pressure is 4.65 cubic feet, and it is expected that 0.9 of the theoretical mean effective pressure will be realized. Find the weight of steam used per I.H.P. per hour. *Ans.* 22.47 pounds.

9. A simple expansive engine with jacketed cylinder uses steam of 90 pounds absolute pressure, and exhausts against an absolute back pressure of 18 pounds. The clearance of the cylinder is 10 per cent, and steam is cut off at one-quarter stroke. The weight of a cubic foot of steam at 90 pounds pressure is 0.2044 pound, the latent heat is 893 units, and the total heat is 1184.4 units. Using a mean pressure factor of 0.77, it is required to find: (a) The weight of steam used in the cylinder per I.H.P. per hour. (b) The weight of steam liquefied in the jacket per I.H.P. per hour, and its equivalent from a feed-water temperature of 132°. (c) The efficiency of the engine. *Ans.* (a) 29.54; (b) 2.13; 1.75; (c) 7.5 per cent.

10. With a jet condenser the temperature of the exhaust steam is 160°, of the injection water 60°, and of the mixture in the condenser 110°. Find the weight of injection water per pound of exhaust steam.

Ans. 21 pounds.

11. With a surface condenser the temperature of the exhaust steam is 170°, and of the condenser 126°. The injection water enters at a temperature of 60° and is discharged at 85°. Find the weight of injection water per pound of exhaust steam. *Ans.* 41.56 pounds.

CHAPTER XIV

DESIGN OF SIMPLE AND COMPOUND ENGINES

185. Engine Design. — The design of an engine of a particular type is based on expected results derived from a study of the best examples of that type in successful practice, the designs of which are known. For example, with a given style of valve it is known that clearance may be confined to a certain minimum percentage of the cylinder volume for each type of engine; also for a given initial pressure the terminal pressure must be so chosen that the consequent approximate steam consumption per unit of power may be regarded as economical for the type. For any design the stroke of the piston and the piston speed are predetermined.

186. Size of Cylinder for a Given Power. — The diameter of the cylinder of an engine depends upon the piston speed and the mean effective pressure permissible in obtaining the given power.

The mean pressure depends upon the initial pressure and the ratio of expansion and may be calculated by means of the formula $p_m = \frac{p_1 (1 + \log_e r)}{r}$ (see page 254). The mean pressure thus obtained makes no allowance for wire-drawing, release, and compression, so that a factor designed to cover these losses must be applied to get the mean pressure to be expected, and from this must be deducted the back pressure in order to get the mean effective pressure.

The M.E.P. derived theoretically is never realized in practice, because there are always causes which make the mean effective pressure found from the indicator diagram less than that due to the initial pressure and ratio of expansion used in the design of the engine. The principal causes which make the actual M.E.P. less than that theoretically due to the initial pressure and ratio of expansion are: Wire-drawing, liquefaction in the cylinder, release of the steam before the piston arrives at the end of the stroke, clearance, compression, back pressure, and *drop* in stage-expansion engines.

These causes have not the same effect in all cases, but experience has shown that a factor depending upon the type and working conditions of the engine may be applied to the theoretical M.E.P. obtained in any case and the result be taken as the M.E.P. to be expected. There is substantial agreement among authorities as to these M.E.P. factors and they may be taken for the different types of engines as follows:

For slow-running pumping engines, from 0.95 to 0.98; for high-speed stationary engines, from 0.90 to 0.95; for two-stage-expansion stationary engines, from 0.75 to 0.85; for triple-expansion engines of the mercantile marine, from 0.60 to 0.65; for triple-expansion engines of war vessels, from 0.55 to 0.60.

187. Piston Speed. — The speed of the piston is determined from considerations of convenience, the type of engine, and the nature of the work to be done. It depends, of course, upon the length of the stroke and the number of revolutions, and is expressed in feet per minute. The mean piston speed is equal to the number of revolutions per minute multiplied by twice the length of the stroke in feet. Experience has shown that a good piston speed for slow pumping engines is from 125 to 175 feet per minute; for high-speed stationary engines, from 500 to 650 feet; for marine engines, from 750 to 1050 feet; for locomotives, from 1000 to 1200 feet; and for torpedo-boat engines a piston speed as high as 1200 feet has been attained.

188. Stroke. — There is no standard rule for determining the stroke of an engine, but the local conditions as to space is sometimes the controlling influence in determining it. Strokes of stationary engines vary from 8 inches to 60 inches, and those of marine engines from 18 inches to 50 inches.

189. The Indicator Diagram in Preliminary Design. — In the design of a steam engine to produce a required power it is customary to predetermine the stroke of the piston, the piston speed, and the initial steam pressure. From this data it is possible to determine by means of the properties of the indicator diagram the steam consumption per I.H.P per hour to be expected from the engine and thus get an intelligent idea of the efficiency of the design. It may be found necessary to change some of the predetermined data in order to secure greater efficiency, but when satisfied in this particular the resulting preliminary diagram furnishes the data for the design of the valve and for the determination of the cylinder diameter.

Example I. — It is desired to design a high-speed non-condensing engine to develop 50 I.H.P. while running at the rate of 375 revolutions per minute. It is predetermined that the stroke of the piston shall be 10 inches, and the initial absolute steam pressure in the valve chest 105 pounds per square inch. It is assumed that a careful design of the steam and exhaust passages will confine the clearance to 2.5 per cent of the volume of the cylinder, and to secure smooth running of the engine at the intended speed a final absolute pressure of compression of 80 pounds will be desirable; it is also fairly assumed that a back pressure of 18 pounds (3 pounds above the atmospheric pressure) will be maintained. It is required to find by means of a preliminary indicator diagram the point of cut-off, the point of exhaust closure, and the diameter of the cylinder.

Solution. — Let

L = length of stroke,

a = part of stroke completed up to cut-off,

b = distance of point of exhaust closure from end of stroke,

c = clearance in percentage of cylinder volume,
p_1 = absolute initial pressure in pounds per square inch,
p_2 = absolute terminal pressure in pounds per square inch,
p_3 = absolute back pressure in pounds per square inch,
p_c = absolute terminal pressure of compression in pounds per square inch,
p_e = M.E.P. = mean effective pressure in pounds per square inch,
r = ratio of expansion,
r_c = ratio of compression,
v_1 = volume of steam admitted to cylinder and clearance up to cut-off,
v_2 = volume of steam in cylinder and clearance at end of stroke.

Taking the steam consumption per unit of power as a measure of the economy of the engine, and regarding the terminal pressure p_2 as an exponent of the steam consumption, we will first determine the terminal pressure to be used in the design. This may be done with sufficient accuracy by means of the empirical formula $Q = \frac{34 p_2 + 23}{\text{M.E.P.}}$, in which Q is the number of pounds of steam consumed per I.H.P. per hour.

Assuming first a terminal pressure of 30 pounds absolute, we will have $r = \frac{105}{30} = 3.5$, the $\log_e$ of which is 1.2528, and we shall have

$$p_m = p_2 (1 + \log_e r) = 30 \times 2.2528 = 67.584 \text{ pounds.}$$

$$\text{Theoretical M.E.P.} = (67.584 - 18)\, 0.9 = 44.63 \text{ pounds.}$$

Then $$Q = \frac{34 \times 30 + 23}{44.63} = 23.37 \text{ pounds}$$

of steam per I.H.P. per hour with a terminal pressure of 30 pounds.

We will try a terminal pressure of 28 pounds. Then $r = \frac{105}{28} = 3.75$, the log. of which is 1.3218, and we shall have $p_m = 28 \times 2.3218 = 65$ pounds.

$$\text{Theoretical M.E.P.} = (65 - 18)\,0.9 = 42.3 \text{ pounds.}$$

Then $$Q = \frac{34 \times 28 + 23}{42.3} = 23 \text{ pounds}$$

of steam per I.H.P. per hour with a terminal pressure of 28 pounds.

As will be shown later, a theoretical consumption of 23 pounds of steam per I.H.P. per hour indicates an actual consumption of about 30 pounds, a result that experience has shown to be fairly economical for the simple expansive non-condensing type of engine, so we will take the terminal pressure p_2 as 28 pounds.

In Fig. 139 assume for convenience that OO', 3 inches in length, represents the volume of the cylinder, and denote it by L. Then the clearance volume will be represented by $OO'' = 0.025\,L = 0.025 \times 3 = 0.075$ inch to the same scale that 3 inches represents the volume of the cylinder. To a pressure scale of 1 inch = 40 pounds erect $O''E$ perpendicular to $O''O'$ and make it $\frac{105}{40}$ inches in length to represent the absolute initial pressure of 105 pounds. Then $O''O'$ and $O''E$ are the lines of ***no pressure*** and ***no volume.***

The expansion being assumed hyperbolic, we shall have $p_1v_1 = p_2v_2$ in which v_1 is the initial volume of steam of pressure p_1, and v_2 the terminal volume of pressure p_2. That is, $105\,v_1 = 28 \times 3.075$, whence $v_1 = 0.82$ inch, which includes the clearance. On a parallel to $O''O'$ through E lay off EG, making it 0.82 inch in length. Erect OF perpendicular to $O''O'$. Then, $EF = c$ and $FG = a = 0.82 - 0.025 = 0.745$ inch, and G is the point of cut-off; that is, the cut-off takes place at $\frac{10}{3} \times 0.745 = 2.483$ inches from commencement of stroke.

At O' erect $O'H$ perpendicular to $O''O'$ making it $\frac{28}{40}$ inch in length to represent the terminal pressure of 28 pounds. Construct the hyperbolic expansion curve GH.

The volume of steam of p_3 pressure remaining in the cylinder at the moment of the exhaust closure on the return stroke must be compressed into the clearance space to a pressure $p_c = 80$

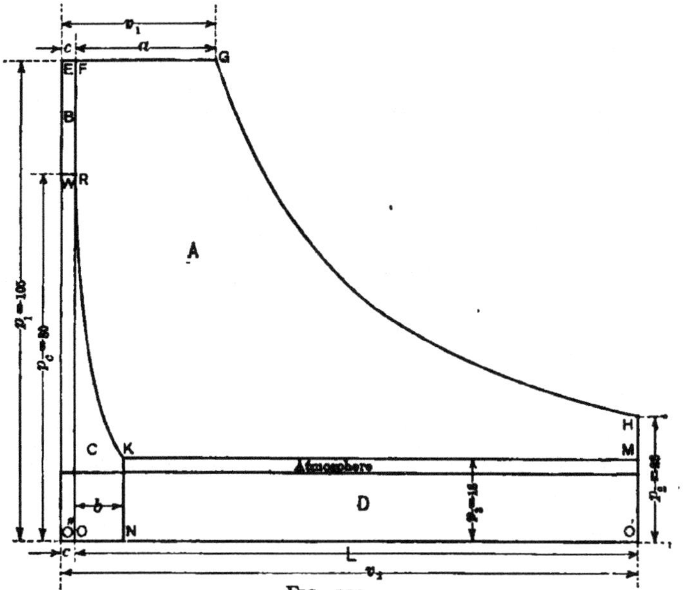

FIG. 139.

pounds. Then we shall have $p_3v_3 = p_cv_c$, or $18\ v_3 = 80 \times 0.075$, whence $v_3 = 0.333$ inch, which includes the clearance; hence $b = 0.333 - 0.075 = 0.258$. That is, the exhaust closes at $0.258 \times \frac{10}{3} = 0.86$ inch from end of stroke.

Make $ON = b = 0.258$ inch. Then $MK = L - b = 3 - 0.258 = 2.742$ inches. $r = \frac{105}{28} = 3.75$, the log. of which is 1.3218; and $r_c = \frac{80}{18} = 4.44$, the log. of which is 1.4907.

Denote the area $FGHMKR$ by A, the area $EFRW$ by B, the area $WRKNO''$ by C, and area $KMO'N$ by D.

$$\text{Area } (A + B + C + D) = \frac{p_1 v_1 (1 + \log_e r)}{40}$$

$$= \frac{105 \times 0.82 \times 2.3218}{40} = 4.998 \text{ square inches.}$$

$$\text{Area } B = \frac{c\,(p_1 - p_c)}{40} = \frac{0.075\,(105 - 80)}{40} = 0.0469 \text{ square inch.}$$

$$\text{Area } C = \frac{p_c v_c (1 + \log_e r_c)}{40} = \frac{80 \times 0.075 \times 2.4907}{40}$$

$$= 0.3736 \text{ square inch.}$$

$$\text{Area } D = \frac{p_3 (L - b)}{40} = \frac{18 \times 2.742}{40} = 1.2339 \text{ square inches.}$$

$$\text{Area } A = \text{Area } (A + B + C + D) - \text{Area } (B + C + D)$$

$$= 4.998 - (0.0469 + 0.3736 + 1.2339) = 3.3436 \text{ sq. ins.}$$

Length of mean ordinate of Area A

$$= \frac{\text{Area } A}{L} = \frac{3.3436}{3} = 1.1145 \text{ inches.}$$

M.E.P. = 1.1145 × 40 = 44.58 pounds, the pressure scale being 40 pounds to the inch.

This calculation has eliminated the losses from compression, clearance, and back pressure, so that a mean pressure factor of 0.95 will be used to allow for losses from wire-drawing and release, giving 44.58 × 0.95 = 42.351 pounds as the M.E.P. to be expected.

Theoretical p_m from the expansion curve to the zero lines of pressure and volume $= p_2 (1 + \log_e r) \doteq 28 \times 2.3218 = 65$ pounds.

Theoretical M.E.P. = 65 − 18 = 47 pounds, which does not include the losses from compression, clearance, wire-drawing and release, so that a mean pressure factor of 0.9 will be used, giving 47 × 0.9 = 42.3 as the M.E.P. to be expected. The mean of these two pressures is 42.325 pounds, which will be used in determining the diameter of the cylinder.

$$\text{Area of piston} = \frac{50 \times 33{,}000}{2\,p_e LN} = \frac{50 \times 33{,}000 \times 12}{2 \times 42.325 \times 10 \times 375}$$
$$= 62.376 \text{ square inches.}$$

Adding 0.884 square inch for the half-section area of a 1.5-inch piston rod, we have 63.26 square inches as the net piston area.

$$\text{Diameter of cylinder} = \sqrt{\frac{63.26}{0.7854}} = 8.975, \text{ say, } 9 \text{ inches.}$$

The mean pressure factors used in the above calculations were determined as follows: It was assumed that the shaded areas *FGN* and *THM* of Fig. 140 would represent losses from wire-

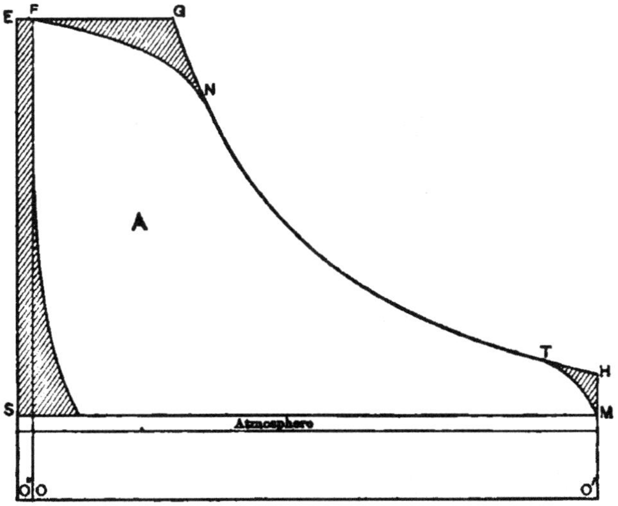

FIG. 140.

drawing and release respectively, and as planimeter measurements showed these areas to be 0.165 square inch, which is 5 per cent of the area *A*, it follows that the M.E.P. factor of 0.95 would make the proper allowance for losses from wire-drawing and release. Similarly, the theoretical M.E.P. above the back pressure of 18 pounds made no allowance for the losses from compression, clearance, wire-drawing, and release. These losses are represented by all the shaded areas of Fig. 140, and their measurement shows them to be a trifle more than 10 per cent of the area *EGHMS*, hence the M.E.P. factor of 0.9 was used.

The properties of the indicator diagram could have been employed in the solution of the problem of Example I without the aid of a scale drawing, thus:

Draw the diagram of Fig. 141, regardless of scale.

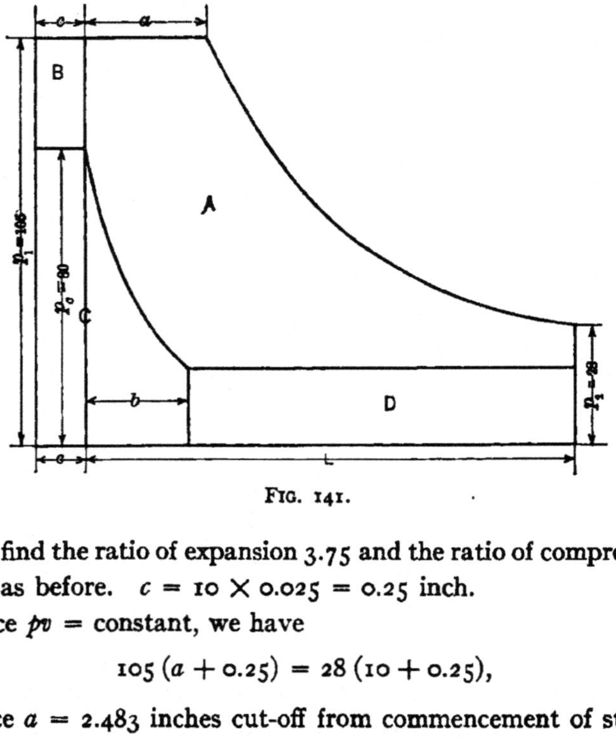

FIG. 141.

We find the ratio of expansion 3.75 and the ratio of compression 4.444 as before. $c = 10 \times 0.025 = 0.25$ inch.

Since pv = constant, we have

$$105\,(a + 0.25) = 28\,(10 + 0.25),$$

whence $a = 2.483$ inches cut-off from commencement of stroke.

And $$80 \times 0.25 = 18\,(b + 0.25),$$

whence $b = 0.861$ inch; that is, the exhaust closes at 0.861 inch from end of return stroke.

The areas expressed in work units are as follows:

$$\text{Area }(A + B + C + D) = p_1 v_1 (1 + \log_e r)$$
$$= 105\,(2.483 + 0.25)\,(1 + 1.3218)$$
$$= 666.275 \text{ inch pounds.}$$

$$\text{Area } B = c\,(p_1 - p_c) = 0.25\,(105 - 80)$$
$$= 6.25 \text{ inch pounds.}$$

$$\text{Area } C = p_c v_c (1 + \log_e r_c) = 80 \times 0.25\,(1 + 1.492)$$
$$= 49.84 \text{ inch pounds.}$$

$$\text{Area } D = p_8\,(10 - b) = 18\,(10 - 0.861)$$
$$= 164.502 \text{ inch pounds.}$$

$$\text{Area } A = \text{Area }(A + B + C + D) - \text{Area }(B + C + D)$$
$$= 666.275 - (6.25 + 49.84 + 164.502)$$
$$= 445.683 \text{ inch pounds.}$$

$$\text{M.E.P.} = \frac{\text{Area } A}{L} = \frac{445.683}{10} = 44.5683 \text{ pounds, as before.}$$

The disadvantage of this method is that the theoretical indicator diagram of the engine is not presented for inspection. In all designs the diagram should be drawn to scale, and this shorter method used as a check to results.

190. Steam Consumption per I.H.P. per Hour. — Engines of different sizes and types do not use the same weight of steam in the development of a given power, but each type, with proper design and working under known conditions, may be expected not to exceed a certain maximum steam consumption per unit of power. The question of steam consumption, while not an exact measure of the efficiency of an engine, is of prime importance in engine design, as it is the leading factor in the determination of the boiler capacity required in any particular case.

Suppose, for example, in the design of a boiler to supply steam for a given engine, it is assumed that 20 pounds of coal will be consumed on each square foot of grate surface per hour and that each pound of coal will evaporate 8 pounds of water. We shall then have $20 \times 8 = 160$ pounds of water evaporated per hour per square foot of grate surface. If the ratio of grate surface to heating surface be 1 to 32, this would mean an evaporation of $\frac{160}{32} = 5$ pounds of water per square foot of heating surface per hour; and if the weight of the boiler, including its water, be taken at the fair estimate of 25 pounds per square foot of heating surface, we would have $\frac{25}{5} = 5$ pounds weight of boiler for each pound of water evaporated per hour. In other words, for each pound of steam saved at the engine per I.H.P. per hour there would be a saving of 5 pounds in boiler weight, which, for 200 I.H.P., would mean a saving of 1000 pounds. In marine practice, where the I.H.P. runs into thousands, the saving of a pound of steam per I.H.P. per hour by proper engine design means the saving of from 20 to 40 tons in boiler weight, increasing by that amount the cargo capacity of a merchant ship or, in the case of a war vessel, permitting an increase in the weight of armor or of an increase in its steaming radius by giving it a greater coal-carrying capacity.

Taking the steam consumption per I.H.P. as the measure of the performance of an engine, independently of the boiler, a comparison of this consumption with that of the ideal engine working between the given extremes of temperature t_1 and t_2 is a measure of the economy of the engine.

The efficiency of the ideal engine, as we have seen, is $\frac{t_1 - t_2}{t_1 + 461°}$, and since the production of one I.H.P. per minute requires the expenditure of $\frac{33000}{778} = 42.42$ thermal units, it follows that the minimum expenditure in the ideal engine to produce one I.H.P. will be

$$42.42 \div \frac{t_1 - t_2}{t_1 + 461°} = \frac{42.42\,(t_1 + 461°)}{t_1 - t_2}$$

thermal units per minute.

If H_w denotes the units of heat required to produce one pound of steam from the temperature t_f of the feed water and at the temperature t_1 of the steam, and if Q denotes the number of pounds of steam per I.H.P. per hour, then, for the ideal engine, we shall have

$$Q = \frac{60\,[42.42\,(t_1 + 461^\circ)]}{H_w\,(t_1 - t_2)}$$

$$= \frac{60\,[42.42\,(t_1 + 461^\circ)]}{[1091.7 + 0.305\,(t_1 - 32^\circ) - (t_f - 32^\circ)]\,[t_1 - t_2]}.$$

The values of Q for condensing and non-condensing engines given in the table below were computed by the formula just given, the results showing a marked gain in efficiency due to the use of the condenser. With the condensing engine the feed water was assumed to have been taken from the condenser at a temperature of 140° and with the non-condensing engine it was assumed that the feed water was heated to a temperature of 222°.

STEAM CONSUMPTION OF THE IDEAL ENGINE PER I.H.P. PER HOUR.

Initial absolute pressure, pounds per square inch.	t_1	Values of Q, condensing engine.	Values of Q, non-condensing engine.
70	303°	11.18 lbs.	24.38 lbs.
90	320°	10.33 lbs.	20.52 lbs.
110	335°	9.65 lbs.	18.02 lbs.
130	347°	9.20 lbs.	....
150	358°	8.83 lbs.	....
180	373°	8.37 lbs.	....
250	401°	7.67 lbs.	....

Since the relative efficiencies of non-condensing and condensing engines do not vary much from 70 per cent and 52 per cent of the ideal efficiencies respectively (see Art. 178), we may expect a consumption of $\frac{20.52}{0.7} = 29.3$ pounds of steam per I.H.P. per hour for the ordinary non-condensing engine, and a consumption of $\frac{7.67}{0.52} = 14.7$ pounds for the triple-expansion type of engine commonly used. Assuming the boiler to evaporate 8.5 pounds of water per pound of coal, these figures would indicate a coal consumption of 3.45 pounds of coal per I.H.P. per hour for the non-condensing engine, and 1.73 pounds for the triple-expansion condensing engine. These results agree closely with those obtained in actual practice.

191. Measurement of Steam Consumption from the Indicator Diagram. — The volume of the cylinder is that cylindrical volume whose diameter is the diameter of the piston and whose altitude is the length of the stroke, and the clearance volume consists of the volume of the steam passage from the valve chest to the cylinder and of that space between the piston and the cylinder head when the piston is at the end of its stroke. The volume of steam available for use at any point of the stroke is therefore equal to the volume displaced by the piston up to that point plus the volume of the clearance space. If the point be taken at cut-off, or at any point after cut-off, and the volume be multiplied by the specific weight (weight per cubic foot) of the steam at the absolute pressure at the point, the product will be the weight of the steam in pounds; deducting from this the weight of steam retained in the cylinder by the closure of the exhaust on the return stroke and which is compressed into the clearance space, we get the weight of steam used per stroke. If the weight of steam expended per stroke be multiplied by the number of strokes per hour and the product be divided by the indicated horse-power of the engine, the quotient will be the number of pounds of steam used per I.H.P. per hour.

Since the clearance volume is always expressed in terms of percentage of the cylinder volume, it follows that all the quantities involved in the measurement of the volume of steam supplied to the cylinder have the cross-section area of the cylinder as a common factor, so that this factor may be omitted and each of the volume quantities be represented on the indicator diagram by its linear dimension.

To estimate the steam consumption per I.H.P. per hour of an engine from the diagram used in its preliminary design we proceed as follows:

Let A denote the area of the piston in square inches and let a, b, c, and L of Fig. 142 be expressed in feet. At the moment

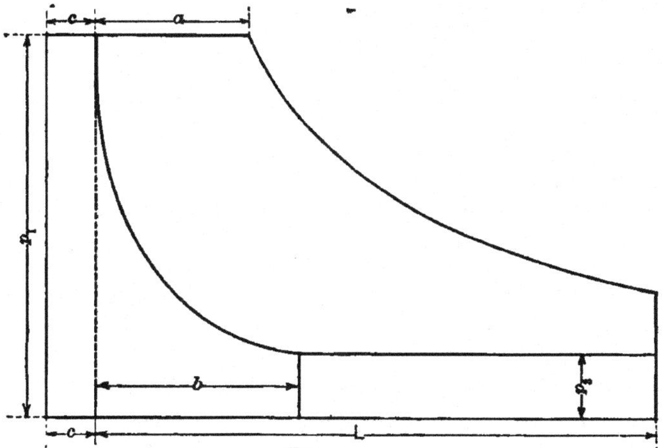

FIG. 142.

of cut-off the volume of steam at pressure p_1 within the cylinder and clearance is $\frac{A\,(a+c)}{144}$ cubic feet, and if w_1 denotes the weight of a cubic foot of steam at pressure p_1 the weight of this steam is $\frac{A\,(a+c)\,w_1}{144}$ pounds. At the moment of exhaust closure the weight of steam at p_3 pressure within the cylinder and clearance is $\frac{A\,(b+c)\,w_3}{144}$ pounds, w_3 denoting the weight of a cubic foot of steam at pressure p_3. This latter weight of steam is not allowed to escape in the exhaust but is compressed into the clearance space and used again during the next stroke of the engine, so that $\frac{A\,(a+c)\,w_1}{144} - \frac{A\,(b+c)\,w_3}{144}$ is the weight of steam used by the engine per stroke. If N denotes the number of revolutions of the engine per minute, then

$$60 \times 2N\left[\frac{A(a+c)w_1}{144} - \frac{A(b+c)w_2}{144}\right] = \frac{60 \times 2NA}{144}[(a+c)w_1 - (b+c)w_2]$$

is the weight of steam used per hour, and

$$\frac{120NA}{144}[(a+c)w_1 - (b+c)w_2] \div \frac{2p_eLAN}{33,000} = \frac{13,750}{p_e}\left[\frac{(a+c)w_1}{L} - \frac{(b+c)w_2}{L}\right]$$

is the expression for the number of pounds of steam used by the engine per hour per I.H.P.

The volume fractions $\frac{a+c}{L}$ and $\frac{b+c}{L}$ are constant as long as a, b, c, and L are measured in the same units, so if these volume fractions be denoted by x_1 and x_2 respectively, we shall have

$$\text{Pounds of steam per I.H.P. per hour} = \frac{13,750}{p_e}(x_1w_1 - x_2w_2),$$

an expression general in its application and entirely independent of the size and speed of the engine. That is, if the indicator diagram of an engine be given and its pressure scale be known, an approximation of the steam consumption of the engine per I.H.P. per hour may be made by means of the formula just given. The points of cut-off and exhaust closure, the clearance c, and the mean effective pressure p_e may be found from the diagram very approximately, and then, since the length of the diagram is directly proportional to L, the volume fractions $\frac{a+c}{L} = x_1$ and $\frac{b+c}{L} = x_2$ are readily found. The pressures p_1 and p_2, of course, are measured from the diagram and the corresponding weights per cubic foot w_1 and w_2 are taken from a table of the properties of steam.

In the case of the engine of Example I we found from Fig. 139, partly reproduced in Fig. 143, that $x_1 = \frac{a+c}{L} = \frac{0.82}{3} = 0.273$, and $x_3 = \frac{b+c}{L} = \frac{0.333}{3} = 0.111$, and from the table of the properties of steam we find w_1 and w_3 corresponding to pressures

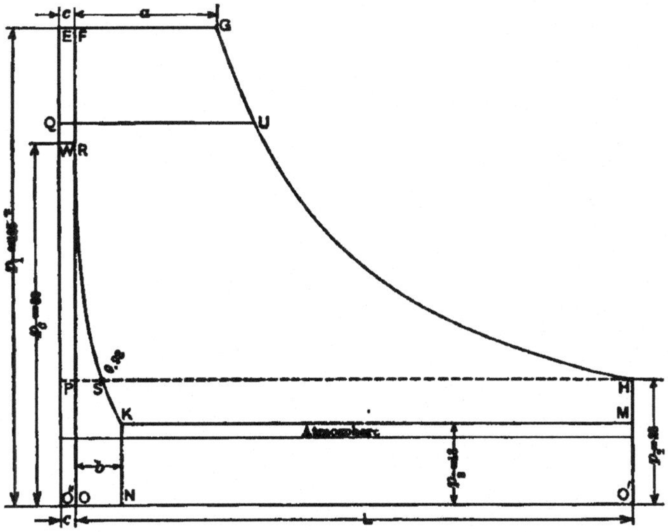

FIG. 143.

of 105 pounds and 18 pounds to be 0.238 pound and 0.04547 pound respectively.

Then, for the engine of the example, we shall have

Steam per I.H.P. per hour

$$= \frac{13,750}{42.325}(0.273 \times 0.238 - 0.111 \times 0.04547)$$
$$= 19.47 \text{ pounds.}$$

The measurement was made at the point of cut-off, but it might have been made at any other point of the stroke, as, for example, at U where the pressure measures 84 pounds. The ratio $\frac{QU}{OO'}$ is found to be 0.341 and the weight of a cubic foot of steam at 84 pounds pressure is 0.193 pound. Then

Steam per I.H.P. per hour

$$= \frac{13,750}{42.325}(0.341 \times 0.193 - 0.111 \times 0.04547)$$

$$= 19.74 \text{ pounds.}$$

The consumption of steam per I.H.P. thus obtained is based on the weight of steam shown by the indicator diagram to have left the cylinder in the exhaust, but it does not include the waste of steam due to liquefaction in the cylinder, radiation and other causes of loss. The result though is valuable as a means of comparing one engine with another when working under like conditions, or of comparing the performance of the same engine working under different conditions. The consumption of steam shown by indicator diagrams is always less than the quantity actually passing through the cylinder, the conditions of actual practice being such that from 10 per cent to 40 per cent, depending upon the type of engine, should be added to obtain results approximately correct.

Since there is always water in the cylinder at the point of cut-off, and since more or less of the water in the cylinder due to liquefaction is re-evaporated into steam as the piston advances, the quantity of steam in the cylinder varies during the stroke. In stage-expansion engines the steam used in the L.P. cylinder first passes through the H.P. and I.P. cylinders, and consequently the steam consumption of the H.P. cylinder will be the measure of the consumption of the whole engine. If the measure be taken also from the diagrams of the intermediate and low-pressure cylinders, as it should be for purposes of comparison, it will be found that the result from the high-pressure diagram will be the greatest. The difference in the results may be taken as a fair measure of the loss in transmission.

If the measurement of the steam consumption of the engine of Example I be taken at H, Fig. 143, the termination of the expansion curve, the volume of steam then in the cylinder and clearance at the terminal pressure $O'H$ is $PH = O''O'$. If K be the point in the back-pressure line where the exhaust closes and compression begins, then $O''N$ is the volume of steam of pressure NK that is retained in the cylinder for compression and does not escape. This volume when compressed to the pressure $O'H$ is PS, therefore $\frac{SH}{O''O'}$ is the percentage of the volume of the steam of terminal pressure found in the cylinder and clearance at the end of the stroke that escapes in the exhaust. If we denote the ratio $\frac{SH}{O''O'}$ by x_2, the weight of a cubic foot of steam of the theoretical terminal pressure by w_2, and a percentage factor by K, then experience has shown that the actual steam consumption is given very approximately by the expression

$$\text{Pounds of steam per I.H.P. per hour} = \frac{13{,}750\, w_2 x_2}{p_e}(1 + K).$$

The values for K are well defined for different types of engines and may be taken as follows:

For single-cylinder non-condensing engines, 40 per cent; for single-cylinder condensing engines, 30 per cent; for non-condensing compound engines, 25 per cent; for condensing compound engines, 20 per cent; and for triple-expansion engines, 10 per cent.

Referring to Fig. 143, the ratio $\frac{SH}{O''O'}$ is found by measurement to be 0.93, and the weight of a cubic foot of steam of the theoretical terminal pressure of 28 pounds is found from the table to be 0.069 pound. Then for the single-cylinder non-condensing engine of Example I, we shall have

Steam per I.H.P. per hour

$$= \frac{13{,}750 \times 1.4 \times 0.069 \times 0.93}{42.325} = 29.19 \text{ pounds.}$$

Example II. The Preliminary Design of Stage-expansion Engines. — It is desired to design a non-condensing cross-compound engine to develop 160 I.H.P. It is predetermined that the absolute initial steam pressure shall be 150 pounds, the stroke 14 inches, the revolutions per minute 260, and that the volumetric cylinder ratio $\frac{\text{L.P.}}{\text{H.P.}}$ shall be 3.65.

It is fairly assumed that an absolute back pressure of 18 pounds will be maintained, and that a careful design of the valve and steam passages will confine the H.P. clearance to 5 per cent and that of the L.P. cylinder to 3 per cent.

Since the power of a stage-expansion engine is measured by the size of the low-pressure cylinder, its dimensions are first to be determined, and in doing so we regard the terminal pressure in that cylinder as an exponent of the steam consumption. The first step will be to decide what the L.P. terminal pressure shall be in order that the engine when designed shall be economical in the use of steam.

Assume a trial terminal pressure of 26 pounds in the L.P. cylinder. We shall then have $\frac{150}{26} = 5.77$ as the ratio of expansion, assuming the whole of the expansion to take place in the L.P. cylinder. Then

$$p_m = p_2 (1 + \log_e r) = 26 \times 2.7527 = 71.57 \text{ pounds.}$$

$$\text{Expected M.E.P.} = (71.57 - 18)\, 0.85 = 45.53 \text{ pounds.}$$

Then $$Q = \frac{34 p_2 + 23}{\text{M.E.P.}} = \frac{34 \times 26 + 23}{45.53} = 19.92 \text{ pounds}$$

of steam per I.H.P. per hour, the terminal pressure in the L.P. cylinder being 26 pounds. This result indicates an actual consumption of $19.92 \times 1.25 = 24.9$ pounds (see page 327), which is rather large, so a terminal pressure of 23 pounds will be tried. Thus

$$r = \frac{150}{23} = 6.522, \text{ the } \log_e \text{ of which is } 1.875,$$

$$p_m = 23 \times 2.875 = 66.125.$$

Expected M.E.P. = (66.125 − 18) 0.85 = 40.9 pounds.

$$Q = \frac{34 \times 23 + 23}{40.9} = 19.68 \text{ pounds}$$

of steam per I.H.P. per hour, indicating an actual consumption of 19.68 × 1.25 = 24.6 pounds. This result may be regarded as fairly good for the non-condensing compound type of engine, so a terminal pressure in the L.P. cylinder of 23 pounds will be adopted.

Assume for convenience that OO', Fig. 144, 4 inches in length, represents the stroke volume of the L.P. cylinder of the required engine, and denote it by L. Then the clearance volume of the L.P. cylinder will be represented by $OO'' = 4 \times 0.03 = 0.12$ inch and will be denoted by k. To a pressure scale of 1 inch = 40 pounds erect $O''E$ perpendicular to $O''O'$, making it $\frac{150}{40}$ inches in length to represent the absolute initial pressure of 150 pounds. Then $O''O'$ and $O''E$ are the lines of *no pressure* and *no volume* respectively. Erect $O'H$ perpendicular to $O''O'$ and make it $\frac{23}{40}$ inch long to represent the terminal pressure of 23 pounds. On $O'H$ take the point M so that $O'M$ measures $\frac{18}{40}$ inch to represent the back pressure of 18 pounds. Then MU' parallel to OO' will be the exhaust line, or line of back pressure. Assuming the exhaust to close at 85 per cent of the stroke, make MK equal to $\frac{85}{100}$ of the stroke OO'.

The expansion curve GH may now be constructed, its terminal point H and rectangular asymptotes $O''E$ and $O''O'$ being known. Construct also a part KT of the compression curve.

In Art. 148, page 243, it was shown that, for the total ratio of expansion for stage-expansion engines, $r = \frac{\phi(1 + c)}{a + c}$. We shall then have

$$6.522 = \frac{3.65\,(1 + 0.05)}{a + 0.05}, \text{ whence } a = 0.5376.$$

That is, the cut-off in the H.P. cylinder takes place at 53.76 per cent of stroke.

The volume of the H.P. cylinder will be represented by

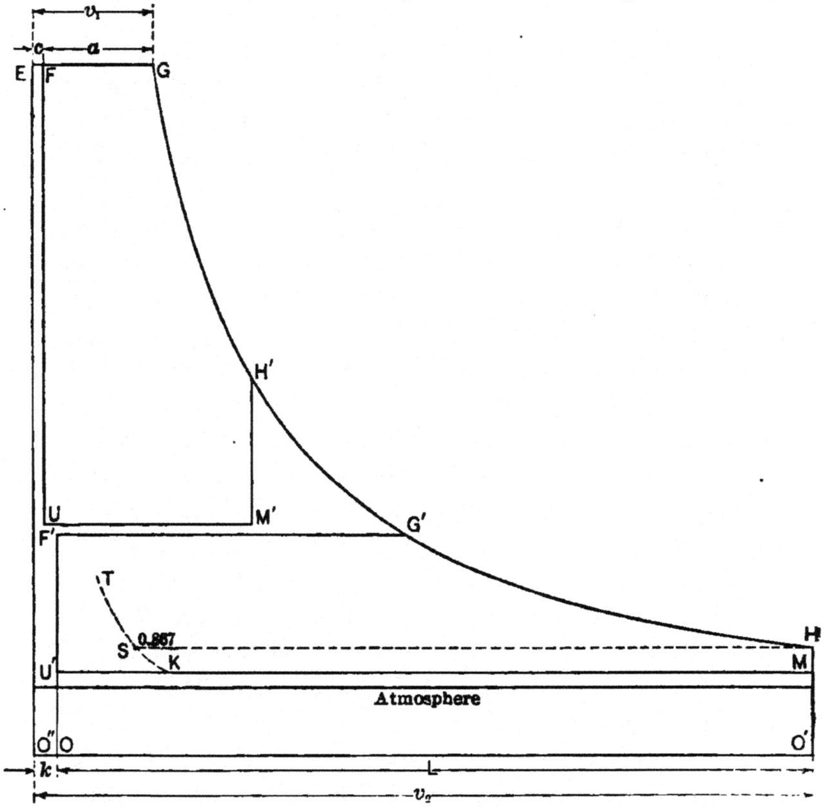

FIG. 144.

$\frac{4}{3.65}$ = 1.096 inches to the same scale that 4 inches represents the volume of the L.P. cylinder. The H.P. clearance volume will be represented by $1.096 \times 0.05 = 0.0548$ inch, and $1.096 \times 0.5376 = 0.5892$ inch represents the part of the H.P. stroke completed up to cut-off. Then

$$v_1 = a + c = 0.5892 + 0.0548 = 0.644 \text{ inch} = EG,$$

proving the accuracy of the construction of the expansion curve.

Denoting the ratio of expansion in the H.P. cylinder by r_1, we shall have

$$r_1 = \frac{1.096 + 0.0548}{0.5892 + 0.0548} = 1.79, \text{ or } r_1 = \frac{1 + 0.05}{0.5376 + 0.05} = 1.79.$$

As the receiver pressure is controlled by the cut-off in the L.P. cylinder, we will try a receiver pressure of 48 pounds. For equality of power in the cylinders the pressure multiplied by the volume of steam given to the H.P. cylinder must equal the pressure multiplied by the volume given to the L.P. cylinder; that is

$$150\,(0.5892 + 0.0548) = 48\,x, \text{ whence } x = 2 \text{ inches,}$$

which represents the volume of steam in the L.P. cylinder and clearance when cut-off takes place. Denoting the ratio of expansion in the L.P. cylinder by r_2 we have

$$r_2 = \frac{4.12}{2} = 2.06.$$

Theoretical p_m in H.P. cylinder

$$= \frac{150\,(1 + \log_e 1.79)}{1.79} = \frac{150 \times 1.5822}{1.79} = 132.58 \text{ pounds.}$$

Assuming that friction in the exhaust passages of the H.P. cylinder will occasion a back pressure of 2 pounds we shall have

Theoretical M.E.P. in H.P. cylinder $= 132.58 - 50 = 82.58$ pounds.

Using a mean pressure factor of 0.85 to allow for losses due to wire-drawing, release, and compression, we have

Expected M.E.P. in H.P. cylinder $= 82.58 \times 0.85 = 70.19$ pounds.

M.E.P. in H.P. cylinder referred to L.P. piston $= \frac{70.19}{3.65} = 19.23$ pounds.

Theoretical p_m in L.P. cylinder

$$= \frac{48\,(1 + \log_e 2.06)}{2.06} = \frac{48 \times 1.7227}{2.06} = 40.14 \text{ pounds.}$$

Expected M.E.P. in L.P. cylinder = (40.14 − 18) 0.85 = 18.82 pounds.

For equality of power in the cylinders the M.E.P. in the L.P. cylinder and the M.E.P. in the H.P. cylinder referred to the L.P. piston should be equal. In this instance there is a difference in these pressures of only 19.23 − 18.82 = 0.41 pound, insuring the practically equal division of the power between the cylinders. Had there been a difference of upwards of 0.5 pound in these pressures it would have been necessary to choose another receiver pressure to bring them nearer equality. The choice of the receiver pressure to equalize the mean effective pressures referred to the L.P. piston is based on these considerations:

An increase in the receiver pressure will increase the back pressure on the H.P. piston and therefore decrease the M.E.P. in the H.P. cylinder, whereas the same increase in receiver pressure will increase the M.E.P. in the L.P. cylinder by reason of the consequent increase of the initial pressure in the L.P. cylinder.

The total calculated M.E.P. referred to the L.P. piston is then 19.23 + 18.82 = 38.05 pounds.

Returning to Fig. 144, lay off the distance $EF = c = 0.0548$ inch to represent the clearance volume of the H.P. cylinder. Draw FU parallel to EO'', the point U being $\frac{50}{40}$ inches from $O''O'$ to correspond to the back pressure of 50 pounds on the H.P. piston. Lay off UM' parallel to $O''O'$, making it 1.096 inches long to represent the volume of the H.P. cylinder. At M' erect a perpendicular to UM', producing it to intersect the expansion curve at H'. The area $FGH'M'U$ is the theoretical indicator diagram for the H.P. cylinder. As a test of the accuracy of the construction the point H' is $\frac{83.8}{40}$ inches from $O''O'$, as it should be to correspond to the terminal pressure $\frac{p_1}{r_1} = \frac{150}{1.79}$ $= 83.8$ pounds in the H.P. cylinder.

For the theoretical diagram of the L.P. cylinder, erect OF' perpendicular to $O''O'$, making it $\frac{48}{40}$ inches in length to correspond to the receiver pressure of 48 pounds. From F' draw a parallel to $O''O'$, producing it to its intersection G' with the expansion curve. Then the area $F'G'HMU'$ is the theoretical indicator diagram for the L.P. cylinder. As a check to the accuracy of this construction, the line $F'G'$ measures $2 - 0.12 = 1.88$ inches, as it should do to represent the volume of the L.P. cylinder up to the point of cut-off.

The gap between the diagrams at $H'G'$ illustrates the loss from drop between the cylinders.

The areas of the H.P. and L.P. diagrams are found by planimeter measurements to be 2.23 and 2.24 square inches respectively, but only 0.85 of each of these areas may be expected to be realized, owing to the losses from wire-drawing, release, and compression.

To reduce the expected area of the H.P. diagram to pressure, we have $\frac{2.23 \times 0.85}{1.096} = 1.73$ inches as the length of the mean ordinate, so that the M.E.P. from the H.P. diagram is $1.73 \times 40 = 69.2$ pounds, the pressure scale being 40 pounds to the inch. When referred to the L.P. piston the M.E.P. from the H.P. diagram becomes $\frac{69.2}{3.65} = 18.96$ pounds per square inch. The area of the L.P. diagram when reduced to pressure becomes $\frac{2.24 \times 0.85 \times 40}{4} = 19.04$ pounds per square inch, so that the total M.E.P., obtained from the diagrams and referred to the L.P. piston, is $18.96 + 19.04 = 38$ pounds per square inch.

The expected M.E.P. due to the initial pressure of 150 pounds and the total ratio of expansion of 6.522 is

$$\left[\frac{150\,(1 + \log_e 6.522)}{6.522} - 18\right]0.85 = 40.9 \text{ pounds.}$$

The total M.E.P. obtained by calculation was 38.05 pounds, and the mean of the three mean effective pressures is

$$\frac{38 + 40.9 + 38.05}{3} = 38.98 \text{ pounds.}$$

This M.E.P. will be used in determining the diameter of the L.P. cylinder.

$$\text{Area of L.P. piston} = \frac{160 \times 33{,}000}{2\,p_e LN} = \frac{160 \times 33{,}000 \times 12}{2 \times 38.98 \times 14 \times 260}$$
$$= 223.28 \text{ square inches.}$$

Adding 1.99 square inches for the half-section area of a 2.25-inch piston rod, we have 225.27 square inches as the net area of the L.P. piston.

$$\text{Diameter of L.P. cylinder} = \sqrt{\frac{225.27}{0.7854}} = 16.94, \text{ say, } 17 \text{ inches.}$$

We have

$$\frac{\text{Volume L.P. cylinder + volume clearance}}{\text{Volume H.P. cylinder + volume clearance}}$$
$$= \frac{\text{L.P. piston area } (1 + k)}{\text{H.P. piston area } (1 + c)} = 3.65,$$

whence

$$\text{Area of H.P. piston} = \frac{223.28 \times 1.03}{3.65 \times 1.05} = 60 \text{ square inches.}$$

$$\text{Net area of H.P. piston} = 60 + 1.99 = 61.99 \text{ square inches.}$$

$$\text{Diameter of H.P. cylinder} = \sqrt{\frac{61.99}{0.7854}} = 8.884, \text{ say, } 9 \text{ inches.}$$

From H, Fig. 144, draw a parallel to $O''O'$ to its intersection S with the compression curve KT. Then HS represents the volume of steam of p_2 pressure expended per stroke. The ratio $\frac{SH}{O''O'}$ is found by measurement to be 0.867, and the weight of a cubic foot of steam at 23 pounds pressure is found from the the steam table to be 0.05676 pound. Then

Steam per I.H.P. per hour

$$= \frac{13,750 \times 0.867 \times 0.05676 \times 1.25}{38.98} = 21.7 \text{ pounds.}$$

The temperatures of steam at pressures of 150 pounds and of 23 pounds are 358.5° and 235.5° respectively, so that we shall have

$$\text{Ideal efficiency} = \frac{t_1 - t_2}{t_1 + 461} = \frac{358.5 - 235.5}{358.5 + 461} = 0.15, \text{ or } 15 \text{ per cent.}$$

It is reasonable to suppose that the boiler will evaporate 8.5 pounds of water per pound of coal, which would mean the expenditure of $\frac{21.7}{8.5} = 2.553$ pounds of coal per I.H.P. per hour.

Assuming the thermal value of the fuel to be 14,500 B.t.u. per pound, we would have

Thermal efficiency of engine and boiler

$$= \frac{42.42 \times 60}{14,500 \times 2.553} = 0.06875, \text{ or } 6.875 \text{ per cent.}$$

These results are satisfactory, so that an engine with a high-pressure cylinder of 9 inches diameter and low-pressure cylinder of 17 inches diameter will fulfil the required conditions, provided the valves are properly designed.

According to the calculations the cut-off in the H.P. cylinder must take place at $14 \times 0.5376 = 7.526$ inches from commencement of stroke, and in the L.P. cylinder the cut-off is to take place at $14 \times \frac{2 - 0.12}{4} = 6.58$ inches from commencement of stroke in order to maintain a receiver pressure of 48 pounds. The design of the valves must provide for these cut-offs and also for a low-pressure exhaust closure at 85 per cent of the stroke.

The points of cut-off might have been found in this manner:

With a stroke of 14 inches and a ratio of expansion in the H.P. cylinder of 1.79 and clearance of 5 per cent, we will have

$$1.79 = \frac{14 + 14 \times 0.05}{a_1 + 14 \times 0.05}, \text{ whence } a_1 = 7.51 \text{ inches.}$$

In like manner, for a ratio of expansion of 2.06 in the L.P. cylinder and clearance of 3 per cent, we will have

$$2.06 = \frac{14 + 14 \times 0.03}{a_2 + 14 \times 0.03}, \text{ whence } a_2 = 6.58 \text{ inches.}$$

Example III. — It is desired to design a condensing compound engine to develop 160 I.H.P. It is predetermined that the absolute initial pressure at the valve chest shall be 150 pounds, the stroke 14 inches, the number of revolutions per minute 260, and the cylinder ratio $\frac{\text{L.P.}}{\text{H.P.}} = 3.65$. It is fairly assumed that a back pressure of 3 pounds will be maintained in the condenser, and that the clearance of the H.P. cylinder will be confined to 5 per cent and that of the L.P. cylinder to 3 per cent.

As in Example II, we shall first decide upon the terminal pressure in the L.P. cylinder.

Assume a trial terminal pressure of 12 pounds in the L.P. cylinder. Then $\frac{150}{12} = 12.5$ = ratio of expansion, assuming the whole of the expansion to take place in the L.P. cylinder. Then

$$p_m = p_2 (1 + \log_e r) = 12 (1 + 2.5262) = 42.314 \text{ pounds.}$$

$$\text{Expected M.E.P.} = (42.314 - 3)\, 0.85 = 33.42 \text{ pounds.}$$

$$Q = \frac{34 p_2 + 23}{\text{M.E.P.}} = \frac{34 \times 12 + 23}{33.42} = 12.89 \text{ pounds}$$

of steam per I.H.P. per hour, the terminal pressure in the L.P. cylinder being 12 pounds. This result indicates an actual consumption of $12.89 \times 1.2 = 15.47$ pounds. With the boiler furnishing superheated steam of 150 pounds pressure it is reasonable to expect an expansion to a terminal pressure of 10 pounds, so we will try a L.P. terminal pressure of 10 pounds. Then

$$r = \frac{150}{10} = 15, \text{ the } \log_e \text{ of which is } 2.708.$$

$$p_m = 10 \times 3.708 = 37.08 \text{ pounds.}$$

$$\text{Expected M.E.P.} = (37.08 - 3)\,0.85 = 28.97 \text{ pounds.}$$

$$Q = \frac{34 \times 10 + 23}{28.97} = 12.53 \text{ pounds}$$

of steam, indicating an actual consumption of $12.53 \times 1.2 = 15$ pounds per I.H.P. per hour. This result is fairly favorable for the type of engine, so a terminal pressure of 10 pounds in the L.P. cylinder will be adopted.

Proceeding as in Example II, let OO', Fig. 145, 4 inches in length, represent the stroke volume of the L.P. cylinder of the required engine, and denote it by L. Then the clearance volume of the L.P. cylinder will be represented by $OO'' = 4 \times 0.03 = 0.12$ inch, and will be denoted by k. To a pressure scale of 1 inch = 40 pounds erect $O''E$ perpendicular to $O''O'$, making it $\frac{150}{40}$ inches in length to represent the absolute initial pressure of 150 pounds. Then $O''O'$ and $O''E$ are the lines of no pressure and no volume respectively. Erect $O'H$ perpendicular to $O''O'$ and make it $\frac{10}{40}$ inch long to represent the terminal pressure of 10 pounds in the L.P. cylinder. On $O'H$ take the point M so that $O'M$ measures $\frac{8}{40}$ inch to represent the back pressure of 3 pounds. Then MU' parallel to OO' will be the exhaust line, or line of back pressure. Assuming the exhaust to close at 85 per cent of the stroke, make MK equal to $\frac{85}{100}$ of the stroke OO'.

The expansion curve GH may now be constructed, its terminal point H and rectangular asymptotes $O''E$ and $O''O'$ being known. Construct also a part KT of the compression curve.

As in Example II, we have

$$15 = \frac{3.65\,(1 + 0.05)}{a + 0.05}, \text{ whence } a = 0.2055,$$

that is, the cut-off in the H.P. cylinder takes place at 20.55 per cent of stroke.

The volume of the H.P. cylinder will be represented by

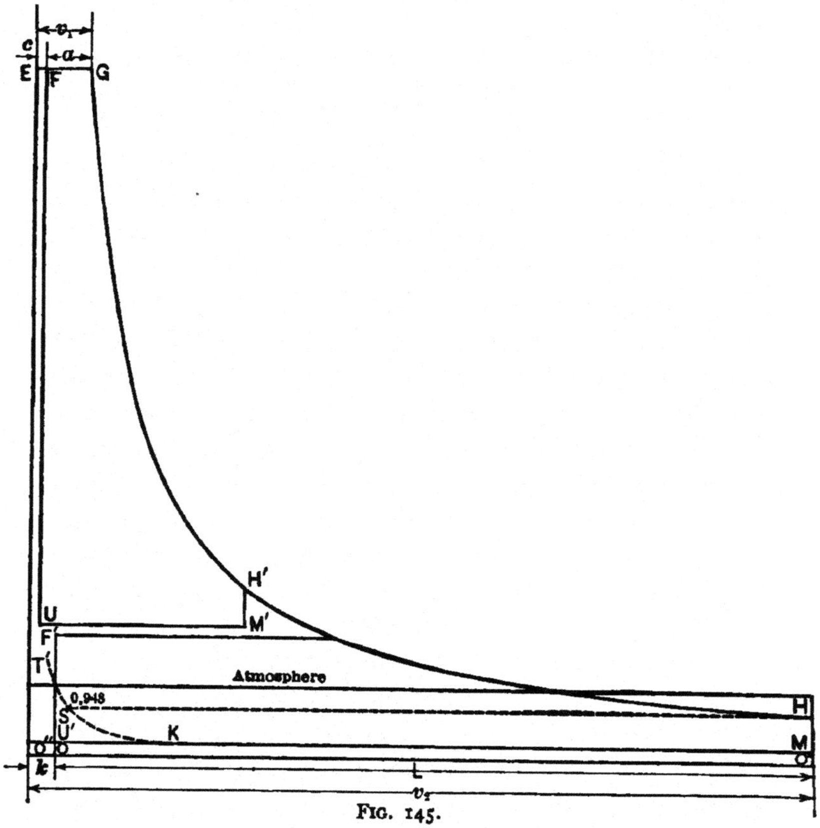

Fig. 145.

$\frac{4}{3.65} = 1.096$ inches to the same scale that 4 inches represents the volume of the L.P. cylinder. The clearance of the H.P. cylinder will be represented by $1.096 \times 0.05 = 0.0548$ inch, and the portion of the H.P. stroke completed up to cut-off by $1.096 \times 0.2055 = 0.2252$ inch. Then

$$v_1 = a + c = 0.2252 + 0.0548 = 0.28 = EG,$$

proving the correctness of the construction of the expansion curve.

For the ratio of expansion in the H.P. cylinder we have

$$r_1 = \frac{1.096 + 0.0548}{0.2252 + 0.0548} = 4.11, \text{ the } \log_e \text{ of which is } 1.4134.$$

We will try a receiver pressure of 28 pounds. Then

$$150\,(0.2252 + 0.0548) = 28\,x, \text{ whence } x = 1.5 \text{ inches,}$$

which represents the volume of steam in the L.P. cylinder and clearance at cut-off. For the ratio of expansion in the L.P. cylinder we have

$$r_2 = \frac{4.12}{1.5} = 2.747, \text{ the } \log_e \text{ of which is } 1.01.$$

Theoretical p_m in H.P. cylinder

$$= \frac{p_1\,(1 + \log_e r_1)}{r_1} = \frac{150 \times 2.4134}{4.11} = 88.08 \text{ pounds.}$$

Assuming that friction in the exhaust passages of the H.P. cylinder will occasion a back pressure of 2 pounds, we shall have

Theoretical M.E.P. in H.P. cylinder $= 88.08 - 30 = 58.08$ pounds.

Using a mean pressure factor of 0.85, we have

M.E.P. to be expected in H.P. cylinder

$$= 58.08 \times 0.85 = 49.36 \text{ pounds.}$$

M.E.P. in H.P. cylinder referred to L.P. piston

$$= \frac{49.36}{3.65} = 13.52 \text{ pounds.}$$

Theoretical p_m in L.P. cylinder

$$= \frac{28\,(1 + \log_e 2.747)}{2.747} = \frac{28 \times 2.01}{2.747} = 20.48 \text{ pounds.}$$

M.E.P. to be expected in L.P. cylinder

$$= (20.48 - 3)\,0.85 = 14.85 \text{ pounds.}$$

The expected mean effective pressures differ too greatly for an equality of power in the two cylinders, so we will try a receiver pressure of 26 pounds. Then

$$150\,(0.2252 + 0.0548) = 26\,x, \text{ whence } x = 1.615 \text{ inches.}$$

$$r_2 = \frac{4.12}{1.615} = 2.55, \text{ the } \log_e \text{ of which is } 0.9361.$$

Theoretical p_m in L.P. cylinder

$$= \frac{26\,(1 + \log_e 2.55)}{2.55} = \frac{26 \times 1.9361}{2.55} = 19.74 \text{ pounds}$$

Expected M.E.P. in L.P. cylinder

$$= (19.74 - 3)\,0.85 = 14.22 \text{ pounds.}$$

Expected M.E.P. in H.P. cylinder referred to L.P. piston

$$= \frac{(88.08 - 28)\,0.85}{3.65} = 13.99 \text{ pounds.}$$

These mean effective pressures are near enough equal to insure a practical equality of power in the two cylinders. The total M.E.P. obtained by calculation and referred to the L.P. piston is $14.22 + 13.99 = 28.21$ pounds.

Returning to Fig. 145, lay off the distance $EF = c = 0.0548$ inch to represent the clearance volume of the H.P. cylinder. Draw FU parallel to EO'', the point U being $\frac{28}{40}$ inch from $O''O'$ to correspond to the back pressure of 28 pounds in the H.P. cylinder. Lay off UM' parallel to $O''O'$, making it 1.096 inches long to represent the volume of the H.P. cylinder. At M' erect a perpendicular to UM' to intersect the expansion curve at H'. The area $FGH'M'U$ is the theoretical indicator diagram for the H.P. cylinder. As a test of the accuracy of the construction the point H' is $\frac{36.5}{40}$ inches from $O''O'$, as it should be to correspond to the terminal pressure of $\frac{p_1}{r_1} = \frac{150}{4.11} = 36.5$ pounds in the H.P. cylinder.

For the theoretical indicator diagram of the L.P. cylinder, erect OF' perpendicular to $O''O'$ and make it $\frac{26}{40}$ inch in length to correspond to the receiver pressure of 26 pounds. From F' draw a parallel to $O''O'$, producing it to its intersection G' with the expansion curve. Then $F'G'HMU'$ is the theoretical indicator diagram for the L.P. cylinder. As a check to the accuracy of the construction the line $F'G'$ measures $1.615 - 0.12 = 1.495$ inches, as it should do to represent the volume of the L.P. cylinder up to the point of cut-off.

The gap in the diagram at $H'G'$ illustrates the loss from drop between the cylinders.

The areas of the H.P. and L.P. diagrams are found by planimeter measurements to be 1.55 and 1.59 square inches respectively, but only 0.85 of each of these areas is expected to be realized owing to the losses from wire-drawing, release, and compression.

To reduce the area of the H.P. diagram to pressure, we have $\frac{1.55 \times 0.85}{1.096} = 1.2021$ inches as the length of the mean ordinate of the diagram, so that the M.E.P. from the H.P. diagram is $1.2021 \times 40 = 48.084$ pounds which, when reduced to the L.P. piston area, becomes $\frac{48.084}{3.65} = 13.17$ pounds. The area of the L.P. diagram when reduced to pressure becomes

$$\frac{1.59 \times 0.85 \times 40}{4} = 13.515 \text{ pounds,}$$

so that the total mean effective pressure obtained from the diagrams and referred to the L.P. piston is $13.17 + 13.515 = 26.685$ pounds.

The expected M.E.P. due to the initial pressure of 150 pounds and the total ratio of expansion of 15 is

$$\left[\frac{150\,(1 + \log_e 15)}{15} - 3\right]0.85 = 28.968 \text{ pounds.}$$

The mean of the three mean effective pressures that have been found, and which will be used in determining the diameter of the L.P. cylinder, is

$$\frac{28.21 + 26.685 + 28.968}{3} = 27.954 \text{ pounds.}$$

$$\text{Area of L.P. piston} = \frac{160 \times 33{,}000}{2\, p_e L N}$$

$$= \frac{160 \times 33{,}000 \times 12}{2 \times 27.954 \times 14 \times 260} = 311.34 \text{ square inches.}$$

Adding 1.99 square inches for the half-section area of a 2.25-inch rod, we have 313.33 square inches as the net area of the piston of the L.P. cylinder.

$$\text{Diameter of L.P. cylinder} = \sqrt{\frac{313.33}{0.7854}} = 19.97 \text{ inches, say 20 inches.}$$

We have

$$\frac{\text{Volume L.P. cylinder + volume clearance}}{\text{Volume H.P. cylinder + volume clearance}}$$

$$= \frac{\text{L.P. piston area } (1 + k)}{\text{H.P. piston area } (1 + c)} = 3.65,$$

whence

$$\text{Area of H.P. piston} = \frac{311.34 \times 1.03}{3.65 \times 1.05} = 83.674 \text{ square inches.}$$

Net area of H.P. piston = 83.674 + 1.99 = 85.664 square inches.

$$\text{Diameter of H.P. cylinder} = \sqrt{\frac{85.664}{0.7854}} = 10.44 \text{ inches, say}$$

10.5 inches.

From H, Fig. 145, draw a parallel to $O''O'$ to its intersection S with the compression curve KT. Then HS represents the volume of p_2 steam expended per stroke. The ratio $\frac{SH}{O''O'}$ is found by measurement to be 0.948, and the weight of a cubic foot of steam at 10 pounds pressure is 0.02606 pound. Then

Steam per I.H.P. per hour

$$= \frac{13,750 \times 0.948 \times 0.02606 \times 1.2}{27.954} = 14.58 \text{ pounds.}$$

The temperatures of steam at pressures of 150 pounds and of 10 pounds are 358.5° and 193.2° respectively, so that we shall have

$$\text{Ideal efficiency} = \frac{358.5 - 193.2}{358.5 + 461} = 0.2017, \text{ or } 20.17 \text{ per cent.}$$

With an evaporation of 8.5 pounds of water per pound of coal, we would have an expenditure of $\frac{14.58}{8.5} = 1.71$ pounds of coal per I.H.P. per hour.

Assuming the thermal value of the fuel to be 14,500 B.t.u. per pound, we would have

Thermal efficiency of engine and boiler

$$= \frac{42.42 \times 60}{14,500 \times 1.71} = 0.1027, \text{ or } 10.27 \text{ per cent.}$$

These results are satisfactory, so that an engine with a high-pressure cylinder of 10.5 inches diameter and a low-pressure cylinder of 20 inches diameter will fulfil the required conditions, provided the valves are properly designed.

According to the calculations the cut-off in the H.P. cylinder must take place at $14 \times 0.2055 = 2.88$ inches from commencement of stroke, and in the L.P. cylinder the cut-off is to take place at $14 \times \frac{1.615 - 0.12}{4} = 5.233$ inches from commencement of stroke in order to maintain a receiver pressure of 26 pounds. The design of the valves must provide for these cut-offs and also for a low-pressure exhaust closure at 85 per cent of the stroke.

In comparison with the non-condensing compound engine of Example II there is a decrease in steam consumption of

$$\frac{21.64 - 14.58}{21.64} = 0.326, \text{ or of } 32.6 \text{ per cent}$$

in favor of the condensing engine.

Example IV. — Required the cylinder diameter of a simple high-speed engine to develop 60 I.H.P. The piston speed is to be 650 feet per minute, the absolute initial steam pressure 95 pounds, the stroke 12 inches, the back pressure 3 pounds above the atmosphere, and the clearance 4 per cent.

Solution. — It is desirable with this type of engine to maintain a terminal pressure of at least 26 pounds, but before adopting that pressure in the design the probable steam consumption will be investigated.

$$r = \tfrac{95}{26} = 3.654, \text{ the log}_e \text{ of which is } 1.294.$$

$$\text{Theoretical } p_m = \frac{95\,(1 + 1.294)}{3.654} = 59.64 \text{ pounds.}$$

Using a mean pressure factor of 0.9, we have

$$\text{Expected M.E.P.} = (59.64 - 18)\,0.9 = 37.476 \text{ pounds.}$$

$$Q = \frac{34 \times 26 + 23}{37.476} = 24.2 \text{ pounds}$$

of steam per I.H.P. per hour. This indicates an actual consumption of $24.2 \times 1.4 = 33.88$ pounds, which is not excessive for the type of engine.

$$\text{Area of piston} = \frac{60 \times 33{,}000}{37.476 \times 650} = 81.285 \text{ square inches.}$$

Adding 1.036 square inches for the half-section area of a $1\frac{5}{8}$-inch rod, we have 82.321 square inches as the net piston area.

$$\text{Diameter of cylinder} = \sqrt{\frac{82.321}{0.7854}} = 10.24, \text{ say } 10.25 \text{ inches.}$$

Example V. — Find the cylinder dimensions of a triple-expansion engine to develop 6000 I.H.P. for a vessel of the mercantile marine. Piston speed, 1000 feet per minute; stroke, 45 inches; initial absolute steam pressure, 250 pounds; absolute back pressure 3 pounds; and the estimated clearances of the H.P., I.P., and L.P. cylinders are 15 per cent, 14 per cent, and 12 per cent respectively.

Solution. — Assume the volumetric cylinder ratio $\frac{\text{L.P.}}{\text{H.P.}}$ to be 10, and the mean pressure factor 0.65.

Assuming a terminal pressure in L.P. cylinder of 15 pounds we will have $\frac{250}{15} = 16.67$ as the number of expansions.

$$\text{Theoretical } p_m = 15(1 + \log_e 16.67) = 15 \times 3.81 = 57.15 \text{ pounds.}$$

$$\text{Expected M.E.P.} = (57.15 - 3)\, 0.65 = 35.2 \text{ pounds.}$$

$$Q = \frac{34\,p_2 + 23}{\text{M.E.P.}} = \frac{34 \times 15 + 23}{35.2} = 15.14 \text{ pounds}$$

of steam per I.H.P. per hour, a result not favorable for the type of engine, so a terminal pressure of 14 pounds will be tried.

$$\text{Ratio of expansion} = \tfrac{250}{14} = 17.86.$$

$$\text{Theoretical } p_m = 14(1 + \log_e 17.86) = 14 \times 3.879 = 54.306 \text{ pounds.}$$

$$\text{Expected M.E.P.} = (54.306 - 3)\, 0.65 = 33.349 \text{ pounds.}$$

$$Q = \frac{34 \times 14 + 23}{33.349} = 14.96 \text{ pounds}$$

of steam per I.H.P. per hour, the terminal pressure in the L.P. cylinder being 14 pounds. This result may be regarded as favorable, so a terminal pressure of 14 pounds will be adopted.

$$\text{Area of L.P. piston} = \frac{6000 \times 33{,}000}{33.349 \times 1000} = 5937 \text{ square inches.}$$

Adding 14.14 square inches for the half-section area of a 6-inch piston rod, we have 5951.14 square inches as the net area of the L.P. piston.

$$\text{Diameter of L.P. cylinder} = \sqrt{\frac{5951.14}{0.7854}} = 87 \text{ inches.}$$

A cylinder 87 inches in diameter is objectionable both on the score of weight and athwartship space occupied. It is desirable then to have two low-pressure cylinders, each of cross-section area of $\frac{5937}{2} = 2968.5$ square inches. This arrangement will give the engine four cranks, securing a better balance and more uniform crank effort.

Adding 14.14 square inches, we have 2982.64 square inches as the net piston area of each L.P. cylinder.

$$\text{Diameter of each L.P. cylinder} = \sqrt{\frac{2982.64}{0.7854}} = 61.63 \text{ inches.}$$

It is not desirable to have the diameter of the L.P. cylinders a fractional number, and as a slight change in the diameter of such large cylinders will not materially alter results, we will make the diameter of each L.P. cylinder 62 inches. The net piston area for each cylinder will then be 3019 square inches and a nominal area $3019 - 14.14 = 3004.86$ square inches.

We have

$$\frac{\text{Volume L.P. cylinder} + \text{its clearance volume}}{\text{Volume H.P. cylinder} + \text{its clearance volume}} = 10,$$

whence

$$\text{Section area of H.P. cylinder} = \frac{3004.86 \times 1.12 \times 2}{10 \times 1.15}$$
$$= 585.3 \text{ square inches.}$$

Net area of H.P. piston $= 585.3 + 14.14 = 599.44$ square inches.

$$\text{Diameter of H.P. cylinder} = \sqrt{\frac{599.44}{0.7854}} = 27.62, \text{ say 28 inches.}$$

Making the H.P. cylinder 28 inches in diameter its net area will be 615.75 square inches.

Making the I.P. cylinder 2.5 times the H.P. cylinder, we have

$$\text{Diameter of I.P. cylinder} = \sqrt{(28)^2 \times 2.5} = 44.26, \text{ say 44 inches.}$$

The net area of the I.P. piston is then 1520.5 square inches, and the nominal area 1520.5 − 14.14 = 1506.36 square inches. We have

$$r = 17.86 = \frac{\phi(1 + c)}{a + c} = \frac{10(1 + 0.15)}{a + 0.15},$$

whence $a = 0.498$, which is the fraction of the H.P. stroke completed when cut-off occurs.

The volume of the H.P. cylinder, including clearance, up to cut-off is 0.498 + 0.15 = 0.648 expressed in percentage of stroke.

Following the practice of the Navy Bureau of Steam Engineering, we shall have

Section area of I.P. cylinder

$$= \frac{585.3 \times 0.648 \times \sqrt{17.86}}{1.14} = 1406 \text{ square inches.}$$

Adding 14.14 square inches we have 1420.14 square inches as the net area of the I.P. piston.

$$\text{Diameter of I.P. cylinder} = \sqrt{\frac{1420.14}{0.7854}} = 42.51 \text{ inches.}$$

Making the diameter of the I.P. cylinder 43 inches, its net piston area will be 1452 square inches.

The stroke having been predetermined to be 45 inches, the engine will have to make $\frac{1000}{3.75 \times 2} = 133$ revolutions per minute. The engine will then be

$$\frac{28'' \times 43'' \times 2 \text{ of } 62''}{45'' \text{ stroke}} \times 250 \text{ pounds pressure} \times 133 \text{ r.p.m.}$$
$$= 6000 \text{ I.H.P.}$$

Section area of cylinders: H.P., 616 square inches; I.P., 1452 square inches; L.P., 3019 × 2 = 6038 square inches.

Deducting 14 square inches from each area for the rod, we have Net piston areas: H.P., 602 square inches; I.P., 1438 square inches; L.P., 6010 square inches.

Volumetric section areas of cylinders: H.P., $602 \times 1.15 = 692$ square inches; I.P., $1438 \times 1.14 = 1639$ square inches; L.P., $6010 \times 1.12 = 6731$ square inches.

Ratios of net piston areas referred to L.P. cylinder:

$$1 : 4.179 : 9.983.$$

Ratios of volumetric cylinder areas referred to L.P. cylinder:

$$1 : 4.107 : 9.723.$$

In order to obtain the most uniform crank effort for this engine the work done in each cylinder should be the same. Regarding the M.E.P. as the exponent of the power developed, we have, in this instance, 33.349 as the total work of the engine expressed in terms of M.E.P. referred to the piston of a single cylinder having a volume equal to that of the two L.P. cylinders. Then, for equality of work, the M.E.P. in each cylinder must be $\frac{33.349}{4} = 8.337$ pounds referred to the single L.P. cylinder. Therefore the M.E.P. in the H.P. cylinder should be $9.983 \times 8.337 = 83.23$ pounds, and in the I.P. cylinder the M.E.P. should be $4.179 \times 8.337 = 34.84$ pounds. The ratio of the single L.P. cylinder to each of the two L.P. cylinders is 2, so that the M.E.P. in each L.P. cylinder should be $8.337 \times 2 = 16.674$ pounds. We would then have

I.H.P. in H.P. cylinder

$$= \frac{83.23 \times 602 \times 1000}{33{,}000} = \frac{83.23 \times 602}{33} = 1518.$$

$$\text{I.H.P. in I.P. cylinder} = \frac{34.84 \times 1438}{33} = 1518.$$

$$\text{I.H.P. in each L.P. cylinder} = \frac{16.674 \times 3005}{33} = 1518.$$

This makes a total of 6072 I.H.P. for the four cylinders, an excess of 72 I.H.P. over that required.

It is the usual practice in distributing the power in triple-expansion engines with two L.P. cylinders to assign one-third the power each to the H.P. and I.P. cylinders and one-sixth to each of the L.P. cylinders. Such an arrangement gives very unequal loads on the crank pins, which can be avoided by determining, as above, the mean effective pressures required in each cylinder to produce the same load on each crank, and then arrange the points of cut-off and receiver pressures accordingly.

To avoid fractions in this instance, and without material error, assume the stroke volume ratios of the cylinders, referred to the L.P. cylinder, to be 1 : 4 : 10. Then, if 10 inches represents the stroke volume of the L.P. cylinder, the stroke volume of the H.P. cylinder will be represented by 1 inch, and that of the I.P. cylinder by $\frac{10}{4} = 2.5$ inches. The H.P. clearance volume will be represented by 0.15 inch, and that of the I.P. cylinder by $2.5 \times 0.14 = 0.35$ inch.

$$\text{Total ratio of expansion} = \frac{p_1}{p_2} = \frac{250}{14} = 17.86.$$

$$17.86 = \frac{10\,(1 + 0.15)}{a + 0.15}, \text{ whence } a = 0.494.$$

That is, the cut-off in the H.P. cylinder is at 49.4 per cent of stroke.

$$a + c = 0.494 + 0.15 = 0.644 \text{ inch.}$$

Denoting the ratio of expansion in the H.P. cylinder by r_1, we have

$$r_1 = \frac{1.15}{0.644} = 1.786, \text{ the } \log_e \text{ of which is } 0.5793.$$

$$\text{Theoretical } p_m \text{ in H.P. cylinder} = \frac{250\,(1 + 0.5793)}{1.786} = 221 \text{ pounds.}$$

Denoting the pressure in the first receiver by p_r', we have

$$(221 - p_r')\,0.65 = 83.23, \text{ whence } p_r' = 93 \text{ pounds.}$$

For equality of power between the H.P. and I.P. cylinders, we have

$$250 \times 0.644 = 93\,x, \text{ whence } x = 1.73 \text{ inches.}$$

$$\text{Cut-off in I.P. cylinder} = \frac{1.73 - 0.35}{2.5} = 0.552.$$

That is, the cut-off in the I.P. cylinder is at 55.2 per cent of stroke.

Denoting the ratio of expansion in the I.P. cylinder by r_2, we have

$$r_2 = \frac{2.5 + 0.35}{1.73} = 1.647, \text{ the } \log_e \text{ of which is } 0.499.$$

$$\text{Theoretical } p_m \text{ in I.P. cylinder} = \frac{93 \times 1.499}{1.647} = 84.64 \text{ pounds.}$$

Denoting the pressure in the second receiver by p_r'', we have

$$(84.64 - p_r'')\,0.65 = 34.84, \text{ whence } p_r'' = 31.05 \text{ pounds.}$$

For equality of power between the I.P. and L.P. cylinders, we have

$$93 \times 1.73 = 31.05\,x, \text{ whence } x = 5.18 \text{ inches.}$$

$$\text{Cut-off in L.P. cylinder} = \frac{5.18 - 1.2}{10} = 0.398.$$

That is, the cut-off in the L.P. cylinder is at 39.8 per cent of stroke.

The valves should be designed to effect these cut-offs.

PROBLEMS

1. Stroke of engine, 24 inches; initial absolute pressure of steam at cylinder, 100 pounds, but wire-drawing reduces the pressure to 95 pounds at cut-off. Back pressure, 18 pounds; cut-off, 0.25 stroke; clearance, 10 per cent. Release takes place at 90 per cent of stroke, and exhaust closes after 75 per cent of the return stroke is completed. The expansion and compression being hyperbolic, construct the expected indicator diagram, and find the terminal pressure, p_2; the final pressure of compression, p_c; the mean effective pressure, p_e, by means of ordinates; and the estimated steam consumption per I.H.P. per hour.

Ans. $p_2 = 33.25$; $p_c = 63$; $p_e = 42$; steam per I.H.P. per hour = 29.6 pounds.

2. An 18.5″ × 30″ engine makes 129 revolutions per minute, the consumption of fuel being 5.5 long tons a day. The mean effective pressure is 36 pounds, and 26 pounds of steam are used per I.H.P. per hour. Find the number of pounds of coal burned per I.H.P. per hour, and the number of pounds of water the boiler must evaporate per pound of coal.

Ans. 2.73 pounds of coal per I.H.P. per hour, and 9.524 pounds of water evaporated per pound of coal.

3. An engine, 18.5″ × 30″, makes 129 revolutions per minute. The mean effective pressure is 36.6 pounds per square inch, and the consumption of coal per I.H.P. per hour is 3 pounds. The weight of steam used per stroke is 0.3478 pound. Find the number of pounds of steam used per I.H.P. per hour; the coal consumption (long tons) per day; the number of pounds of water the boiler must evaporate per pound of coal.

Ans. Pounds of steam per I.H.P. per hour, 28; tons of coal per day, 6.18; pounds of water to be evaporated per pound of coal, 9.33.

4. It is desired to design a high-speed non-condensing engine to develop 60 I.H.P. while running at the rate of 350 revolutions per minute. It is predetermined that the stroke of the piston shall be 12 inches, and the initial absolute steam pressure in the valve chest 110 pounds per square inch. It is assumed that a careful design of the steam and exhaust passages will confine the clearance to 2.5 per cent of the volume of the cylinder, and to secure smooth running of the engine at the intended speed a final absolute pressure of compression of 85 pounds will be desirable; it is also fairly assumed that a back pressure of 18 pounds absolute will be maintained. It is required to find by means of a preliminary indicator diagram the point of cut-off, the point of exhaust closure, and the diameter of the cylinder.

5. It is desired to design a non-condensing cross-compound engine to develop 150 I.H.P. It is predetermined that the absolute initial steam pressure shall be 140 pounds, the stroke 14 inches, the revolutions per minute 250, and that the volumetric cylinder ratio $\frac{\text{L.P.}}{\text{H.P.}}$ shall be 3.6. It is fairly assumed that an absolute back pressure of 18 pounds will be maintained, and that a careful design of the valve and steam passages will confine the H.P. clearance to 5 per cent and that of the L.P. cylinder to 3 per cent.

6. It is desired to design a condensing compound engine to develop 150 I.H.P. It is predetermined that the absolute initial pressure in the valve chest shall be 140 pounds, the stroke 14 inches, the number of revolutions per minute 250, and the cylinder ratio $\frac{\text{L.P.}}{\text{H.P.}} = 3.6$. It is fairly assumed that a back pressure of 3 pounds will be maintained in the condenser,

and that the clearance of the H.P. cylinder will be confined to 5 per cent and that of the L.P. cylinder to 3 per cent.

7. Find the cylinder dimensions of an engine to develop 75 horse-power. Piston speed, 600 feet per minute; absolute pressure of steam, 95 pounds; cut-off, 0.3 stroke; back pressure, 2 pounds above the atmosphere; clearance, 5 per cent.

8. Required the dimensions of a compound engine for use in the mercantile marine to develop 1800 I.H.P. Piston speed, 725 feet per minute; absolute initial pressure, 112 pounds; back pressure in the condenser, 2 pounds; cut-off in H.P. cylinder, 0.4 stroke. Estimated clearances: in H.P. cylinder, 10 per cent; in L.P. cylinder, 12 per cent.

9. Find the dimensions of a triple-expansion engine to be used in the mercantile marine, the horse-power to be developed being 4000. Piston speed, 800 feet per minute; initial absolute pressure, 160 pounds; absolute back pressure, 3 pounds; cut-off in H.P. cylinder, 0.6 stroke. The estimated clearances are: 16 per cent for the H.P. cylinder and 14 per cent for the L.P. cylinder.

CHAPTER XV

COMBINING DIAGRAMS OF STAGE-EXPANSION ENGINES

192. Combining Diagrams of Stage-expansion Engines. — Owing to the difference in the initial pressures in the cylinders of stage-expansion engines, springs of different tensions are used in the indicators, and consequently the diagrams of the several cylinders are to different scales.

It is possible and profitable to combine the diagrams of a stage-expansion engine into a single diagram which shall exhibit the changes in pressure and volume the steam undergoes from the moment of its admission into the H.P. cylinder until its final release from the L.P. cylinder. Such a diagram enables a comparison to be made between the work actually accomplished and that theoretically due to the initial pressure and ratio of expansion, under the supposition that the total expansion takes place in the L.P. cylinder. Owing to fundamental defects in the steam engine itself, the results obtained from the combined diagram cannot be more than approximate, but the process of the combination is instructive, and if care be exercised in the interpretation of the results, valuable information concerning the general performance of the engine may be obtained and factors determined which may be of value in subsequent designs.

In the process of combining the diagrams, the pressure scale of all the diagrams is the same, while the length of each must be such as to represent the stroke volume of the cylinder from which it was taken. It is important that the clearance volume of each cylinder, expressed in percentage of stroke volume, be added to the length of each diagram.

The diagrams on page 355 were taken from one of the two triple-expansion engines of the battleship *South Carolina* when on her trial off the Delaware Breakwater, August 24, 1909.

The engines are of the four-cylinder type (two L.P. cylinders), the diameters being 32″, 52″, 72″ and 72″; stroke, 48″; diameter of all piston rods, 7.25″; clearance: H.P., 15.5 per cent; I.P., 12.5 per cent; L.P., 12 per cent; back pressure, 3.5 pounds absolute.

From the data we deduce:

Net piston areas in square inches: H.P., 783.609; I.P., 2103.079; two L.P., 8101.737.

Volumetric cylinder ratios:

$$\frac{\text{L.P.}}{\text{H.P.}}=\frac{8101.737\times 1.12}{783.609\times 1.155}=10;\qquad \frac{\text{L.P.}}{\text{I.P.}}=\frac{8101.737\times 1.12}{2103.079\times 1.125}=3.835.$$

Referred to the L.P. cylinder, the volumetric ratios are 1 to 3.835 to 10.

Referred to the L.P. piston, the ratios of net piston areas are 1 to 3.852 to 10.339.

The first step in the process of combination is to superpose the diagrams from the two ends of each cylinder, and thus obtain the mean diagrams shown on page 356.

Erect perpendiculars to the atmospheric lines, touching the diagrams at each end. Lay off, to the pressure scale of each diagram, the perfect vacuum lines, 14.7 pounds below the respective atmospheric lines. Lay off *ao* equal to the respective cylinder clearance percentage of *ab*, *ab* representing the stroke volume, and erect the clearance lines *oc*. Select the point *x* on each diagram as near after the actual point of cut-off as the eye can detect, and where, it is fair to assume, the expansion is according to Boyle's law. Construct the hyperbolic

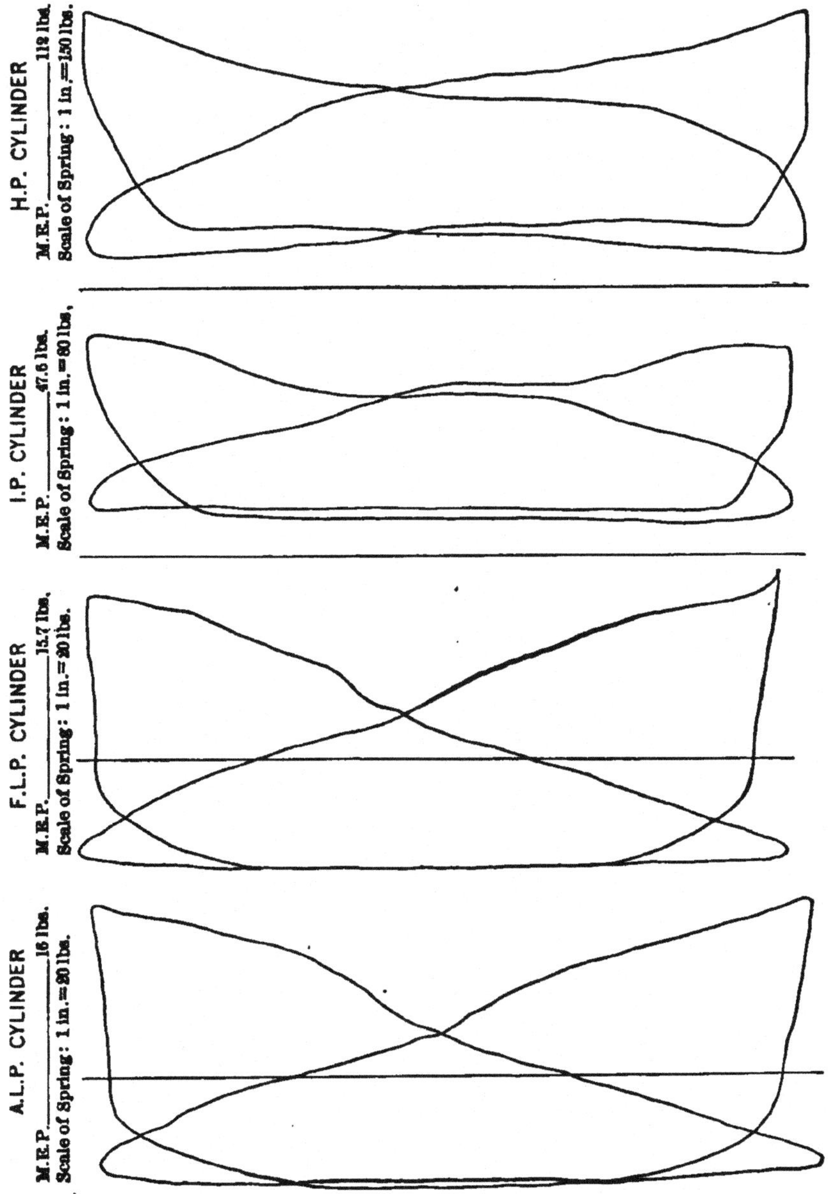
H.P. CYLINDER
M.E.P. 112 lbs.
Scale of Spring: 1 in.=150 lbs.
I.P. CYLINDER
M.E.P. 47.6 lbs.
Scale of Spring: 1 in.=80 lbs.
F.L.P. CYLINDER
M.E.P. 15.7 lbs.
Scale of Spring: 1 in.=20 lbs.
A.L.P. CYLINDER
M.E.P. 16 lbs.
Scale of Spring: 1 in.=20 lbs.

MEAN H.P. DIAGRAM

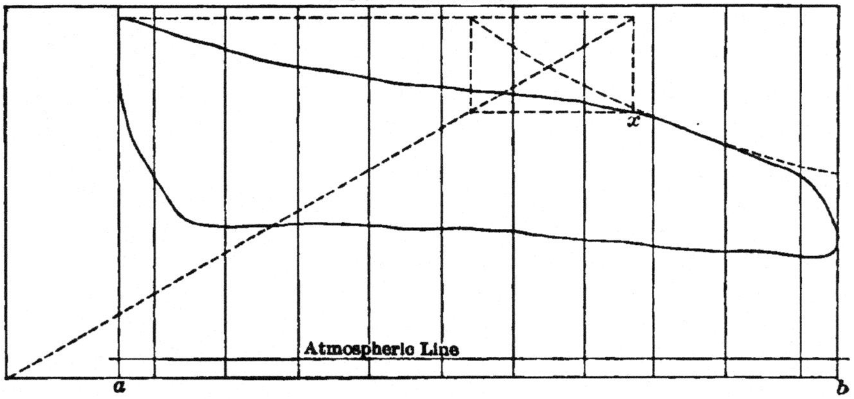

MEAN I. P. DIAGRAM

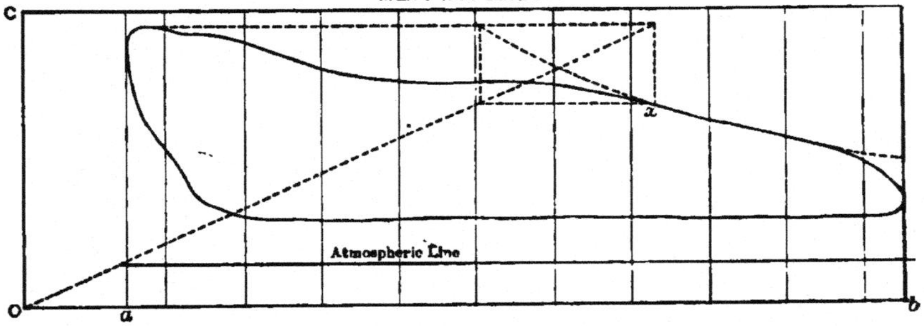

MEAN L.P. DIAGRAM

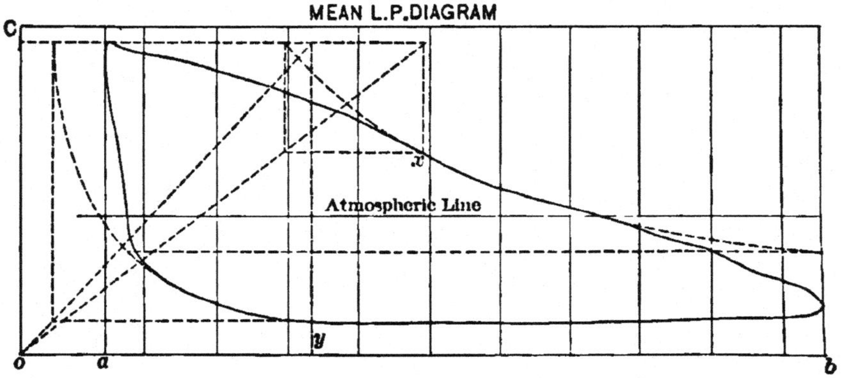

expansion curves for each diagram, from which a fair idea may be obtained as to the character of the actual expansion in each cylinder. Select the point y, as near after the L.P. exhaust closure as the eye can detect, and construct the hyperbolic compression curve. Erect ordinates and obtain the mean effective pressure from each diagram.

Draw the coördinate axes OX and OY, Fig. 146, representing the lines of no pressure and no volume respectively.

Assume arbitrarily a pressure scale of 40 pounds to the inch. If we assume that a line 4 inches in length represents the stroke volume of the L.P. cylinder, then $4 \times 0.12 = 0.48$ inch will represent the clearance volume of the L.P. cylinder. On OX take $AO = 0.48$ inch and $AB = 4$ inches to represent the volumes of the L.P. clearance and the L.P. cylinder respectively.

From the mean diagrams we find the initial pressure in the H.P. cylinder to have been 299 pounds, in the I.P. cylinder 110 pounds, and in the L.P. cylinder 33.5 pounds. On OY take OW, OH, and OP equal respectively to $\frac{299}{40}$ inches, $\frac{110}{40}$ inches, and $\frac{33.5}{40}$ inch to represent these pressures.

Draw WD parallel to OB, making it $\frac{4.48}{10} = 0.448$ inch in length to represent the volume of the H.P. cylinder and its clearance. Make DE equal in length to $\frac{4}{10.339} = 0.387$ inch to represent the stroke volume of the H.P. cylinder. Then $0.448 - 0.387 = 0.061$ inch $= WE$, the volume of the H.P. clearance.

Draw HM parallel to OB, making it $\frac{4.48}{3.835} = 1.168$ inches in length to represent the volume of the I.P. cylinder and its clearance. Make MT equal in length to $\frac{4}{3.852} = 1.038$ inches to represent the stroke volume of the I.P. cylinder. Then 1.168 –

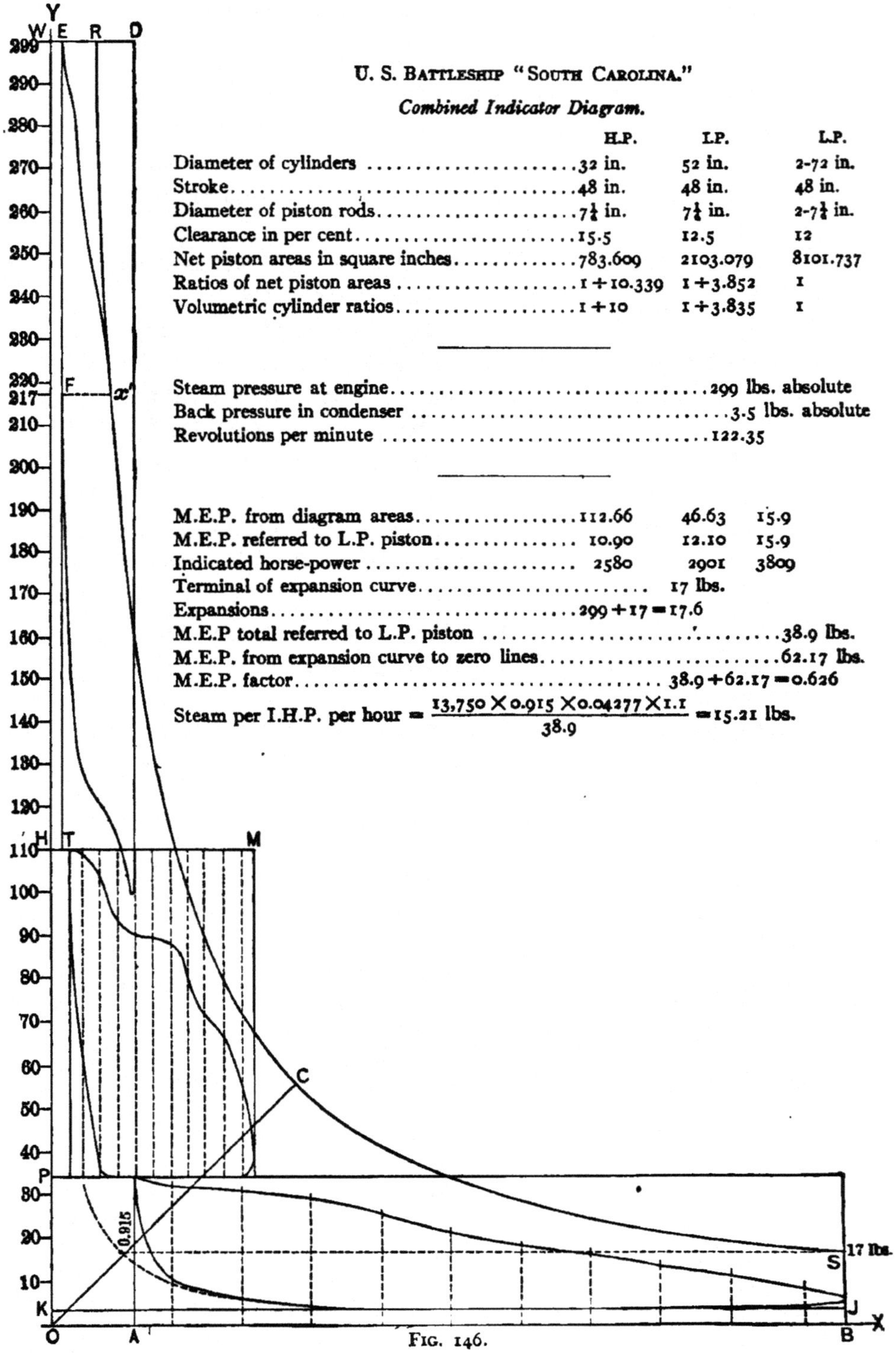

U. S. BATTLESHIP "SOUTH CAROLINA."

Combined Indicator Diagram.

	H.P.	I.P.	L.P.
Diameter of cylinders	32 in.	52 in.	2-72 in.
Stroke	48 in.	48 in.	48 in.
Diameter of piston rods	$7\frac{1}{4}$ in.	$7\frac{1}{4}$ in.	2-$7\frac{1}{4}$ in.
Clearance in per cent	15.5	12.5	12
Net piston areas in square inches	783.609	2103.079	8101.737
Ratios of net piston areas	1 + 10.339	1 + 3.852	1
Volumetric cylinder ratios	1 + 10	1 + 3.835	1

Steam pressure at engine .... 299 lbs. absolute

Back pressure in condenser .... 3.5 lbs. absolute

Revolutions per minute 122.35

M.E.P. from diagram areas	112.66	46.63	15.9
M.E.P. referred to L.P. piston	10.90	12.10	15.9
Indicated horse-power	2580	2901	3809

Terminal of expansion curve .... 17 lbs.

Expansions .... 299 + 17 = 17.6

M.E.P total referred to L.P. piston .... 38.9 lbs.

M.E.P. from expansion curve to zero lines .... 62.17 lbs.

M.E.P. factor .... 38.9 + 62.17 = 0.626

$$\text{Steam per I.H.P. per hour} = \frac{13{,}750 \times 0.915 \times 0.04277 \times 1.1}{38.9} = 15.21 \text{ lbs.}$$

FIG. 146.

1.038 = 0.13 inch = HT, the clearance volume of the I.P. cylinder.

Divide AB, TM, and ED each into ten equal parts and draw ordinates midway between the points of division (the ordinates of the H.P. diagram have been omitted for clearness). Transfer to these ordinates the pressures obtained from the corresponding ordinates of the mean diagrams on page 356, and through the points thus obtained draw the transferred diagrams of Fig. 146. These diagrams are those of the H.P., I.P., and L.P. cylinders to a pressure scale of 40 pounds to the inch and of lengths corresponding to the respective ratios of the net piston areas.

From the mean diagram of the H.P. cylinder, page 356, it is found by measurement that the cut-off at x occurred at 71.5 per cent of the stroke, the pressure at the point x being 217 pounds. At the 217 pound pressure mark on OY draw a parallel to OB to its intersection x' with the transferred H.P. diagram. Then x' is a point of the theoretical expansion curve and Fx' should measure $ED \times 0.715 = 0.387 \times 0.715 = 0.277$ inch.

Having the point x' the hyperbolic expansion curve $Rx'S$ may now be constructed and the terminal pressure BS will be found to be 17 pounds, the ratio of expansion therefore being $\frac{299}{17} = 17.6$.

The accuracy of the construction of the expansion curve may be tested by applying to the curve the equation $xy = \frac{a^2}{2}$ of the rectangular hyperbola; thus, $BS = \frac{\overline{OC}^2}{2 \times OB}$. Also, from the general equation $p_1v_1 = p_2v_2$, we have $BS = \frac{OW \times WR}{OB}$.

The area $WRCSBO$ included between the expansion curve and the zero lines of pressure and volume is that due the initial pressure of 299 pounds and the ratio of expansion 17.6, supposing the whole of the expansion to take place in the L.P. cylinder. By planimeter measurement this area is found to be 7.37 square inches. The areas of the transferred diagrams are by measurement: H.P., 1.09 square inches; I.P., 1.21 square inches; L.P., 1.59 square inches, the total being 3.89 square inches.

The ratio $\frac{3.89}{7.37} = 0.528$ is a factor illustrating the losses due to wire-drawing, drop, release, compression, and clearance. It affords also an insight into the difference between the results obtained from the working conditions and those to be expected from the same degree of expansion in a single cylinder under theoretical conditions.

M.E.P. from H.P. diagram area $= \frac{1.09 \times 40}{0.387} = 112.66$ pounds.

M.E.P. from I.P. diagram area $= \frac{1.21 \times 40}{1.038} = 46.63$ pounds.

M.E.P. from L.P. diagram area $= \frac{1.59 \times 40}{4} = 15.9$ pounds.

High pressure M.E.P. referred to L.P. piston $= \frac{112.66}{10.339} = 10.9$ pounds.

Intermediate pressure M.E.P. referred to L.P. piston $= \frac{46.63}{3.852}$ $= 12.1$ pounds.

Total M.E.P. referred to L.P. piston $= 10.9 + 12.1 + 15.9$ $= 38.9$ pounds.

Theoretical p_m from expansion curve to zero lines of pressure and volume $= 17\ (1 + \log_e 17.6) = 17 \times 3.863 = 65.67$ pounds.

Theoretical M.E.P. $= 65.67 - 3.5 = 62.17$ pounds.

The ratio $\frac{38.90}{62.17} = 0.626$ is a mean pressure factor which may be used in determining the size of cylinder for this type of engine to produce a given power, the ratio of expansion and piston speed being predetermined.

Steam per I.H.P. per hour

$$= \frac{13{,}750 \times 0.915 \times 0.04277 \times 1.1}{38.9} = 15.21 \text{ pounds.}$$

Assuming an evaporation of 8 pounds of water per pound of coal, we have

$$\text{Coal per I.H.P. per hour} = \frac{15.91}{8} = 1.99 \text{ pounds.}$$

These results of steam and coal consumption are not economical for the triple-expansion type of marine engine. The engines of war vessels are designed, however, to run economically at the ordinary cruising speeds, the question of economy being of no consideration when driven to full power in emergencies. In this instance the *South Carolina* was on her trial trip and was forced to the development of 9290 I.H.P., an increase of 12.6 per cent over the power for which the engines were designed.

The steam consumption per I.H.P. per hour, shown by the mean indicator diagrams on page 356, when measured at cut-off in each cylinder, is as follows:

For the H.P. cylinder, steam per I.H.P. per hour

$$= \frac{13{,}750\,(0.88 \times 0.441 - 0.26 \times 0.282)}{10.339 \times 38.9} = 10.77 \text{ pounds.}$$

For the I.P. cylinder, steam per I.H.P. per hour

$$= \frac{13{,}750\,(0.812 \times 0.1822 - 0.38 \times 0.085)}{3.852 \times 38.9} = 10.6 \text{ pounds.}$$

For the L.P. cylinder, steam per I.H.P. per hour

$$= \frac{13{,}750\,(0.566 \times 0.055 - 0.398 \times 0.0098)}{38.9} = 9.62 \text{ pounds.}$$

As an illustration of the use of the mean pressure factor of 0.626 found above in the design of triple-expansion engines of war vessels, we will take the case of the battleship *South Carolina*, each of whose engines was to develop 8250 I.H.P. at 125 r.p.m.

and absolute initial pressure at the H.P. cylinder of 265 pounds. The stroke was predetermined to be 4 feet, the back pressure to be 3.5 pounds, and the ratio of expansion to be 18.

M.E.P. from expansion curve to zero lines of pressure and volume $= \frac{265\,(1 + \log_e 18)}{18} = \frac{265 \times 3.89}{18} = 57.22$ pounds.

M.E.P. to be expected $= (57.22 - 3.5)\,0.626 = 33.63$ pounds.

Denoting the area of the L.P. piston by A, we shall have

$$A = \frac{8250 \times 33{,}000}{33.63 \times 4 \times 250} = 8095 \text{ square inches.}$$

The diameter corresponding to this area is 104 inches, so it is desirable to have two L.P. cylinders, each of a cross-section area of $\frac{8095}{2} = 4048$ square inches. Adding 20.64 square inches for the half-section area of a 7.25-inch rod, we have $4048 + 20.64 = 4068.64$ square inches as the net area of each L.P. piston. Then

$$\text{Diameter of each L.P. cylinder} = \sqrt{\frac{4068.64}{0.7854}} = 71.97 \text{ inches.}$$

The diameter used was 72 inches.

CHAPTER XVI

THE ZEUNER VALVE DIAGRAM

193. **Valve Design.** — The essential features and functions of the slide valve have been considered in Chapter V, and it is now the intention to consider, with a degree of particularity sufficient for practical purposes, the main features of its design.

No detail of the steam engine is of more importance than the correct design of its valve gear, which includes the valve and the mechanism that gives it motion. A defective design may occasion a loss in fuel as great as 20 per cent, and cause an unevenness in the motion of the engine with a resulting wear and tear in the working parts that may lead to serious casualty.

The angularities of the eccentric and connecting rods introduce such irregularities into the motion of the valve and of the piston that mathematical formulæ showing their relative positions during the distribution of the steam in the cylinder are too abstruse and complicated for practical use. Of the graphic methods devised to simplify the problem, that of Zeuner is as simple and complete as any. It permits the angularity of the connecting-rod to be taken into account, though not that of the eccentric rod; but since the length of the eccentric rod is great compared with the throw of the eccentric its angularity introduces an inappreciable irregularity, and the motion given to the valve is assumed to be harmonic. The rule and compass are the only instruments needed in the construction of the diagram, and its accuracy depends in no small degree upon the skill of the draftsman.

Take *AB*, Fig. 147, equal to the travel of the valve, and bisect it at *O*. Let *O* represent the center of the shaft, *OA* the direction of the crank when on the dead center, and *OE* the direction of the center line of the eccentric, the angular advance *DOE* being denoted by θ. On *AB* as a diameter describe a circle. The intersection, *E*, of this circle with *OE* will represent the center of the eccentric, since the half-travel of the valve is equal to the distance between the center of the shaft and the center of the eccentric.

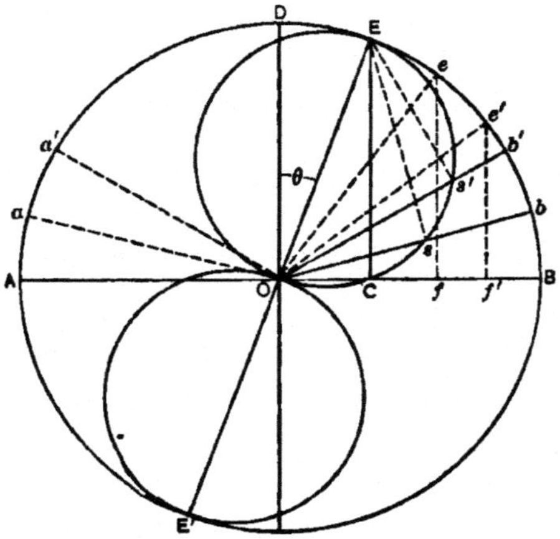

FIG. 147.

The motion of the valve being assumed harmonic the perpendicular *EC* let fall from *E* upon *AB* gives *OC* as ***the distance the valve has moved from its mid-position.*** As the shaft rotates the crank will assume the directions *Oa*, *Oa'*, etc., and the eccentric the positions *Oe*, *Oe'*, etc.; and if from *e* and *e'* perpendiculars be let fall upon *AB*, the distances *Of* and *Of'* intercepted between the center of the circle and the feet of the perpendiculars will represent the distances the valve has moved from its central position when the crank has the directions *Oa* and *Oa'* respectively.

Instead of the crank and eccentric moving in their circular paths, suppose them to remain fixed and the radius *OB* to revolve in the opposite direction and assume the positions *Ob* and *Ob'*, corresponding to *Oa* and *Oa'*. From *E* let fall perpendiculars *Es* and *Es'* on *Ob* and *Ob'*. By similar triangles it is easily shown that *Os* and *Os'* are equal respectively to *Of* and *Of'*; therefore

Os and *Os′* represent equally well the movements of the valve from its central position when the crank assumes the directions *Oa* and *Oa′*. Since the angles *OsE* and *Os′E* are right angles, the locus of the points *s* and *s′* is the circumference of a circle described on *OE* as a diameter.

If upon *OE* as a diameter a circle be described, the radii vectors *Os* and *Os′* give the distances the valve has moved from its central position for the angular positions *Ob* and *Ob′* of the crank, it being remembered that for the eccentric positions *Oe* and *Oe′* the real crank positions would be *Oa* and *Oa′*. This artifice is usual and convenient in the construction of the Zeuner diagram and contributes to its simplicity.

A precisely similar construction and reasoning will show that a circle described on a diameter *OE′*, equal to and diametrically opposite to *OE*, so that *EE′* is a straight line, will give, by means of the crank intercepts, the movement of the valve during the return stroke.

We now have a diagram which shows the distance the valve has moved from its central position for any position of the crank, and therefore the corresponding position of the piston may be found. It will be observed that the distance the valve has moved from its middle position is the important feature of the Zeuner diagram.

194. Angular Advance of the Eccentric. — The position of the eccentric radius when the crank is on the dead center may be found as follows:

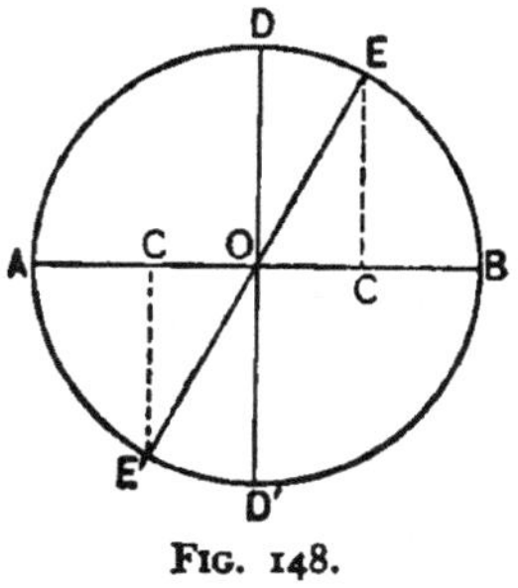

FIG. 148.

Describe a circle *AEBE′*, Fig. 148, with the throw of the eccentric as a radius. Let *OA* be the direction of the crank when on the dead center. From *O* set off in a direction away from the crank the distance *OC* equal to the lap plus the lead. Draw the vertical *CE*, cutting the circle at E. Then OE will be the position of the eccentric radius when the crank is on the dead center, and the angle DOE is the ***angular advance of the eccentric.*** This follows from the fact that the valve has moved from its central position a distance equal to the lap plus the lead when the crank is on the dead center. The above construction for the angular advance holds when the valve takes steam at its outside edges; but if, as is not infrequently the case, the valve takes steam at its inside edges, the lap plus the lead must be laid off from O towards the crank, and OE' will then be the position of the eccentric, and $D'OE'$ the angle of advance.

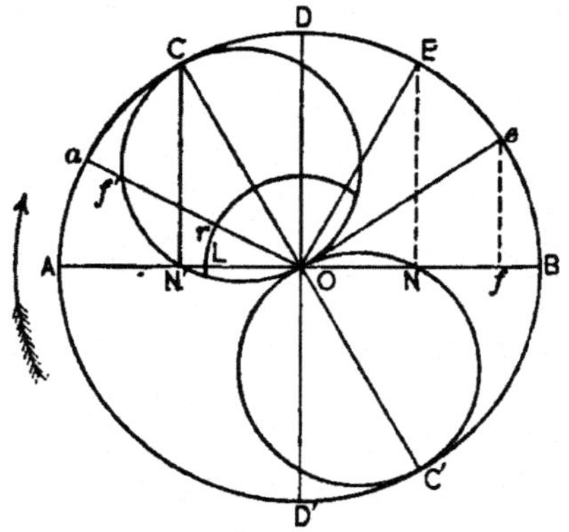

FIG. 149.

With a radius equal to the throw of the eccentric or the half travel of the valve describe the circle AEC', Fig. 149. Let OA be the direction of the crank when on the dead center, and OE the direction of and equal to the throw of the eccentric. Draw the diameter DD' perpendicular to OA. From E drop a perpendicular to AB intersecting it at N. Then ON is the distance the valve has moved from its mid-position when the crank is on the dead center. From O set off ON' equal to ON, and draw $N'C$ perpendicular to AB. Draw the diameter CC'. The angle COD = angle DOE = angular advance of the eccentric. If the crank move to Oa the eccentric will take the position Oe, such that the angles AOa and EOe are equal. Drop ef perpendicular to AB. Then Of will be the distance the valve has moved from its central position for the crank position Oa. On Oa set off $Of' = Of$. From what has been shown in Fig. 147, all points such as N' and f' will lie in the circumference of two circles described on the diameters OC and OC', each equal to the half travel of the valve, and make an angle with the perpendicular to the line of dead centers equal to the angular advance of the eccentric measured in the direction opposite to that of the motion of the crank.

If with O as a center and a radius OL equal to the outside lap of the valve a circle be described, the distance the port is open for the admission of steam for any crank position, as Oa, is given by the intercepted part rf'. This will be well understood from the fact that the opening of the port is equal to the movement of the valve from its central position minus the lap.

Since ON' is the movement of the valve from its central position when the crank is on the dead center A, it must be equal to the lap plus the lead, and since OL is the lap $N'L$ must, therefore, be the lead.

We may now construct a complete diagram, and in order that it be not too complicated it will be drawn for the steam and exhaust on only one side of the piston, the side remote from the shaft, and the lap, lead, etc., determined by the diagram will be for the end of the valve remote from the shaft.

The stroke of the piston from the head end to the crank end of the cylinder is known as the *forward* stroke (top stroke in vertical engines), and from the crank to the head end, the *return* stroke (bottom stroke in vertical engines). The diagram will then give the dimensions of the end of the valve to obtain the admission and cut-off of the steam for the forward stroke of the piston, and for the release and compression of the same steam on the return stroke.

With a center O, Fig. 150, and radius OA equal to the half travel of the valve, describe a circle. AB being the line of dead centers, the diameter DD' is drawn perpendicular to it and the angle DOE, equal to the angular advance, is laid off from DO in a direction opposite to that of the motion of the crank as indicated by the arrow. On the radii OE and OE' as diameters describe the circles OyE and OXE'. For the stroke under consideration the first of these circles is known as the primary valve circle, or steam circle, and the second as the secondary valve circle, or exhaust circle.

With a radius OL equal to the steam lap describe the arc sLs'. Since the admission and cut-off of the steam must take place when the distance of the valve from its central position is equal

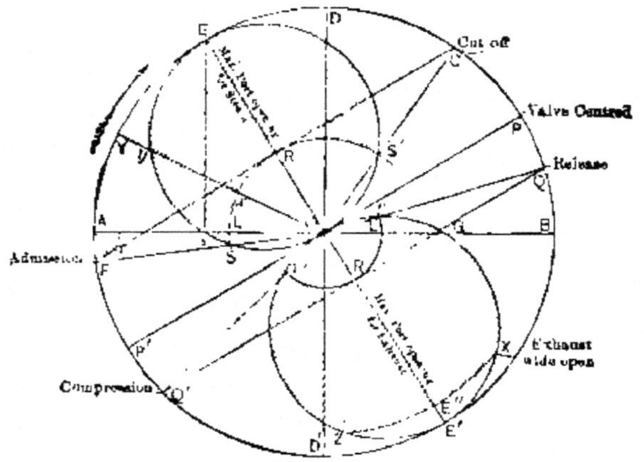

FIG. 150.

to the lap it follows that s and s', the intersections of the lap circle with the steam circle, give the crank positions OF and OC for admission and cut-off respectively. In like manner release and compression occur when the distance of the valve from its central position is equal to the inside lap. If then with O as a center and a radius OL' equal to the inside lap, an arc be described, its intersections e and e' with the exhaust circle will give the crank positions OQ and OQ' for release and compression respectively.

Should the exhaust lap be negative, which is not infrequently the case for the top stroke of vertical engines, the intersections of the exhaust lap circle with the steam circle must be taken to get the crank positions for release and compression. This will become evident after a consideration of the fact that the intercept on the crank position by the steam or the exhaust circle, in every instance, gives the distance of the valve from its central position. The valve is therefore centered when the intercept is zero, that is, when the crank is in the position *OP* or *OP'*, tangents to both circles; *PP'* is therefore perpendicular to *EE'*. Now if the exhaust lap is positive the crank will have to revolve beyond *OP* before the port opens to release and will be intersected by the exhaust circle. If, however, the exhaust lap be negative release will have taken place before the valve is centered, or before the crank reaches the position *OP*, and the intercept will therefore be made by the steam circle.

For any position of the crank, as *OY*, *Oy* is the distance of the valve from its central position, and the intercepted portion *xy* between the lap and the steam circles is the amount the port is open for the admission of steam. The intercepts in the hatched area of the steam circle show port openings to steam and therefore *LG* is the measure of the lead.

From the relative positions of the crank and eccentric and the harmonic motion of the valve it is seen that from the crank positions *OF* to *OE* the opening of the port to steam continually increases, at first quickly and finally slowly; from *OE* to *OC* the valve is closing the port, at first slowly and finally quickly.

The intercepts in the hatched area of the exhaust circle show openings of the port to exhaust. The port will, of course, be wide open to exhaust when the valve has moved from its central position a distance equal to the exhaust lap plus the width of the port, and any further distance the throw of the eccentric may cause the valve to move will result in *overtravel*. If with *O* as a center and a radius equal to the exhaust lap plus the width of the port an arc *XZ* be drawn the port will be shown wide open to exhaust during the period from X to Z and $E'E''$ will show the overtravel.

To draw the diagram for the return stroke the exhaust circle for the forward stroke becomes the steam circle for the return stroke and *vice versa*.

If the steam and exhaust laps are the same for both strokes the lap circles must be completed so as to intersect the opposite circles and thus give the corresponding events of the return stroke. Frequently they are different for the two strokes, in which case the correct radii must be used.

This is essentially the Zeuner valve diagram and by its application the numerous problems of the slide valve may be solved. By varying some of the points the alterations in the others may easily be found, and by assuming certain elements the remainder may be determined.

Some confusion exists at times as to what is meant by the different names given to the strokes of the piston. It should be remembered that the stroke known as *forward*, *outward*, *top*, *head*, and *down* is that in which the piston moves toward the shaft; and the stroke known as *return*, *inward*, *bottom*, *crank*, and *up* is that in which the piston moves away from the shaft.

195. Geometric Features of the Diagram. — There are four geometric features of the diagram which are useful in the solution of problems.

1. The line joining the points of admission and cut-off is tangent to the lap circle.

In Fig. 151, let OF and OC be the crank positions, respectively, for admission and cut-off, and let $N'LN$ be an arc of the lap circle. Join FC, cutting EE' at some point M. It is required to prove that M is the point of tangency of FC and the arc $N'LN$. The triangle FOC is isosceles, and since ME is the greatest opening of the port, it is midway between the admission and cut-off of the steam; therefore EE' bisects the vertical angle FOC and is perpendicular to the base FC and bisects it at M. Draw EN, and in the right triangles EON and COM we shall have the hypotenuse CO equal to the hypotenuse EO. The acute angles at E and at C are equal because their sides are respectively perpendicular to each other; therefore the triangles are equal and $ON = OM$. But ON is the lap; therefore if, with O as a center and ON as a radius, the arc $N'LN$ be described it will be tangent to FC at M.

A similar construction and reasoning will show that a line joining the point of release Q with the point of exhaust closure, or of compression Q', will be tangent to the exhaust lap circle.

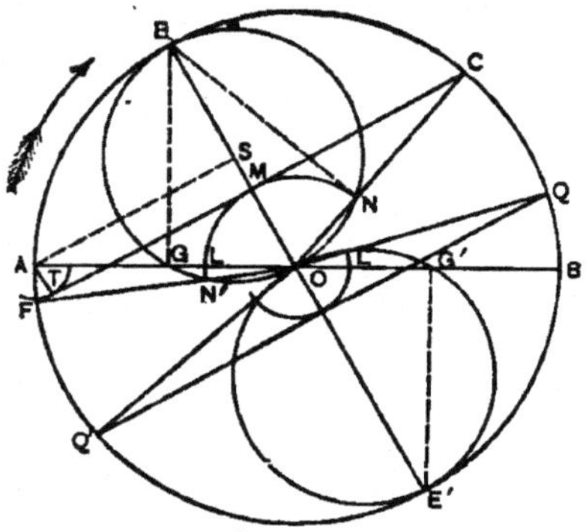

FIG. 151.

2. If from the dead point as a center an arc be described that will be tangent to the line joining the points of admission and cut-off the radius of this circle will be equal to the lead.

With dead point A, Fig. 151, as a center describe an arc tangent to FC; then its radius AT will be equal to the lead. Through A draw AS parallel to FC, and therefore perpendicular to EE'. The right triangles ASO and EGO are equal, and $GO = SO$. But $LO = MO$, therefore $GL = SM = AT =$ the lead.

3. If the points of admission and cut-off and maximum opening of the port to steam were given, or if the points of release and compression and the width of port were given (taking the maximum opening of port to exhaust as the width of port), there would not be sufficient data to construct the diagram. We can, however, determine the travel of the valve from the data given in either case and then construct the diagram, thus:

With center O, Fig. 152, describe the circle EQQ' with a diameter taken at random of 2.5 inches. Let Q and Q' be the points respectively of release and compression. These points would be given by the angles BOQ and AOQ', or by some definite part of the stroke, now represented indefinitely by AB. Join QQ', and from what has preceded we know that QQ' will be tangent to the exhaust lap circle, and that EE', the perpendicular to QQ' through O, will give the position of the eccentric, and indefinitely

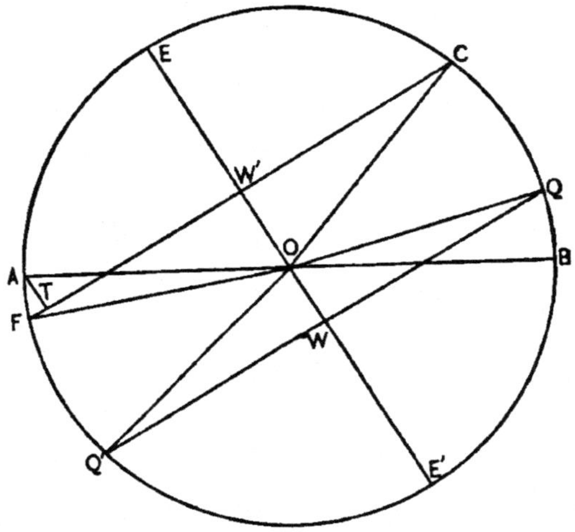

FIG. 152.

the diameters of the steam and exhaust circles. WE' will then represent to some scale the maximum opening to exhaust, which is assumed to be the width of the port, given in this case as 1.5 inches. We would then have the proportion:

$$\frac{\text{Travel of valve}}{\text{Width of port}} = \frac{EE'}{WE'} = \frac{2.5}{\frac{15}{16}}, \quad \text{or} \quad \frac{\text{Travel}}{1.5} = \frac{2.5}{\frac{15}{16}}$$

$$\text{Travel} = \frac{2.5 \times 1.5 \times 16}{15} = 4 \text{ inches.}$$

The practical method of determining the travel would be as follows:

Draw *ab* and *ac*, Fig. 153, making any convenient angle with each other. Lay off $ad = WE'$ of Fig. 152, which, by actual measurement, is $\frac{15}{16}$ inch. This represents the width of the port, in this case 1.5 inches. Lay off $ae = 1.5$ inches and join *de*. Our scale is now complete. Lay off on *ac* the distance $af = EE'$ of Fig. 152, which is 2.5 inches, and it will represent the travel of the valve. From *f* draw a parallel to *de* to its inter-

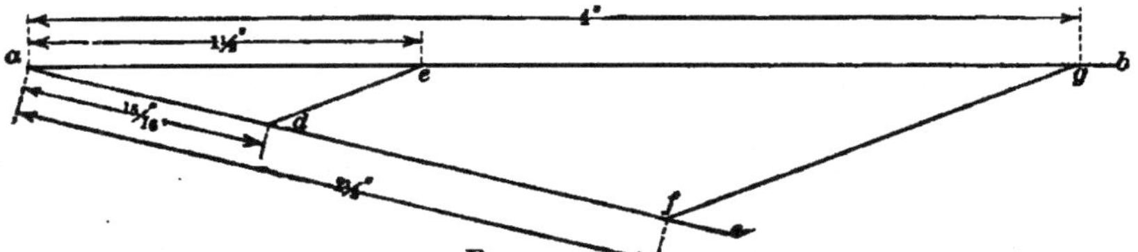

FIG. 153.

section *g* with *ab*; then *ag*, which measures 4 inches, is the travel of the valve. The diagram may now be completed.

If the points of admission and cut-off, *F* and *C*, Fig. 152, and $W'E$ the maximum opening of the port to steam were given, the travel of the valve could be determined in a similar manner. We would have

$$\frac{\text{Travel of valve}}{\text{Max. opening of steam port}} = \frac{EE'}{W'E}.$$

Should the point of admission, the point of cut-off, and the lead be given, the travel could be found from the proportion

$$\frac{\text{Travel of valve}}{\text{Given lead}} = \frac{EE'}{AT}.$$

4. Since *EGO*, Fig. 151, is a semicircle and *O* one extremity of its diameter, if at the extremity *G* of the chord *OG* (equal to the lap plus the lead) a perpendicular be erected it will meet the circumference at *E*, the other extremity of the diameter.

196. Information from the Diagram. — An examination of Fig. 150 reveals twelve features of the diagram that give important information, viz.:

1. Admission at F.
2. Cut-off at C.
3. Release at Q.
4. Compression at Q'.
5. Steam lap, OL.
6. Exhaust lap, OL'.
7. Steam lead, $LG = AT$.
8. Exhaust lead, $L'G'$.
9. Travel of valve, EE'.
10. Maximum opening of port to steam, RE.
11. Maximum opening of port to exhaust, $R'E''$.
12. Angular advance of the eccentric, DOE.

Points 1, 2, and 4 are frequently given in angular units, and these, with some other point in linear units, are sufficient to construct the diagram. These points may be given as occurring at certain points of the stroke, and in order to find the corresponding crank positions the circle described on EE' (equal travel of the valve) as a diameter may be taken to represent the stroke of the piston on a new scale; or another circle described with O as a center and to any convenient scale may be taken as the crank circle.

Points 1, 2, 5, 7, and 10 belong to the steam side of the diagram, and points 3, 4, 6, 8, and 11 to the exhaust side, points 9 and 12 being common to both.

Three points are sufficient data to construct either the steam side or the exhaust side alone, but for the complete diagram four points must be given, and one or two of these must belong to a different side than the others. One of the given points must be in linear units, for if only angles were given the linear dimensions of the valve might be made anything we please so long as they bear a certain ratio to each other.

If the width of the port in the face of the cylinder be given as a part of the data, it may be taken as the maximum opening of the port to exhaust, since the opening to exhaust will be greater than the opening to steam, and the port is made only sufficiently wide to allow a full opening.

If the cut-off be one of the given points it is understood to be the mean cut-off. The obliquity of the connecting-rod occasions a longer cut-off on the stroke toward the shaft than on the stroke away from it and a compromise must be made to secure the desired mean cut-off. If, for example, the mean cut-off were given as 0.6 of the stroke we may determine the crank positions at cut-off as follows: On any diameter representing the stroke,

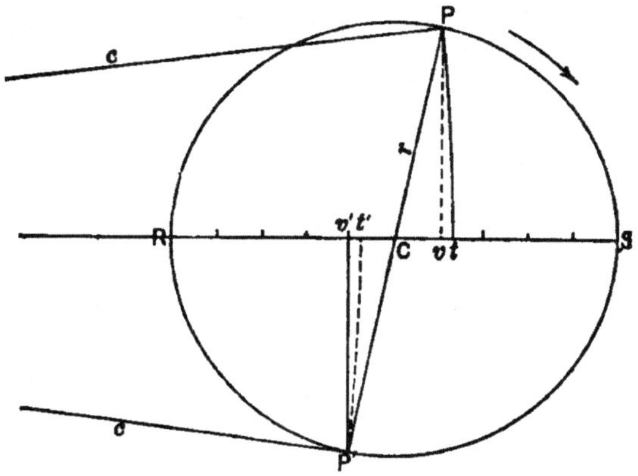

FIG. 154.

as *RS*, Fig. 154, describe a circle. Take *Rv* equal to 0.6 of *RS* and erect a vertical at *v*, intersecting the circumference at *P*. Then *CP* is the crank position for cut-off on the forward stroke, and $vt = \frac{r^2 \sin^2 \theta}{2c}$ (see page 129) is the error in piston displacement due to the angularity of the connecting-rod. Similarly, *CP'* is the crank position for the return stroke, and *v't'* the error in piston displacement.

Figure 155 is a complete diagram for the valve of the high-pressure cylinder of a triple-expansion marine engine. The

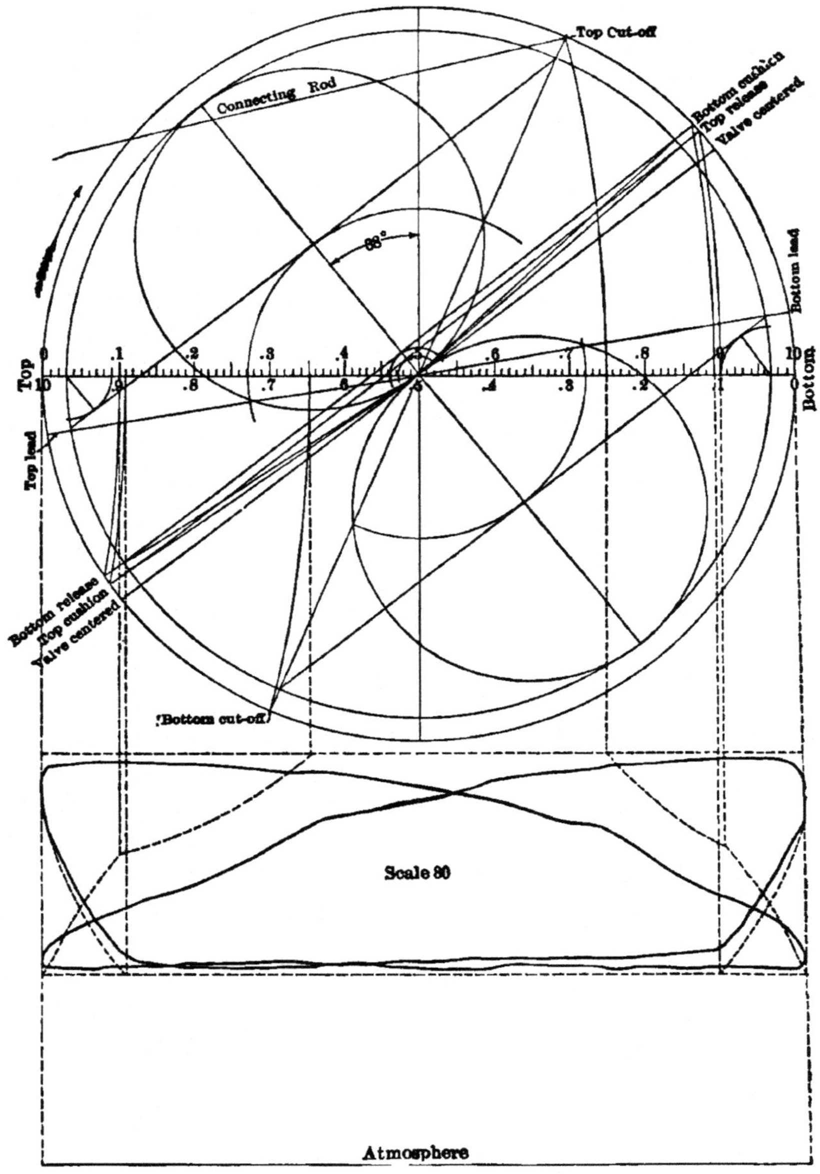

Top Cut-off
Connecting Rod
Bottom cushion
Top release
Valve centered
88°
Bottom lead
Top
Bottom
Top lead
Bottom release
Top cushion
Valve centered
Bottom cut-off
Scale 80
Atmosphere

FIG. 155.

theoretical indicator diagrams are shown dotted and the diagrams that may reasonably be expected appear in full lines.

PROBLEMS

1. Travel of valve, 4.25 inches; lead angle, 6°; angle at cut-off, 105°. Find the lap, lead, and the angular advance of the eccentric. Scale, full size. *Ans.* Lap, 1 3/32 inches; lead, 3/16 inch; angular advance, 40° 30′.

2. Stroke of engine, 3 feet; length of connecting-rod, 5 feet; steam lap, 1 inch; exhaust lap, 1/4 inch; width of port, 2 inches; overtravel of the exhaust port, 1/8 inch. Compression begins when the piston is one-ninth of the stroke from the end. Find the travel of the valve, angular advance, lead, and the point of cut-off expressed as a decimal of the stroke. Scale of stroke, 1/8, and of diagram, full size.

Ans. Travel, 4.75 inches; lead, 11/32 inch; angular advance, 8°; cut-off, 84.5 per cent.

3. Crank angle at cut-off, 105°; lead angle, 8°; maximum port opening to steam, 7/8 inch. Find the travel of the valve, the lap, the lead, and the angular advance. *Ans.* Travel, 3.875 inches; lap, 1 1/16 inches; lead, 7/32 inch; angular advance, 42°.

4. Cut-off, 0.7 stroke; lap, 1 inch; lead, 3/16 inch. Show that the angular advance is 36° 25′ 28″, and that the travel of the valve is 4 inches.

5. Travel of valve, 5 inches; width of port, 2 inches; opening of port to steam, 1.25 inches; exhaust lap, 3/8 inch; lead, 3/16 inch. Find the angular advance, lap, point of cut-off, point of release, point of compression, and overtravel.

Ans. Angular advance, 35° 6′; lap, 1.25 inches; cut-off, 0.71; release, 94.5 per cent; compression, 86 per cent; overtravel, 1/8 inch.

CHAPTER XVII

ENTROPY

197. Entropy. — In a pressure-volume diagram, such as that of the indicator, the energy expended, expressed in foot pounds, is represented by the area included by the lines of the diagram, the rectangular coördinates of which are pressure and volume.

In Fig. 156 let the horizontal through O be the line of absolute zero of temperature, and let A denote the thermodynamic state point of a substance at the absolute temperature T_2. If a quantity of heat Q be added to the substance such that B denotes its state point at the absolute temperature T_1, then we may represent the quantity of heat units added by the area $aABb$, and the same area would represent the same quantity of heat taken from the substance in cooling it from T_1 to T_2. If this area be divided by its mean vertical dimension, the quotient will be the horizontal dimension ab, to which the name *entropy* has been given, and the lines inclosing the area is a *temperature-entropy* diagram, the abscissas of which are entropy and the ordinates absolute temperatures.

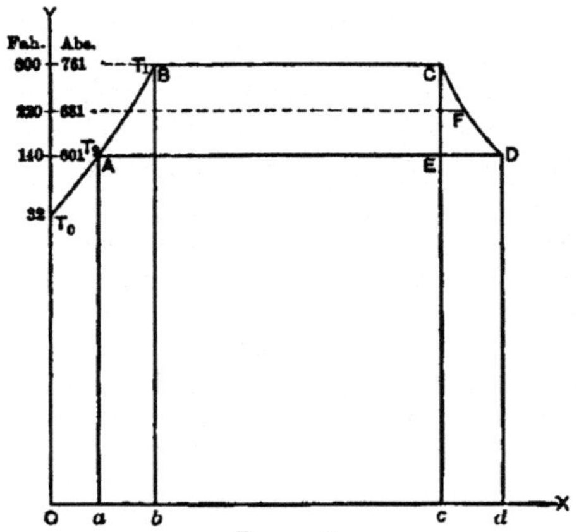

FIG. 156.

Taking O as the zero of entropy, Oa is the entropy ϕ_2 of the substance at its state point A and Ob its entropy ϕ_1 at its state point B, so that $\phi_1 - \phi_2 = Ob - Oa = ab$ is the change in entropy of the substance due to the addition of Q units of heat.

If $d\phi$ represents an indefinitely small change in entropy due to the addition of an indefinitely small quantity of heat dQ, then the temperature T may be regarded as constant during the very short time required to add dQ, and we shall have $d\phi = \frac{dQ}{T} = \frac{wc\,dT}{T}$, in which w is the weight of the substance considered and c its specific heat. We shall then have

$$\text{Entropy} = \phi = \phi_1 - \phi_2 = wc\int_{T_2}^{T_1}\frac{dT}{T} = wc\log_e\frac{T_1}{T_2},$$

in which T_1 and T_2 are the absolute temperatures of the substance at the commencement and at the end of the change in entropy.

It will be observed from the preceding paragraph that entropy is an abstract ratio, having no physical existence and incapable of being measured by any physical means. It is an indescribable quality of heat, the significance of which is difficult to grasp, but is nevertheless a recognized valuable aid in thermodynamic discussions not within the scope of this book.

The temperature-entropy diagram is specially applicable to the study of the steam turbine, where an indicator diagram is impossible, but a table or chart of the entropy of steam must be available in order to make calculations.

198. Temperature-entropy Diagram for Steam. — Entropy of steam may be reckoned from any point, but it is usual and convenient to consider its zero entropy to be that of water at the freezing-point temperature of 32° F, or 493° absolute, denoted by T_0.

Consider one pound of water at 32° F which it is desired to convert into steam at a temperature of 300° from a feed-water temperature of 140°. Then $T_1 = 761°$, $T_2 = 601°$, and $T_0 = 493°$.

The weight considered being 1 pound, and taking the specific heat of steam at 1, we shall have

$$\text{Entropy of steam} = \log_e \frac{T_1}{T_2} = \log_e T_1 - \log_e T_2.$$

The entropy of the water when heated from T_0 to T_2 is

$$\log_e T_2 - \log_e T_0 = \log_e 601 - \log_e 493 = 0.2;$$

and when heated from T_2 to T_1 the entropy is

$$\log_e 761 - \log_e 601 = 0.23.$$

Take OX and OY, Fig. 156, as the axes of entropy and temperature respectively. To the scale of 0.1 inch = 40°, locate on OY the temperatures 32°, 140°, and 300°; and to the inch scale lay off on OX the entropies $Oa = 0.2$ and $ab = 0.23$. In this manner the points T_2 and T_1 of the *water line* curve $T_0T_2T_1$ are determined. The heat units supplied to the water during its change of entropy from T_0 to T_2 is represented by the area OT_0Aa, and area $aABb$ represents the units supplied during the change from T_2 to T_1.

When the boiling point temperature T_1 is reached we know that any further addition of heat does not change T_1, so that the heat is added at constant temperature during the conversion of the water at T_1 into steam at T_1, the change being isothermal and represented by a parallel to OX through B. The quantity of heat added during this change is obviously the latent heat of steam at the T_1 temperature at 761° or at 300° F, which is 909.5 B.t.u., and this quantity of heat is represented by the rectangular area beneath the isothermal through B, the horizontal dimension, or the entropy, which is $\frac{909.5}{761} = 1.19$, or 1.19 inches to our scale, giving the length of the isothermal BC.

The total entropy of a pound of steam at 300° temperature is thus found to be $0.2 + 0.23 + 1.19 = 1.62 = Oc$.

Now suppose the steam from its state point C of temperature T_1 to expand against a resistance to temperature T_2, receiving from a jacket or otherwise a sufficient amount of heat to keep the steam in a dry, saturated condition; then the curve of expansion will be CD, the point B falling down the water line to A. The area $aADd$ represents the latent heat of steam at T_2 temperature, or at 140° F, hence the entropy AD is $\frac{1013}{601} = 1.68$, or 1.68 inches to our scale, locating the point D. The curve CD is the temperature-entropy curve for dry or saturated steam expansion. Points other than C and D are found in a similar manner. Thus, the latent heat of steam at 220° is 965.2 B.t.u., and $\frac{965.2}{681} = 1.417$ inches to scale, giving the point F.

If the steam at its state point D and temperature T_2 now have its heat abstracted by a condenser so that it exists in the liquid state of temperature T_2, the change in entropy will be DA.

Then, the total heat supplied is represented by the area $aABCDd$, and the heat rejected to the condenser by the area $aADd$, so that the area $ABCD$ represents the heat converted into work, and the efficiency is $\frac{ABCD}{aABCDd}$.

If the steam at its state point C and temperature T_1 had expanded adiabatically to temperature T_2, there could then be no area under the expansion line, since there would have been no transference of heat and therefore no change in entropy. The expansion line would then be along the adiabatic Cc (vertical line of constant entropy) and would be represented by CE when the temperature falls to T_2, B again falling to A. During the adiabatic expansion some condensation takes place due to the work done in overcoming the resistance, so that at the state point E the original pound of water exists as a mixture of steam and water at the temperature T_2. If now the steam in the mixture have its latent heat abstracted by a condenser, reducing the mixture to the liquid state at T_2, the change in entropy will be EA.

The total heat supplied in this instance is represented by the area $aABCc$, and the heat rejected to the condenser by the area $aAEc$, leaving the area $ABCE$ to represent the heat converted into work.

The use of the jacket in keeping the steam dry during expansion occasioned the performance of the additional work represented by the area ECD.

In the first instance when the steam expanded from C to D, receiving heat from the jacket to prevent any condensation, we had at D a pound of dry steam at temperature T_2, the latent heat of which was represented by the area $aADd$. In the second instance when the steam expanded adiabatically from C to E, we had at E a pound of a mixture of steam and water at T_2, and the latent heat of the steam of this mixture was represented by the area $aAEc$. The difference of these areas, which is area $cEDd$, must then represent the latent heat of the steam liquefied during the adiabatic expansion, and since all these areas are proportional to their horizontal dimensions, we shall have AE representing the proportion of steam and ED the proportion of water in the mixture at state point E. The dryness fraction of the steam at the end of the adiabatic expansion will then be $\frac{AE}{AD}$, and the moisture fraction will be $\frac{ED}{AD}$.

CHAPTER XVIII

THE STEAM TURBINE

199. Turbines. — The word *turbine* in mechanics signifies some form of wheel which revolves on its axis under the impelling force of a fluid, such as water, air or steam, thus furnishing the motive power for the operation of machinery. The most familiar example is that of the water turbine, in which a wheel is made to revolve by means of the impulse or of the reaction of a stream of water. The most recent turbine development is that in which the heat energy of steam is the agent employed in producing a heat engine that rivals the reciprocating steam engine in economical performance.

200. The Steam Turbine. — In the steam turbine the heat energy of the steam is converted into kinetic energy and is transferred in that form to the rotating parts of the turbine. The steam is admitted in the form of a jet, and if the operation of the turbine depends upon the direct impact of the jet it is known as an *impulse* turbine; and if the operation is due to the reaction of the jet it is known as a *reaction* turbine. Because of their excessive speed, turbines of the purely reaction type are not used, but the reaction principle, in combination with an impulse action, is employed in the type known as *impulse-reaction* turbines.

In the simplest form of the impulse turbine, such as that of De Laval, the steam jet issues at a high velocity from fixed diverging nozzles and impinges on *buckets* or *blades* mounted in the periphery of a rotating wheel. During the passage of the steam through the nozzles, and before impinging on the buckets, it expands from the initial pressure to the pressure at which it is exhausted, whether to the atmosphere or to a condenser, so that the steam while in contact with the moving buckets is at the constant pressure of the exhaust. During the expansion in the nozzles the pressure energy of the steam is transformed into kinetic energy, the transformation resulting in an enormous increase in the volume and velocity of the steam.

In the impulse-reaction turbine, such as that of Parsons, there are alternate rings of fixed and movable blades, the fixed blades projecting radially inward from the enveloping casing, while the movable blades project radially outward from the surface of a cylindrical drum mounted on the shaft of the turbine. The steam enters the first ring of fixed blades at initial pressure and while passing through these blades it expands, its pressure falling, and in thus expanding it does work upon itself and attains a velocity. From the fixed blades the steam is deflected so as to impinge on the adjacent ring of movable blades, giving up to them the energy of its velocity, causing the drum to revolve. This action of the steam is purely impulse. In the passage of the steam through the moving blades it expands still further, the expansion being accompanied with a further fall in pressure and acquirement of velocity, and as its direction of movement is changed it exerts a push on the moving blades as it passes to the next ring of fixed blades. This action of the steam on the moving blades is purely reaction. This cycle of operations is repeated a number of times until the exhaust pressure is reached.

The steam turbine, though still in the process of development, has already a wide field of usefulness. It is successfully employed in the propulsion of steamships, and is extensively applied to belt and rope transmission, and to the operation of electric generators, blowers, air compressors and centrifugal pumps. It is particularly suitable to the last named application, as the speeds required for the best efficiency of centrifugal pumps are readily obtained.

The loss from friction in the steam turbine, with the exception of that in the shaft bearings, is due to the passage of the steam through the buckets, and as steam velocities of from 2500 to 4000 feet per second are not uncommon the friction generated may be very great. Particularly is this the case with wet steam, the entrained water not only increasing the friction but also exerting an erosive action on the buckets. For these reasons the use of superheated steam with the steam turbine is a necessity if it is to be operated at its maximum efficiency.

In all turbines the steam is received at high pressure and small volume per unit of weight and discharged at a greatly increased volume and reduced pressure. The changes are directly due to the expansion of the steam, and the manner in which the expansion is effected constitutes the main difference in types.

201. The De Laval Steam Turbines.—One of the earliest and most successful steam turbines is that of De Laval. It is of the impulse type and, as now manufactured, has a very broad field of application.

De Laval turbines are classified under four types, viz.:

Class A. — The steam is expanded completely to the terminal pressure in one set of nozzles and impinges against a single row of buckets or blades, carried by a single wheel. The speed is reduced by means of a helical pinion and gear to suit the driven machine. This is known as the De Laval Single-stage Geared Turbine.

Class B. — The steam is expanded completely to the lower pressure limit in a single set of nozzles, as in Class A, and the impulse of the steam is taken by a single row of moving blades, but no reduction gear is employed. This is known as the De Laval Single-stage Impulse Turbine.

Class C. — The steam is expanded, as in Classes A and B, in a single set of nozzles, but the velocity of the steam is abstracted by two rows of moving blades, each row mounted on a separate wheel with one row of stationary guide blades in between. This is known as the De Laval Two-stage Impulse Turbine.

Class D. — The steam is expanded through successive sets of nozzles, with corresponding pressure steps, the velocity produced in each stage being expended upon a corresponding row of moving blades. The driven machine may be directly connected or driven by means of a gear according to speed requirements. This type is known as the De Laval Multi-stage Turbine.

The Class A turbines in units varying from 7 H.P. to 600 H.P. are adapted to all conditions requiring moderate speed and high steam economy, particularly under condensing conditions.

The Class B type, now abandoned and superseded by the Class C type, was adapted to special work where economy was of little importance.

The Class C type in units of from 1 H.P. to 750 H.P. is intended to cover all non-condensing conditions where high speeds of from 2400 r.p.m. to 4000 r.p.m. can be used and direct connection made to small generators, centrifugal pumps, blowers, small machinery, etc. With the reduction gear it can be adapted for conditions requiring low speed, so that the principal distinction between Class A and Class C De Laval turbines is that the former is designed for condensing and the latter for non-condensing conditions.

The Class D multi-stage turbine is used for all conditions where horse-power is above the range of Classes A and C, and is used both with and without gear, depending on the nature of the machine to be driven.

A representation of the wheel and divergent nozzle of the De Laval turbine is shown in Fig. 157. The casing, through which the nozzles enter, and which incloses the wheel, is removed, as are also the outer ends of some of the buckets, in order that the general idea of the course and action of the steam may be clearly illustrated. The nozzles are set at an angle of 20° to the plane of the wheel, and one of them is shown transparent so that its divergent action may be seen.

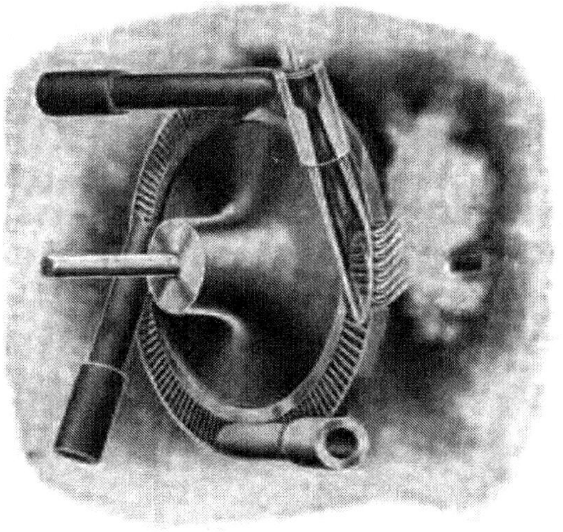

FIG. 157.

As already stated, the complete expansion of the steam takes place during its passage through the nozzle. The steam enters the small end at initial pressure and is discharged to the buckets at the large end at about atmospheric pressure if the turbine is non-condensing, and at a pressure corresponding to the vacuum if condensing. The design of the nozzle is of the first importance, the object being to secure the conversion of the maximum of the heat energy of the steam into kinetic energy, or energy of motion. The cross section of the nozzle is determined by the pressure and quantity of steam to be passed and the amount of heat to be converted into work. The shape of the De Laval nozzle has been determined by calculations based upon the theoretical action of steam during adiabatic expansion and also upon the results of a great number of experiments. The entrance of the nozzle is slightly rounded, followed by a narrow throat. The cross section of the throat determines the quantity of steam that can pass with a given initial pressure. From the throat to the outlet the cross section increases in such a way that, as the steam continues to expand, there is an approximately uniform acceleration in velocity. In the case of water this enlargement of section would be accompanied by a reduction in velocity, but with steam there is a continual gain in velocity and a decrease in density, pressure and temperature. The expansion of the steam in the nozzle takes place within an exceedingly brief period of time, less than $\frac{1}{2000}$ second, so that water particles resulting from expansion, or from damp steam, do not have time to form masses of appreciable weight that would have an erosive effect upon the blades.

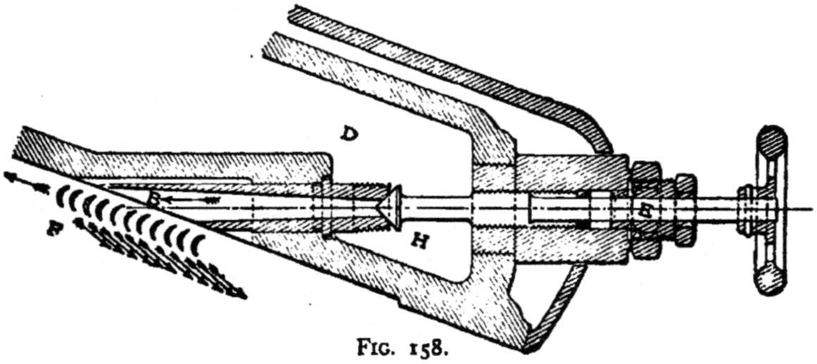

FIG. 158.

A section of the De Laval nozzle and valve is shown in Fig. 158. Steam of initial pressure in chamber *D* enters the nozzle through the valve *H*, and after completing its expansion within the nozzle impinges with great velocity on the buckets *F*. The number of nozzles varies from 1 to 15, according to the power and steam and exhaust conditions of the machine, and any number of them may be put out of action by closing their inlet valves, thus regulating the power required.

It is a condition of efficiency of the De Laval turbine that the steam shall leave the buckets at a minimum velocity, thus securing the maximum transference of energy. If there were no losses within the buckets, and if the direction of motion of the steam were the same as that of the bucket at the moment of impact, the buckets would have to move at one-half the velocity of the steam in order that the steam should be brought to rest by the reversal of the direction of its motion in the bucket. As a matter of fact, it is not found feasible to set the nozzles at an angle less than 20° to the direction of motion of the buckets. Since the velocity of the steam may vary in practice from 2600 to 4400 feet per second, the peripheral speeds of the wheel may vary from about 1300 to 2200 feet per second, which is much too high for driving ordinary machines, so that gearing is used to effect a reduction in the speed.

The speed reduction gear of the De Laval turbine is a distinguishing characteristic. It consists of a double pinion cut directly upon the turbine wheel shaft and of either one or two double-gear wheels mounted on the secondary or driven shaft or shafts. A fine pitch is used for the teeth in order that a large number may be in contact at one time, thus reducing the unit pressure between the surfaces of the teeth and preventing the lubricant from being forced out, and also reducing the danger of breakage.

In the construction of the wheel of the De Laval turbine the problem of successfully resisting the stresses produced by the centrifugal force arising from the high velocity presented itself. As these stresses vary with the square of the velocity, it was evident that a wheel in the form of a flat disk would burst at the normal speed required for the operation of the turbine, so the form of the wheel shown in Fig. 159 was devised to meet the requirements. As shown by the section view, the wheel is heavy at the boss and tapers in section towards the peripheral ring to which the buckets are attached. The centrifugal force is relatively small at the boss and is a maximum at the periphery. The gradual increase in section area from periphery to boss pro-

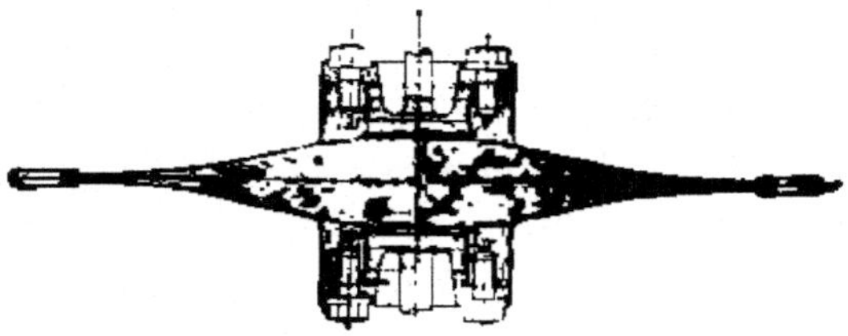

FIG. 159.

vides an excess of material that becomes available for the support of the highly stressed section at the periphery. The buckets are inserted into milled slots in the peripheral ring of the wheel, as shown in Fig. 160, and the centrifugal force due to their velocity loads the ring to an amount equal to the stresses produced in the body of the wheel. The section of the wheel immediately under the ring is purposely made small, and so designed that fracture will take place at that section should the speed approximate double that of the normal. A fracture of that description would break the rim into such small pieces that no damage to the wheel case would result, and, the buckets then gone, the action of the steam on the wheel would immediately cease. The wheels are turned from solid steel forgings, and are carefully polished in order to minimize the loss of power due to friction between the disk and the steam. The wheels of the larger sizes are made solid in body in order to give the maximum of supporting material at

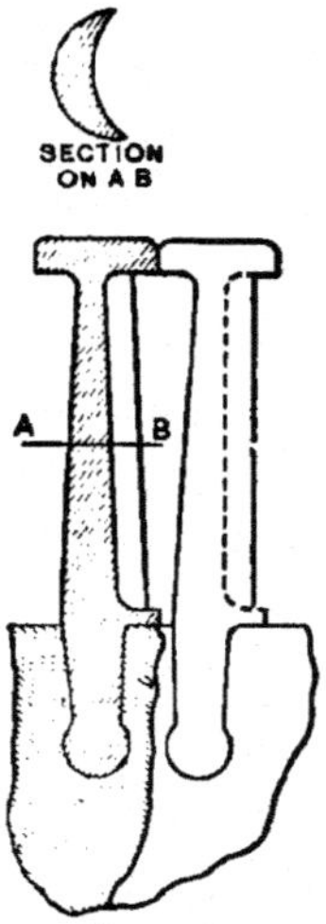

FIG. 160.

the boss, and, in such cases, the shaft is made in two parts and secured to the wheel.

The De Laval buckets are drop forged from nickel steel. They are fully formed in one operation, resulting in a smooth hardened surface which resists erosion better than does a machined surface. Lugs are formed at the outer ends, each lug touching the bucket in front to form a continuous band over the ends, as shown in Fig. 160. The shank of the bucket is finished with parallel sides, except at the inner extremity, where a cylindrical bulb is formed. These shanks fit in radial slots milled in the rim of the turbine wheel, the round part fitting into a hole bored transversely through the rim at the bottom of the slot.

FIG. 161.

The working parts of a De Laval single-geared Class A turbine are shown in Fig. 161. Mounted on the turbine shaft *A* is the turbine wheel *B* with the buckets fitted around its circumference. The pinion *C* is made solid with the shaft, and, when in position, meshes with the teeth of the gear wheel *H* mounted on the power shaft *I*. The pinion bearings, *E* and *D*, and the power-shaft bearings, *J* and *L*, are in two parts as shown. The ball bearing

F bears most of the weight of the wheel and is self-adjusting, being held in its seat by a cap and springs. The packing bushing *G* prevents leaks of air or steam where the shaft passes through the wheel case. These bushings are made of babbitt, lined with a special composition of graphite, so that they do not require lubrication, and the graphite forms a film between the shaft and the bushing that effectually prevents leakage. The packing bushing is split, so that it can easily be removed, and is held in place against a spherical seat by a nut and spring. The spherical seat is made in the form of a floating ring, so that the bushing is free to move in response to any oscillations of the turbine shaft. The governor is shown at *n*. In detail, the governing mechanism consists of two weights mounted on a governor head. The weights are hinged on knife edges and act against a spring, which can be adjusted by a spanner nut. The weights, in moving outward, press upon a collar on a spindle, pushing the latter outward and at the same time compressing the spring. This spindle, in turn, presses against a bell-crank lever operating the throttle valve through which steam is admitted to the steam chest of the turbine.

A horizontal section of a single-geared De Laval turbine is shown in Fig. 162. Steam enters the nozzle chamber *V*, passing thence through the nozzles to act on the buckets of the wheel *C*, and then escapes to exhaust chamber *W*. The wheel is inclosed between the case *A* and cover *B*. The packing bushings are shown at *J* and *K*. The pinion bearings are *P* and *Q*, and *M* is the gear wheel mounted on the power shaft *Y*. The governor *T* is held in the end of the power shaft by a tapered shank. The driven shaft is coupled to the power shaft by means of the flexible coupling *R*.

202. The De Laval Two-stage Impulse Turbine. — In the single-stage De Laval turbine, where the bucket velocity approximates one-half the velocity of the steam that has been expanded

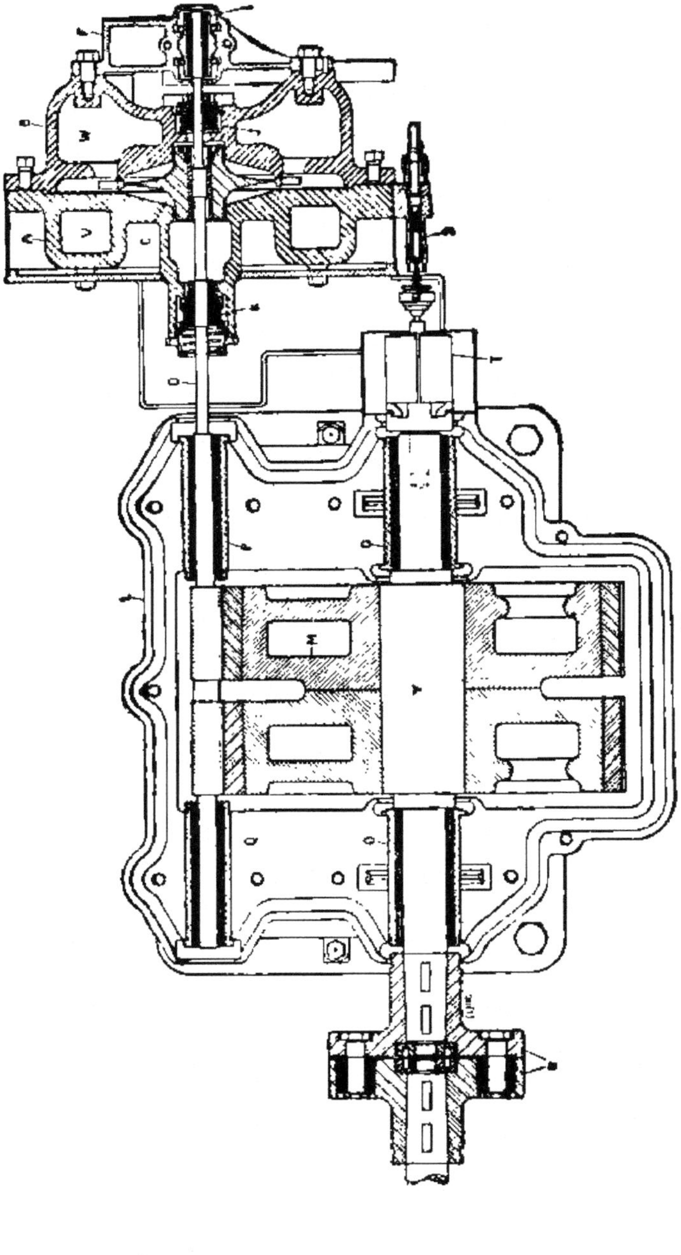

FIG. 162.

from the initial to the terminal pressure in one nozzle, the necessary reduction of the speed to the speed required for the driven machine is obtained by means of gearing. This arrangement has given the highest steam economy for turbines of all sizes up to 500 H.P., and while the use of the gear improves conditions for both the driven and driving machines by permitting each to run at its most advantageous speed, still the expense involved has prevented the general introduction of the geared turbine in many situations where small power is required, such as for driving steam-plant auxiliaries, including pumps, fans, stokers, exciters, etc., and in the operation of small units on shipboard, as ballast pumps, lighting sets, ventilating fans, etc.

In the De Laval Class C two-stage impulse turbine the speed reduction is obtained by having two rows of moving buckets to abstract the velocity of the steam, the two rows being separated by a row of stationary guide vanes similar in form to the buckets and held by dovetail sockets in a steel ring within the casing. This ring is made wide enough to serve as armor to prevent damage in case the wheels should be disrupted by over-speeding. Each row of moving buckets is carried in the periphery of a separate wheel mounted on the shaft.

The steam is expanded to the exhaust pressure, as in the Class A type, and on issuing from the nozzles delivers the first impulse to the first row of moving buckets, thus parting with some of its velocity energy. As it discharges from the first row of buckets the steam is directed by the guide vanes so as to give the second impulse to the second row of buckets, which is accompanied with a further abstraction of velocity energy.

In Fig. 163 is shown a view from above into the turbine case after the rotating parts and bearings have been removed. The individual nozzles by which steam is directed upon the first row of moving buckets and the intermediate stationary guide vanes by which the steam is directed upon the second row of moving buckets are seen in place. In Fig. 164 is shown the complete rotating member and bearings removed from the casing. The diagram, Fig. 165, shows the relative positions of

FIG. 163.

buckets, guide vanes, nozzles and controlling valves. In Fig. 166 is shown the steel armor ring holding the guide vanes which receive the steam discharged from the first row of moving buckets and redirect it upon the second row of buckets.

FIG. 164.

203. The De Laval Multi-stage Turbine. — The types of the De Laval turbine in which the complete expansion of the steam takes place within the nozzles are limited in capacity to about 700 H.P., and in order to increase the range of power without increasing the diameter of the wheel it is necessary to resort to multiple stages in which the steam is expanded through successive sets of nozzles, with corresponding pressure steps, the

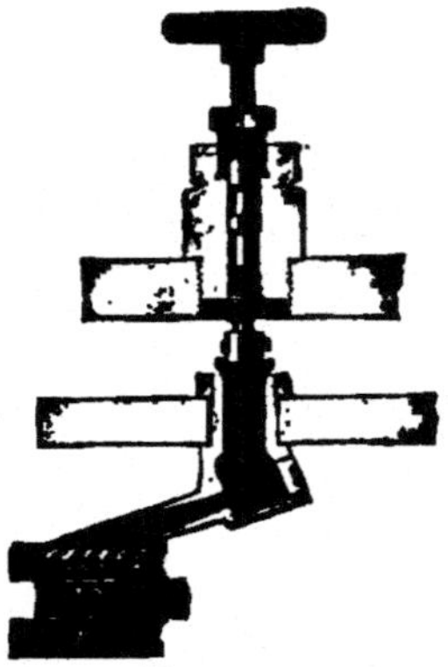

FIG. 165.

FIG. 166.

velocity produced in each stage being expended upon a corresponding row of moving buckets. This arrangement constitutes the Class D type of De Laval multiple-stage turbines which are adaptable to capacities ranging from 700 H.P. to 7000 H.P. The driven machine may be directly connected, or driven by means of a gear, according to speed requirements.

The general arrangement is shown in Fig. 167. The rotating member consists of a heavy shaft upon which is mounted a series of disks, or wheels, each revolving in an independent chamber formed between diaphragms held in a cylindrical casing. Steam is admitted to the steam chest at the right-hand end of the casing, and flows thence through nozzles and impinges upon the buckets of the first wheel. The nozzles of the first stage are formed by tubes carefully bored and reamed and set in the nozzle ring. They occupy only a portion of the circumference of the ring and are controlled by hand-operated valves. These valves are not used for speed regulation but for the admission of more or less steam to meet load changes. The nozzles of the succeeding stages, Fig. 168, are formed between guide-vanes set around the entire periphery of the diaphragms. These guide-vanes are made of nickel bronze, accurately formed in

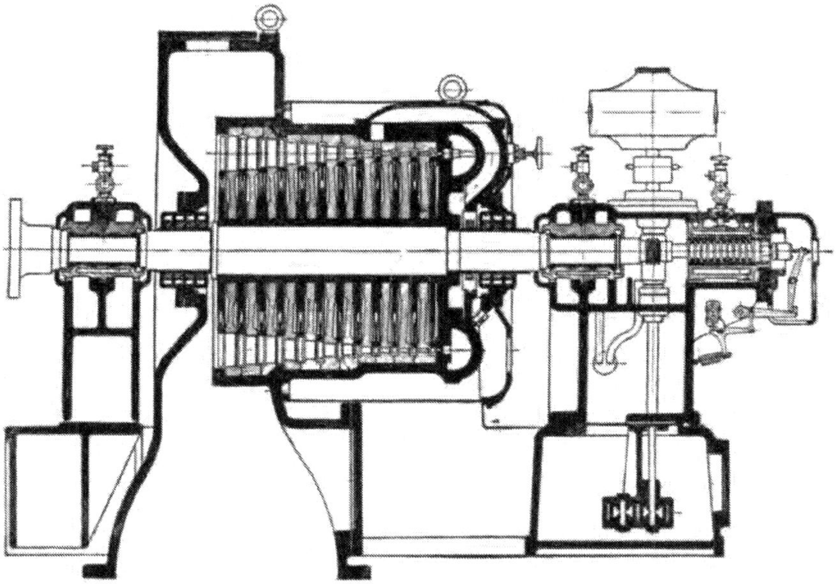

FIG. 167.

dies. They are spaced and located on the rim of the diaphragm by pins and held in place by a steel band shrunk over their tips. Two pins are used for each vane to determine its proper angle, and, therefore, in connection with the shape of the vanes, to fix the contour and cross-sectional area of the nozzles formed between the successive vanes. The steel bands are wider than the vanes and diaphragms, see Fig. 168, and touch one another, forming a continuous cylinder which incloses not only the diaphragms but also the rotating wheels, and provides a complete lining of forged steel for the wheel case, and protecting it from damage if by any chance a wheel should burst. The buckets, or blades, against which the steam impinges are similar to those of Class A, and are secured in the rims of the revolving wheels in the same manner.

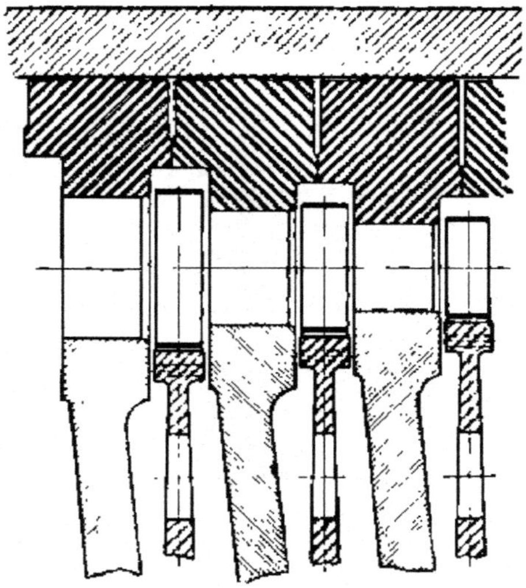

FIG. 168.

The speed reduction is controlled by the number of stages of the expansion. If, for example, the turbine were supplied with steam at pressure p_1 and exhausted at pressure p_2, the heat energy converted into kinetic energy would be $H_1' - H_2'$, under the supposition that the steam expanded adiabatically. Then, if the turbine were designed for equal steam velocity in each stage, the kinetic energy for each of n stages would be $\frac{H_1' - H_2'}{n}$, expressed in B.t.u., the corresponding velocity of which may be found from the Mollier total-heat-entropy diagram. Since for the maximum abstraction of the velocity energy of the steam the velocity of the buckets should be about one-half that of the steam, it follows that the speed of the buckets will be that corresponding to $\frac{H_1' - H_2'}{2n}$, expressed in feet per second.

For example, if steam at 160 pounds pressure and 100° of superheat were supplied to a ten-stage impulse turbine and exhausted at a pressure of 2 pounds, the turbine design providing for an equal steam velocity in each stage, we find from the Mollier diagram that the kinetic energy of the steam in each stage is $\frac{H_1' - H_2'}{10} = \frac{1253 - 947}{10} = 30.6$ B.t.u., the corresponding velocity of which is 1230 feet per second, so that the bucket speed would be about 615 feet per second.

Had the complete expansion taken place in a single nozzle, as in the Class A single-stage De Laval type of turbine, the conversion into kinetic energy of $H_1' - H_2' = 1253 - 947 = 306$ B.t.u. would give to the steam a velocity of 3920 feet per second, giving a bucket velocity of about 1960 feet per second.

In each stage of the expansion from the initial to the exhaust pressure, the steam as it discharges from a row of moving buckets is directed by the guide-vanes to the buckets of the next succeeding wheel, the velocity produced in each stage being expended upon a corresponding row of moving buckets.

The increase in the cross-sectional area of the passages required for the expansion of the steam as it passes through the turbine is gained by lengthening the buckets, reducing the diameters of the wheels correspondingly, and by increasing the bore of the casing, as shown in Fig. 167.

204. The Parsons Turbine. — A well-known type of the impulse-reaction turbine is that of Parsons, in which the steam acts alternately on rings of fixed and movable blades. The fixed blades project radially inward from the enveloping casing, while the movable blades project radially outward from the surface of a hollow cylindrical drum, which is co-axial with the casing and mounted on the shaft of the turbine. The action of the steam on the blades is shown in Fig. 169, the arrows indicating the course of the steam. The entering steam first encounters row 1 of fixed blades, and in passing through them a partial expansion of the steam takes place, accompanied by a pressure fall and a conversion of heat energy into velocity energy. The steam then enters row 2 of moving blades, delivering to them an impulse due to its acquired velocity. In passing

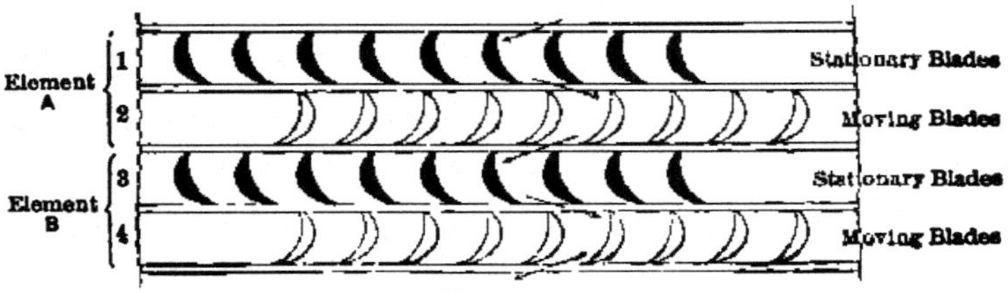

FIG. 169.

through the moving blades a further expansion of the steam takes place, accompanied by another fall in pressure and acquirement of velocity, and in leaving the blades the steam delivers to them a push in the nature of reaction, due to the reluctance of the expanding steam current to having its direction of motion changed. These impulse and reaction forces are the contributing causes of the rotation of the drum or rotor. This cycle of operations of the steam between the alternate fixed and movable blades is repeated until the desired pressure of exhaust is reached.

The passages through the blades are of constantly increasing area corresponding to the increased volume of the steam due to its expansion. This increase in area of the passages is obtained by progressively increasing the length of the blades until the mechanical limit of such increase is reached, at which point the diameters of the drum and casing are abruptly increased, thus allowing for another progression in the heights of the blades.

The expansion of the steam continues, its volume increasing until the height of blades required again becomes excessive, when the drum diameter is again increased and the blade heights reduced on the enlarged diameter. These steps in the expansion are continued until the pressure of the exhaust is reached. As the design of the turbine is such as to provide for an equal steam velocity in each stage of the expansion, it follows that the rotational speed of the rotor is determined by the number of stages in the expansion. The number of stages vary from 50 to 70, the rotational speed of the rotor varying from 500 r.p.m. to 1000 r.p.m. in marine practice, and from 1500 r.p.m. to 5000 r.p.m. in stationary practice. The capacities vary from 1000 H.P. to 10,000 H.P., the initial steam pressure varying from 90 pounds to 175 pounds absolute.

A longitudinal section of the Parsons turbine, as developed in this country by the Westinghouse Machine Company, and known as the Westinghouse-Parsons turbine, is shown in Fig. 170. Steam enters through the governor valve into the space *A* and passes out to the left through the first ring of fixed blades, issuing thence to act on the first ring of movable blades as explained. The steam continues its expansion through successive stages, 60 in this instance, and is finally exhausted through *D* to the condenser. In providing for the passage of the increasing volume of the steam it will be observed that the heights of the blades progressively increase through the three steps *R*, *R*, *R* in the expansion. To prevent an end thrust to the left on the shaft, due to the pressure of the steam as it passes through the turbine, the balance disks *P*, *P*, *P* are provided. There can be no end thrust from the steam at *A*, because the pressures on *P* and on the blades are equal and opposite. The steam pressures to the left on the annular surfaces of the drum where the diameters are increased are balanced by equal pressures to the right on *P*, *P* by means of the balance ports *E*, *E*. Similarly, there is a

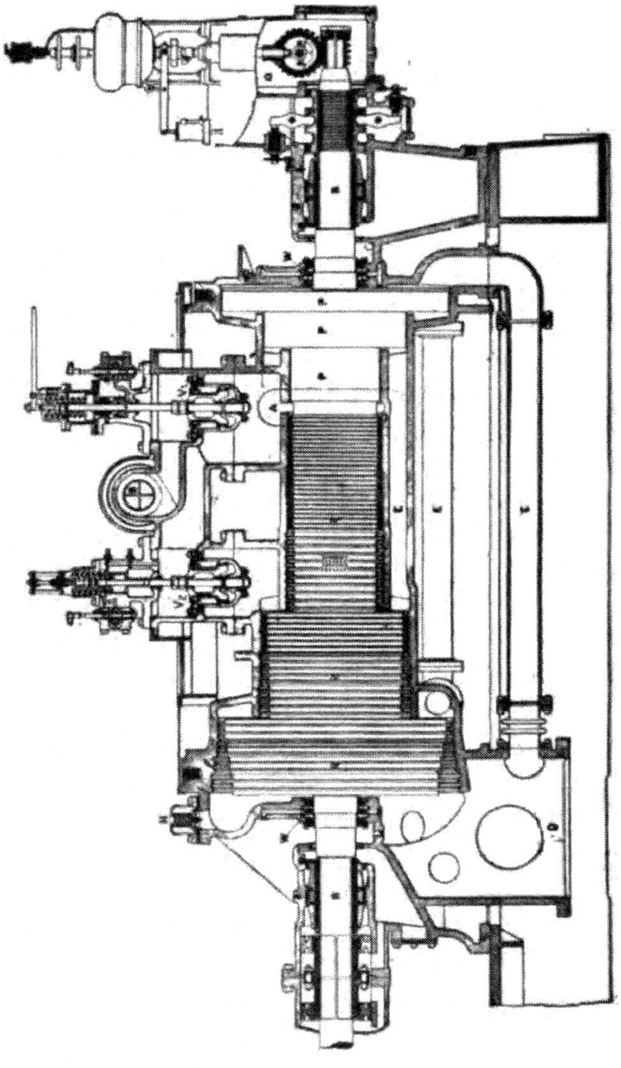

FIG. 170.

balance of the exhaust pressure on the left end of the rotor and on the right side of the largest of the balance disks by means of the lowest balance port *E*. A small thrust bearing is shown on the right which, however, has no thrust to take care of, but serves to maintain the correct adjustment of the balance disks.

The shaft passes through the stuffing-boxes *W*, *W*, which are carefully made to preserve the vacuum by preventing air leaking into the exhaust chamber.

The governor is driven by a screw and worm gearing contained in the chamber *G*, and operates, by suitable connections, the steam valve V_1. The secondary valve V_2 supplies steam for overloads.

205. Westinghouse Turbines for Small Powers. — The field of application of the steam turbine in small powers, already extensive, is still growing, displacing the reciprocating engine in many instances. The Westinghouse Machine Company classifies under four heads the applications of its small power turbine as follows:

1. Direct-current Generators:
   - (a) Excitation.
   - (b) Train lighting.
   - (c) Small industrial plants.
   - (d) Railway signal service.
2. Alternating-current Generators:
   - (a) Lighting and power.
3. Centrifugal Pumps:
   - (a) Condenser auxiliary service.
   - (b) Elevator service.
   - (c) Boiler feeding.
   - (d) Fire service — marine or stationary equipments.
   - (e) Water supply for moderate or high lift.
   - (f) Irrigation and dredging.

4. Centrifugal Blowers or Exhausters:

   (a) Ventilation.
   (b) Blast for furnaces.
   (c) Mechanical draft — forced or induced.
   (d) Compressor service — air or gas, moderate pressures.

The proper type of turbine to be used depends upon local conditions. Generally speaking, high-pressure non-condensing machines should be used where water for condensing purposes is not available and where fuel is cheap, or where the exhaust from the turbine may be utilized in heating boiler feed water, or for other heating purposes. The turbine has a decided advantage over the reciprocating engine, in that its exhaust is free from oil, and, consequently, when used in a feed-water heater of the open type does not contaminate the water going to the boilers.

High-pressure condensing turbines should be used where water for condensing purposes is available and where fuel is comparatively expensive.

Low-pressure turbines furnish a most satisfactory solution of the problem of increasing the capacity of an existing plant where non-condensing reciprocating engines are installed and condensing water is available. By the installation of a low-pressure turbine and condenser under these conditions, upwards of 60 per cent increase in capacity may be obtained without enlarging the boiler plant, the additional power developed depending on the amount of exhaust steam available from the non-condensing units.

The Westinghouse small power turbine consists of a single wheel, on which is mounted a single row of blades, the wheel being inclosed in a horizontally split casing containing the nozzles and reversing chambers.

The moving wheel, or rotor, is a high-grade steel casting, accurately machined and balanced. The blades are inserted in a groove cut in the periphery of the wheel, and held in place by rivets. A section through the bottom half of the casing is shown diagrammatically in Fig. 171.

Steam enters the turbine at the inlet and passes through the first-stage nozzles, where its pressure is reduced by expansion and its velocity increased. Leaving the nozzles it impinges on the moving blades, which are shown in solid section, giving up part of its velocity energy, and thence passes to the first reversing chamber, where its direction is changed and it is made to

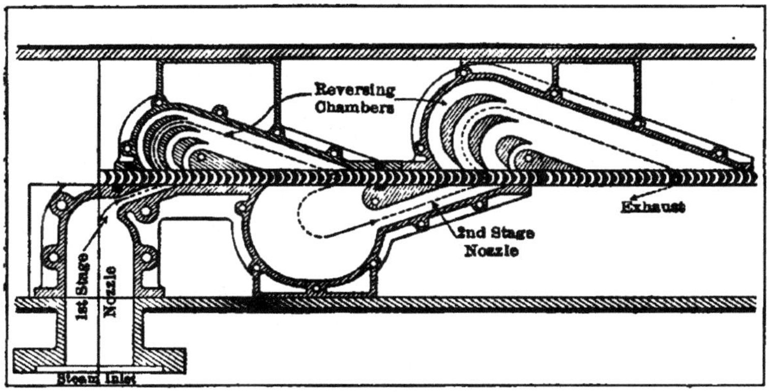

FIG. 171.

pass again through the moving blades, giving up to them the balance of the velocity energy acquired in the first-stage nozzles. The steam then passes through the second-stage nozzles, expanding and acquiring velocity energy, and then repeats the cycle of the first stage, finally passing to the exhaust. The dotted line shows the course of the steam through the two stages.

Turbines of this type of 50 H.P., or less, have but one set of nozzles and one reversing chamber; all others up to 200 H.P. have two stages.

206. The Curtis Turbine. — A well-known type of multi-stage impulse turbine is that invented by C. G. Curtis and built by the General Electric Company. It combines features of both the De Laval and the Parsons types. The expansion of the steam takes place in stationary divergent nozzles, but instead of the steam expanding completely in one set of nozzles, as in the De Laval machine, it expands in stages, each stage being accompanied by a pressure drop. In each stage of the expansion the steam acquires a relatively high velocity and then passes, without further expansion, through movable and stationary buckets to the nozzles of the next stage.

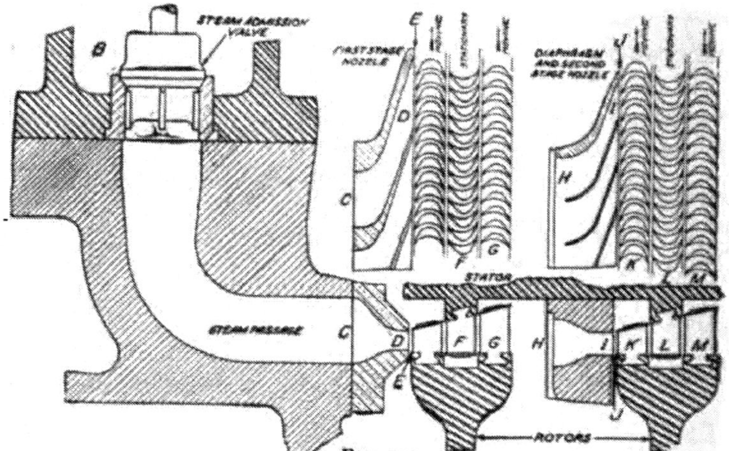

FIG. 172.

The diagrammatic sketch, Fig. 172, shows the course of the steam and the arrangement of the nozzles and buckets of a two-stage Curtis turbine, the nozzles and buckets being shown both in plan and in vertical section. Steam entering the steam chest *B* passes through one or more valves to the bowls *C*. The number of valves open depends upon the load, the action of the valves being controlled by the governor. From the bowls *C* the steam expands through divergent nozzles *D*, losing pressure and acquiring velocity. It then enters the first row of revolving buckets of the first stage at *E*, delivering to them the impulse due to the velocity, and then passes through the stationary buckets *F*, which reverse its direction and redirect it against the second row of revolving buckets *G*.

This constitutes the performance of the steam through one stage. Having entered the first row of buckets at *E* with relatively high velocity, it leaves the row *G* with a relatively low velocity, its velocity energy having been abstracted during the passage. Since the expansion from *C* to *E* in the nozzles *D* covered only a part of the available pressure range, the steam contains a large amount of unexpended energy. The expansion process is, therefore, repeated in a second stage.

The steam having left the buckets *G*, and having had its velocity greatly reduced, reaches a second series of bowls *H*, opening upon a second series of nozzles *I*. Through these nozzles the steam expands from the first-stage pressure to some lower pressure, again acquiring relatively high velocity in its expansion through the nozzles, leaving them at *J* and impinging upon and passing through the moving and stationary buckets *K*, *L* and *M*, precisely as in the first stage, the velocity acquired in the nozzles being expended in passing through the buckets, the steam leaving the second row of moving buckets *M* with relatively low velocity.

This process is continued in the larger machines through several additional stages. As now constructed, the Curtis machines have a single stage in the very small sizes up to six or seven stages in the larger ones.

During the passage of the steam from *E* to *H* the pressure has remained practically constant, so that there is no tendency for the steam to pass otherwise than directly through the buckets, thus avoiding the necessity of maintaining a small clearance between the ends of the revolving buckets and the casing, as is the case in reaction turbines. In practice there is from one to two inches clearance in the Curtis machine. In consequence also of the pressure being the same at *E* and *H* there is no tendency to move the wheel in an axial direction, and hence no need of balance pistons to take up end thrust.

To provide for a greater area of passage as the steam expands from stage to stage, the number and area of the nozzles, as well as the length of the buckets, are increased. The nozzles of the first stage extend around a small portion only of the periphery, so that only those buckets adjacent to the nozzles at any instant are carrying active steam. This applies equally to the stationary row and second revolving row; in fact, the stationary buckets, as built, extend over an arc not much greater than the nozzle arc. In the second stage, however, the nozzle arc becomes longer and wider, thus permitting the flow of steam through a greater number of revolving buckets and necessitating a longer arc of stationary buckets. Finally, when the low-pressure stages are reached, the nozzles and stationary buckets extend around the whole circumference. Greater area for the flow of steam is provided also by increasing the length of the buckets. Usually the buckets of the first or high-pressure stage are something less than an inch in length, while those in the low-pressure stages may be eight or ten inches long.

This lengthening of the buckets from one stage to the other to assist in providing the proper area space for the passage of the expanding steam must not be confused with the difference in size of the two rows in any given stage. The typical relation of moving and stationary buckets in any given stage is shown diagrammatically in Fig. 173, where *A* and *C* are the moving and *B* the stationary buckets. It will be observed that *B* and *C* are progressively longer than *A*, and that the outer ends of the revolving buckets and the base of the stationary buckets form a line diverging from the parallel to the axis. It is erroneous to assume that this divergence of the bucket area is to provide for an increase in volume due to expansion. In fact, the steam enters bucket A at high velocity and leaves bucket C at a comparatively low velocity, and in its passage a continuous drop in velocity has taken place requiring a progressively increasing area of passageway. The buckets are so designed that the product of the area and velocity is constant throughout the passage, the pressure and specific volume of the steam remaining constant.

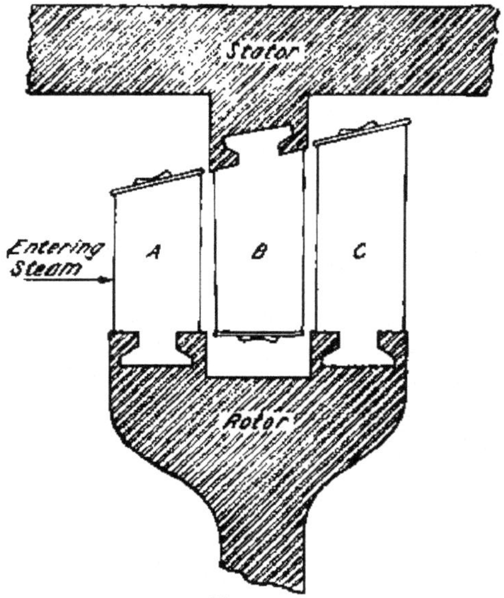

FIG. 173.

In general terms, the Curtis turbine derives its energy from the expansion of steam in a series of steps or stages. Each stage may contain two or more rows of rotating buckets with stationary reversing buckets between them. The impulse given to the moving buckets by the steam and the passage of the steam through the buckets is due to velocity, there being practically no expansion in the stage, the actual expansion being accomplished in the nozzle passages between the stages.

Curtis turbine wheels consist of steel disks carrying the rotating buckets on their peripheries. In some cases the two rows of buckets are carried on separate disks and in others by a single disk. The buckets are held in the periphery by a dovetail-shaped root which fits snugly into a channel of similar section in the rim, as shown in Fig. 173. In Fig. 174 is shown the wheel rim with a row of rims partially inserted. At intervals the dovetail channel is opened for the insertion of buckets as shown at *A*, the openings being afterward filled with a spacing block. After the buckets are assembled, a shroud ring *B*, *B* is riveted to their outer ends. The function of this ring is partly to stiffen the completed row and reduce vibration, but more especially to assist in retaining the steam flow in the bucket space.

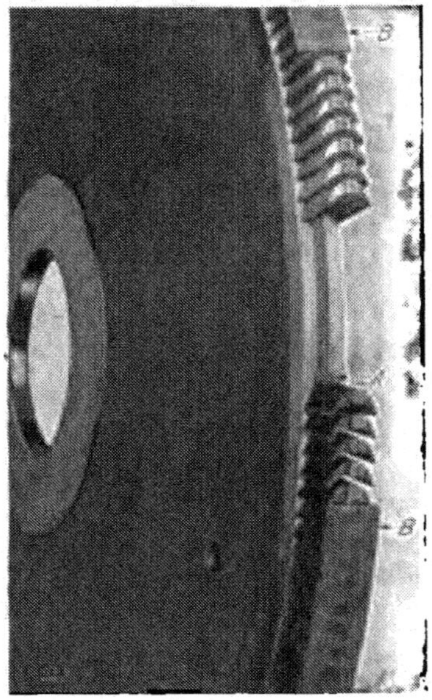

FIG. 174.

A horizontal five-stage Curtis turbine with top half of casing removed is shown in Fig. 175. The revolving wheels are indicated at *A* and the diaphragms between the stages at *B*.

The Curtis multi-stage turbine is well adapted to the development of large powers, installations of capacities as great as 16,000 H.P. being in successful operation.

207. The Curtis Low- and Mixed-pressure Turbines. — For the development of intermediate powers, the General Electric Company has adapted the Curtis turbine to the use of low-pressure steam. In this form they are known as low-pressure and mixed-pressure turbines, and are of the same general construction as the high-pressure machines, but as they work through a smaller pressure range they have fewer stages, and their capacities are governed by varying the pressure of the steam instead of varying the number of working nozzles, as is the case with the high-pressure turbines.

The low-pressure machine is designed to utilize the exhaust steam from reciprocating engines, pumps, air compressors, hoists, etc., in converting it into useful power. The mixed-

FIG. 175.

pressure turbine, in addition, is designed to carry all or any proportion of its rated load on high-pressure steam; special nozzles are provided for this purpose, and are automatically brought into action in case the supply of low-pressure steam is for any reason insufficient for the power required from the turbine.

The special nozzles of the mixed-pressure machines receive their steam direct from the boiler and, after expansion, deliver it to the same wheel as do the low-pressure nozzles. The steam is expanded in these nozzles from high pressure down to the normal pressure of the first stage, and in such expansion acquires a high velocity and contains consequently a great deal of energy — much more than does an equal quantity of low-pressure steam. Mixed-pressure turbines have been adapted to powers as great as 10,000 H.P.

Exhaust steam is supplied to the turbines at slightly more than atmospheric pressure and expanded down to a vacuum, say, of 28 inches. In the case of a plant operating non-condensing, the addition of a low- or mixed-pressure turbine may increase the power more than one-half without additional expenditure of fuel or of enlargement of boiler power. In the case of plants operating with a condenser, the engines may, in general, be operated non-condensing with an increased expenditure of steam and have the exhaust operate a low- or mixed-pressure turbine at a net gain of from 25 per cent to 40 per cent.

These results are possible because of the fact that the turbine is able to expand steam economically to a much greater degree than the reciprocating steam engine. Expansion in the steam engine is limited on account of cylinder condensation, wire-drawing, practical sizes of low-pressure cylinders, and for other reasons.

208. Curtis Turbo-generator Sets. — The direct connection of Curtis turbines to generators of the General Electric Company produce units for the generation of small electric powers that are representative of the best practice in that particular field. These turbo-generator sets are made for capacities of 7, 10, 15, 25, 35, 75, 100, 125, and 300 k.w. direct current and 100 k.w. alternating current, and may be operated with or without a condenser.

All the direct-current generators are driven by overhung turbines consisting of one or two wheels, each wheel having two or three rows of buckets, varying according to size and operating conditions. A 100 k.w., 2400 r.p.m., two-wheel set is shown in Fig. 176. It will be noted that the rotating element

FIG. 176.

of the overhung design is practically a single piece made up of a shaft carrying the armature, commutator and wheels. There are only two bearings, the number that would be required for the armature alone in the case of an ordinary motor or generator, so that the addition of the turbine has not complicated the rotor from an operating standpoint.

209. The Rateau Turbine. — The most prominent among steam turbines, other than those already described, is that of Rateau. It is of the impulse type and consists of a series of disks mounted on the shaft and carrying buckets in their peripheries. Each disk rotates in a compartment of its own formed by diaphragms projecting inward from the casing and extending to a close running fit to the shaft. Each diaphragm is fitted with guide buckets opposite to the buckets of the rotating disks.

Steam enters the buckets of the first diaphragm, where a partial expansion and acquirement of velocity, accompanied by a pressure fall, takes place; thence the steam enters the buckets of the first rotating disk, delivering to them an impulse due to the acquired velocity. Passing from the first set of moving buckets, the steam enters the buckets of the second diaphragm, where a further expansion and acquirement of velocity takes place, thence to the second set of rotating buckets, and so on through successive expansion steps until the pressure of exhaust is reached.

It is, in effect, a multiplication of the De Laval principle, the expansions taking place in the stationary buckets and the pressure remaining constant during the passage of the steam through the moving buckets, thus avoiding end thrust on the shaft.

CHAPTER XIX

TESTING OF ENGINES AND BOILERS

210. The Dynamometer. — In testing an engine it is necessary to determine the brake horse-power (B.H.P.), or the useful power delivered to the shaft independent of the power absorbed in friction in driving the engine itself, and for this purpose some form of dynamometer is used. A form of the absorption dynamometer, known as the Prony brake, is shown in Fig. 177. A

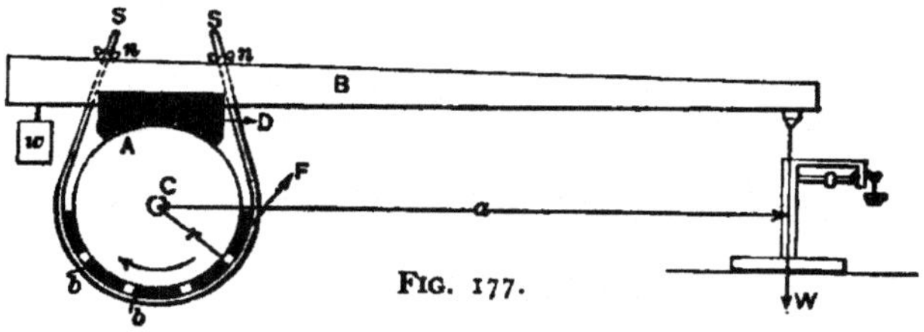

FIG. 177.

drum pulley *A*, keyed to the engine shaft *C*, has a strap *S*, *S* partially encircling it. Bolted to the strap and bearing on the pulley is a series of wooden blocks *b*, *b*. A beam of wood *B* has a wooden shoe *D*, which also bears on the pulley. The weight *w* is adjustable, and serves to balance the preponderance of that part of the beam to the right of the center of the shaft.

If the strap *S*, *S* be tightened about the pulley by means of the nuts *n*, *n* one of two things must happen when the engine starts and the drum revolves in the direction of the arrow: either the grip of the strap must be sufficient to carry the blocks and beam around with the pulley, or the pulley must slip around over the blocks. In practice the nuts are so adjusted that the pulley just slips over the blocks.

The resistance to the carrying of the beam and blocks around with the pulley is the pressure W made to act on a platform scales as shown, so that its amount may be measured in pounds. The resistance to the pulley slipping around in the strap is the friction F.

At the moment of slipping we shall have, by moments about the center of the shaft, $Fr = Wa$, whence $F = \frac{Wa}{r}$, in which r is the radius of the pulley and a the distance from the center of the shaft to the line of pressure on the scales.

For one revolution of the shaft, the work done is

$$F \times 2\pi r = \frac{Wa}{r} \times 2\pi r = 2\pi aW \text{ foot pounds,}$$

when a is measured in feet and W in pounds.

For n revolutions of the shaft the work is $2\pi naW$, and the brake horse-power is $\frac{2\pi naW}{33{,}000}$. If a be taken as $\frac{33}{2\pi} = 5.25$ feet, then the brake horse-power is $\frac{Wn}{1000}$.

In practice the pulley is kept cool by circulating water within it. If both the rubbing surfaces are metallic they should be freely lubricated. An iron pulley rubbing over wooden blocks, as in Fig. 177, requires little lubrication.

211. Objects of Engine and Boiler Tests. — The installation of engine and boiler plants is usually done by contract, and the only means of determining whether the stipulations of the contract have been fulfilled is to test the engine and boiler as a system. Preceding the commencement of such a test, the engine should be run for several hours under the test conditions, and it is very necessary that the conditions remain constant during the test. This is particularly true with respect to the steam pressure. It is equally important that the measuring instruments, such as thermometers, gauges, indicators and scales, should be standardized before the commencement of the test.

Everything pertaining to the cost of the operation and maintenance of a plant enters into its economy, and it is the province of the engineer to obtain the best results at the minimum of cost, and this can be done only by making periodic tests under the working conditions.

The primary object in testing a steam engine is to determine the cost of the power developed. The mean effective pressure, as determined from the indicator diagram, is the exponent of the power, and the cost per horse-power per hour is expressed in thermal units, in pounds of steam, or in pounds of coal.

If the measure is to be in pounds of steam per I.H.P. per hour there are two methods — one for the surface-condensing type of engine and one for the jet-condensing and non-condensing types. With the surface-condensing engine the water resulting from the condensation of the exhaust steam is carefully weighed, allowance being made for all steam used for purposes incidental to the operation of the engine, and correction being made for moisture in the steam. An hour's duration of such a test should be sufficient to obtain accurate results. In testing non-condensing and jet-condensing engines, the weight of steam per I.H.P. per hour is determined by carefully weighing the feed water supplied to the boilers, always making allowance for steam used for purposes other than for driving the engine, such, for example, as for pumping, for jackets, for heating and for calorimeter tests.

When the measure is made in pounds of coal per I.H.P. per hour, the test must be one of the engine and of the boiler combined, in which all the precautions stipulated for a boiler test are to be observed, and the coal carefully weighed. The duration of such trials is from 12 to 24 hours.

Tests may be made to determine the economy derived from the use of such appurtenances as jackets, feed-water heaters, economizers, etc.

The primary object in testing a boiler is to determine the number of pounds of water evaporated per pound of coal, and the conditions of the test should be such that the boiler is not forced beyond what is supposed to be its rated capacity. To force a boiler to its utmost would be a test of its capacity, but since it is far from economical to operate a boiler in that manner, such tests are not of common occurrence.

The Codes of Rules of the American Society of Mechanical Engineers for conducting performance tests of boilers; reciprocating steam engines; steam turbines; complete steam power plants; pumping machinery; complete pumping plants; compressors, blowers, and fans; locomotives; gas producers; gas and oil engines; complete gas power plants; and water wheels are standard and should be followed in all cases.

The Power Test Committee of the Society has submitted its preliminary report on the Codes of 1912 and, though subject to revision, it is not likely that material change will be made in the extracts given below.

Extracts from the Preliminary Report of the Power Test Committee, Codes of 1912, the American Society of Mechanical Engineers, New York

INSTRUCTIONS REGARDING TESTS IN GENERAL

OBJECT

Ascertain the specific object of the test, and keep this in view not only in the work of preparation, but also during the progress of the test, and do not let it be obscured by devoting too close attention to matters of minor importance. Whatever the object of the test may be, accuracy and reliability must underlie the work from beginning to end.

If questions of fulfillment of contract are involved, there should be a clear understanding between all the parties, preferably in writing, as to the operating conditions which should obtain during the trial, and as to the methods of testing to be followed, unless these are already expressed in the contract itself.

Among the many objects of performance tests, the following may be noted:

Determination of capacity and efficiency, and how these compare with standard or guaranteed results.

Comparison of different conditions or methods of operation.

Determination of the cause of either inferior or superior results.

Comparison of different kinds of fuel.

Determination of the effect of changes of design or proportion upon capacity or efficiency, etc.

PREPARATIONS

(*A*) *Dimensions*

Measure the dimensions of the principal parts of the apparatus to be tested, so far as they bear on the objects in view, or determine these from correct working drawings. Notice the general features of the same, both exterior and interior, and make sketches, if needed, to show unusual points of design.

The dimensions of the heating surfaces of boilers and superheaters to be found are those of surfaces in contact with the fire or hot gases. The submerged surfaces in boilers at the mean water level should be considered as water-heating surfaces, and other surfaces which are exposed to the gases as superheating surfaces.

In the case of condensers, feed-water heaters, and the like, the outside surfaces are to be taken. In reheaters and steam jackets, the surfaces to be considered are those exposed to the steam of lower pressure.

The dimensions of engine cylinders should be taken when they are cold, and, if extreme accuracy is required, as in scientific investigations, corrections should be applied to conform to the mean working temperature. If the cylinders are much worn, the average diameter should be found. The clearance of the cylinders may be determined approximately from working drawings of the engine. For accurate work, when practicable, the clearance should be determined by the water measurement method.

(*B*) *Examination of Plant*

Make a thorough examination of the physical condition of all parts of the plant or apparatus which concern the object in view, and record the conditions found, together with any points in the matter of operation which bear thereon.

In boilers, for example, examine for leakage of tubes and riveted or other metal joints. Note the condition of brick furnaces, grates and baffles. Examine brick walls and cleaning doors for air leaks, either by shutting the damper and observing the escaping smoke or by candle-flame test. Determine the condition of heating surfaces with reference to exterior deposits of soot and interior deposits of mud or scale.

See that the steam main is so arranged that condensed and entrained water cannot flow back into the boiler.

Ascertain the interior condition of all steam, air, gas, or water cylinders and the condition of their pistons, and of water plungers and impellers, together with the valves and valve-seats belonging thereto. Locate vacuum leaks in exhaust piping, condensers, packings, etc., using vacuum gauge or candle-flame test. Examine steam, air, gas, or water piping, traps, drip valves, blow-off cocks, safety valves, relief valves, heaters, etc., and make sure that they do not leak. Determine the condition of the blading, nozzles, and valves in steam turbines, and of buckets, guides and draft-tubes in water turbines.

If the object of the test is to determine the highest efficiency or capacity obtainable, any physical defects, or defects of operation, tending to make the result unfavorable, should first be remedied, all fouled parts being cleaned, and the whole put in first-class condition. If, on the other hand, the object is to ascertain the performance under existing conditions, no such preparation is either required or desired.

(*C*) *General Precautions against Leakage*

In steam tests make sure that there is no leakage through blow-offs, drips, etc., or in any steam or water connections of the plant or apparatus undergoing test, which would in any way affect the results. All such connections should be blanked off or satisfactory assurance should be obtained that there is leakage neither out nor in. This is a most important matter, and no assurance should be considered satisfactory unless it is susceptible of absolute demonstration.

(*D*) *Apparatus and Instruments*

Select the apparatus and instruments specified in the Code of Rules applying to the test in hand, locate and install the same, and complete the preparations for the work in view.

The arrangement and location of the testing appliances in every case must be left to the judgment and ingenuity of the engineer in charge, the details being largely dependent upon locality and surroundings. One guiding rule, however, should always be kept in view, viz., see that the apparatus and instruments are substantially reliable, and arrange them in such a way as to obtain correct data.

MISCELLANEOUS INSTRUCTIONS

The person in charge of a test should have the aid of a sufficient number of assistants, so that he may be free to give special attention to any part of the work whenever and wherever it may be required. He should make sure that the instruments and testing apparatus continually give reliable indications, and that the readings are correctly recorded. He should also keep in view, at all points, the operation of the plant or part of the plant under test and see that the operating conditions determined on are maintained and that nothing occurs, either by accident or design, to vitiate the data. This last precaution is especially needed in guarantee tests.

Before a test is undertaken, it is important that the boiler, engine, or other apparatus concerned, shall have been in operation a sufficient length of time to attain working temperatures and proper operating conditions throughout, so that the results of the test may express the true working performance.

It would, for example, be manifestly improper to start a test for determining the maximum efficiency of an externally fired boiler with brick setting until the boiler had been at work a sufficient number of days to dry out thoroughly and heat the brickwork to its working temperature; and likewise improper to begin an engine test for determining the performance under certain prearranged conditions until those conditions had become established by a suitable preliminary run.

An exception should be noted where the object of the test is to obtain the working performance, including the effect of preliminary heating, in which case all the conditions should conform to those of regular service.

In preparation for a test to demonstrate maximum efficiency, it is desirable to run preliminary tests for the purpose of determining the most advantageous conditions.

OPERATING CONDITIONS

In all tests in which the object is to determine the performance under conditions of maximum efficiency, or where it is desired to ascertain the effect of predetermined conditions of operation, all such conditions which have an appreciable effect upon the efficiency should be maintained as nearly uniform during the trial as the limitations of practical work will permit. In a stationary steam plant, for example, where maximum efficiency is the object in view, there should be uniformity in such matters as steam pressure, times of firing, quantity of coal supplied at each firing, thickness of fire, and in other firing operations; also in the rate of supplying the feed water, in the load on the engine or turbine, and in the operating conditions throughout. On the other hand, if the object of the test is to determine the performance under working conditions, no attempt at uniformity is either desired or required unless this uniformity corresponds to the regular practice, and when this is the object the usual working conditions should prevail throughout the trial.

RECORDS

A log of the data should be entered in notebooks or on blank sheets suitably prepared in advance. This should be done in such manner that the test may be divided into hourly periods, or, if necessary, periods of less duration, and the leading data obtained for any one or more periods as desired, thereby showing the degree of uniformity obtained.

The readings of the various instruments and apparatus concerned in the test other than those showing quantities of consumption (such as fuel, water, and gas) should be taken at intervals not exceeding half an hour and entered in the log. Whenever the indications fluctuate, the intervals should be reduced according to the extent of the fluctuation. In the case of smoke observations, for example, it is often necessary to take observations every minute, or still oftener, continuing these throughout the period covering the range of variations.

Make a memorandum of every event connected with the progress of a test, however unnecessary at the time it may appear. A record should be made of the exact time of every such occurrence and the time of taking every weight and every observation. For the purpose of identification the signature of the observer and the date should be affixed to each log sheet or record.

In the simple matter of weighing coal by the barrow-load, or weighing water by the tank-full, which is required in many tests, a series of marks, or tallies, should never be trusted. The time each load is weighed or emptied should be recorded. The weighing of coal should not be delegated to unreliable assistants, and whenever practicable, one or more men should be assigned solely to this work. The same may be said with regard to the weighing of feed water.

DATA REPRESENTED GRAPHICALLY

If it is desired to show the uniformity of the data at a glance the whole log of the trial should be plotted on a chart, using horizontal distances to represent times of observation, and vertical distances on suitable scales to represent various data as recorded.

It is instructive to plot the leading data on such a chart while the test is in progress.

REPORT

The report of a test should represent all the leading facts bearing on the design, dimensions, condition, and operation of the apparatus tested, and should include a description of any other apparatus and auxiliaries concerned, together with such sketches as may be needed for a clear understanding of all points under consideration. It should state clearly the object and character of the test, the methods followed, the conditions maintained, and the conclusions reached, closing with a tabular summary of the principal data and results.

RULES FOR SAMPLING AND DRYING COAL AND ASH AND SAMPLING STEAM

(*A*) *Sampling and Drying Coal*

Select a representative shovelful from each barrow-load as it is drawn from the coal pile or other source of supply, and store the samples in a cool place in a covered metal receptacle. When all the coal has thus been sampled, break up the lumps, thoroughly mix the whole quantity, and finally reduce it by the process of repeated quartering and crushing to a sample weighing about 5 pounds, the largest pieces being about the size of a pea. From this sample two 1-quart air-tight glass fruit jars or other air-tight vessels are to be promptly filled and preserved for subsequent determinations of moisture, calorific value, and chemical composition. These operations should be conducted where the air is cool and free from drafts.

When the sample lot of coal has been reduced by quartering to say 100 pounds, a portion weighing say 15 to 20 pounds should be withdrawn for the purpose of immediate moisture determination. This is placed in a shallow iron pan and dried on the hot iron boiler flue for at least 12 hours, being weighed before and after drying on scales reading to quarter ounces.

The moisture thus determined is approximately reliable for anthracite and semi-bituminous coals, but not for coals containing much inherent moisture. For such coals and for all absolutely reliable determinations the method to be pursued is as follows:

> Take one of the samples contained in the glass jars, and subject it to a thorough air drying, by spreading it in a thin layer and exposing it for several hours to the atmosphere of a warm room, weighing it before and after, thereby determining the quantity of surface moisture it contains. Then crush the whole of it by running it through an ordinary coffee mill or other suitable crusher adjusted so as to produce somewhat coarse grains (less than $\frac{1}{16}$-inch), thoroughly mix the crushed sample, select from it a portion of from 10 to 50 grams (from 0.5 ounce to 2 ounces), weigh it in a balance which will easily show a variation as small as 1 part in 1000, and dry it for one hour in an air or sand bath at a temperature between 240° and 280° F. Weigh it and record the loss, then heat and weigh again until the minimum weight has been reached. The difference between the original and the minimum weight is the moisture in the air-dried coal. The sum of the moisture thus found and that of the surface moisture is the total moisture.

(C) *Sampling Steam*

Construct a sampling pipe or nozzle made of $\frac{1}{2}$-inch iron pipe and insert it in the steam main at a point where the entrained moisture is likely to be most thoroughly mixed. The inner end of the pipe, which should extend nearly across to the opposite side of the main, should be closed and the interior portion perforated with not less than twenty $\frac{1}{8}$-inch holes equally distributed from end to end and preferably drilled in irregular or spiral rows, with the first hole not less than half an inch from the wall of the pipe.

The sampling pipe should not be placed near a point where water may pocket or where such water may affect the amount of moisture contained in the sample. When non-return valves are used, or where there are horizontal connections leading from the boiler to a vertical outlet, water may collect at the lower end of the uptake pipe and be blown upward in a spray which will not be carried away by the steam owing to a lack of velocity. A sample taken from the lower part of this pipe will show a greater amount of moisture than a true sample. With goose-neck connections a small amount of water may collect on the bottom of the pipe near the upper end where the inclination is such that the tendency to flow backward is ordinarily counterbalanced by the flow of steam forward over its surface; but when the velocity momentarily decreases the water flows back to the lower end of the goose-neck and increases the moisture at that point, making it an undesirable location for sampling. In any case it should be borne in mind that with low velocities the tendency is for drops of entrained water to settle to the bottom of the pipe, and to be temporarily broken up into spray whenever an abrupt bend or other disturbance is met.

If it is necessary to attach the sampling nozzle at a point near the end of a long horizontal run, a drip pipe should be provided a short distance in front of the nozzle, preferably at a pocket formed by some fitting and the water running along the bottom of the main drawn off, weighed, and added to the moisture shown by the calorimeter; or, better, a steam separator should be installed at the point noted.

(D) General Location of Sampling Pipe and Thermometer Well

In testing a stationary boiler the sampling pipe should be located as near as practicable to the boiler, and the same is true as regards the thermometer well when the steam is superheated. In an engine or turbine test these locations should be as near as practicable to the throttle valve. In the test of a plant where it is desired to get complete information, especially where the steam main is unusually long, sampling nozzles or thermometer wells should be provided at both points, so as to obtain data at either point as may be required.

In a locomotive the calorimeter should be attached either to the steam dome where it may be connected to the throttle opening, or to the steam passage in the saddle casting on one side.

RULES FOR CONDUCTING EVAPORATIVE TESTS OF BOILERS

OBJECT AND PREPARATIONS

Determine the object, take the dimensions, note the physical conditions, examine for leakages, install the testing appliances, etc., and make preparations for the test according to the instructions contained in the Code.

FUEL

Determine the character of fuel (coal) to be used. For tests of maximum efficiency or capacity of the boiler to compare with other boilers, the coal should be of some kind which is commercially regarded as a standard for the locality where the test is made.

In the Eastern States the standards thus regarded for semi-bituminous coals are Pocahontas (Va. and W. Va.) and New River (W. Va.); for anthracite coals those of the No. 1 buckwheat size, fresh-mined, containing not over 13 per cent ash by analysis; and for bituminous coals, Youghiogheny and Pittsburg coals. In some sections east of the Allegheny Mountains the semi-bituminous Clearfield (Pa.) and Cumberland (Md.) are also considered as standards. These coals when of good quality possess the essentials of excellence, adaptability to various kinds of furnaces, grates, boilers, and methods of firing required, besides being widely distributed and generally accessible in the Eastern market.

There are no special grades of coal mined in the Western States which are widely and generally considered as standards for testing purposes; the best coal obtainable in any particular locality being regarded as the standard of comparison.

A coal selected for maximum efficiency and capacity tests should be the best of its class, and especially free from slagging and unusual clinker-forming impurities.

For guarantee and other tests with a specified coal containing not more than a certain amount of ash and moisture, the coal selected should not be higher in ash and in moisture than the stated amounts, because any increase is liable to reduce the efficiency and capacity more than the equivalent proportion of such increase.

The size of the coal, especially when it is of the anthracite class, should be determined by screening a suitable sample.

APPARATUS AND INSTRUMENTS

The apparatus and instruments required for boiler tests are:

(*a*) Platform scales for weighing coal and ashes.

(*b*) Graduated scales attached to the water glasses.

(*c*) Tanks and platform scales for weighing water (or water meters calibrated in place).

(*d*) Pressure gauges, thermometers, and draft gauges.

(*e*) Calorimeters for determining the calorific value of fuel and the quality of steam.

(*f*) Furnace pyrometers.

(*g*) Gas analyzing apparatus.

OPERATING CONDITIONS

Determine what the operating conditions and method of firing should be to conform to the object in view, and see that they prevail throughout the trial, as nearly as possible.

Where uniformity in the rate of evaporation is required, arrangement can usually be made to dispose of the steam so that this result can be attained. In a single boiler it may be accomplished by discharging steam through a waste pipe and regulating the amount by means of a valve. In a battery of boilers, in which only one is tested, the draft may be regulated on the remaining boilers to meet the varying demands for steam, leaving the test boiler to work under a steady rate of evaporation.

DURATION

The duration of tests to determine the efficiency of a hand-fired boiler should be 10 hours of continuous running or such time as may be required to burn a total of 250 pounds of coal per square foot of grate.

In the case of a boiler using a mechanical stoker, the duration, where practicable, should be at least 24 hours. If the stoker is of a type that permits the quantity and condition of the fuel bed at beginning and end of the test to be accurately estimated, the duration may be reduced to 10 hours, or such time as may be required to burn the above noted total of 250 pounds per square foot.

In commercial tests where the service requires continuous operation night and day, with frequent shifts of firemen, the duration of the test, whether the boilers are hand-fired or stoker-fired, should be at least 24 hours. Likewise in commercial tests, either of a single boiler or of a plant of several boilers which operate regularly a certain number of hours and during the balance of the day the fires are banked, the duration should not be less than 24 hours.

The duration of tests to determine the maximum evaporative capacity of a boiler, without determining the efficiency, should not be less than three hours.

STARTING AND STOPPING

The conditions regarding the temperature of the furnace and boiler, the quantity and quality of the live coal and ash on the grates, the water level, and the steam pressure should be as nearly as possible the same at the end as at the beginning of the test.

To secure the desired equality of conditions with hand-fired boilers, the following method should be employed:

The furnace being well heated by a preliminary run, burn the fire low and thoroughly clean it, leaving enough live coal spread evenly over the grate (from 2 inches to 4 inches ordinarily; from 1 inch to 2 inches for small anthracite coals) to serve as a foundation for the new fire. Note quickly the thickness of the coal bed as nearly as it can be estimated or measured; also the water level, the steam pressure, and the time, and record the latter as the starting time. Fresh coal should then be fired from that weighed for the test, the ashpit thoroughly cleaned, and the regular work of the test proceeded with.

Before the end of the test the fire should again be burned low and cleaned in such a manner as to leave the same amount of live coal on the grate as at the start. When this condition is reached, observe quickly the water level, the steam pressure, and the time, and record the latter as the stopping time. If the water level is not the same as at the beginning, a correction should be made by computation, rather than by feeding additional water after the final readings are taken. Finally remove the ashes and refuse from the ashpit.

In a plant containing several boilers where it is not practicable to clean them simultaneously, the fires should be cleaned one after the other as rapidly as may be, and each one after cleaning charged with enough coal to maintain a thin fire in good working condition. After the last fire is cleaned and in working condition, burn all the fires low (say 4 to 6 inches), note quickly the thickness of each, also the water levels, steam pressure, and time, which last is taken as the starting time. Likewise when the time arrives for closing the test, the fires should be quickly cleaned one by one, and when this work is completed they should all be burned low the same as at the start, and the various observations made as noted.

In the case of a large boiler having several furnace doors requiring the fire to be cleaned in sections one after the other, the above directions pertaining to starting and stopping in a plant of several boilers may be followed.

To obtain the desired equality of conditions of the fire when a mechanical stoker other than a chain grate is used, the procedure should be modified where practicable as follows:

Regulate the coal feed so as to burn the fire to the low condition required for cleaning. Shut off the coal-feeding mechanism and fill the hoppers level full. Clean the ash or dump plate, note quickly the depth and condition of the coal on the grate, the water level, the steam pressure, and the time, and record the latter as the starting time. Then start the coal-feeding mechanism, clean the ashpit, and proceed with the regular work of the test.

When the time arrives for the close of the test, shut off the coal-feeding mechanism, fill the hoppers and burn the fire to the same low point as at the beginning. When this condition is reached, note the water level, the steam pressure, and the time, and record the latter as the stopping time. Finally clean the ash plate and haul the ashes.

In the case of chain grate stokers, the desired operating conditions should be maintained for half an hour before starting a test and for a like period before its close, the height of the throat plate and the speed of the grate being the same during both of these periods.

RECORDS

Half-hourly readings of the instruments are usually sufficient. If there are sudden and wide fluctuations, the readings in such cases should be taken every fifteen minutes, and in some instances oftener.

The coal should be weighed and delivered to the fireman in portions sufficient for one hour's run, thereby ascertaining the degree of uniformity of firing. An ample supply of coal should be maintained at all times, but the quantity on the floor at the end of each hour should be as small as practicable, so that the same may be readily estimated and deducted from the total weight.

The records should be such as to ascertain also the consumption of feed water each hour, and thereby determine the degree of uniformity of evaporation.

QUALITY OF STEAM

If the boiler does not produce superheated steam the percentage of moisture in the steam should be determined by the use of a throttling or separating calorimeter. If the boiler has superheating surface, the temperature of the steam should be determined by the use of a thermometer inserted in a thermometer well.

SAMPLING AND DRYING COAL

During the progress of the test the coal should be regularly sampled for the purpose of analysis and determination of moisture.

ASHES AND REFUSE

The ashes and refuse withdrawn from the furnace and ashpit during the progress of the test and at its close should be weighed so far as possible in a dry state. If wet the amount of moisture should be ascertained and allowed for, a sample being taken and dried for this purpose. This sample may serve also for analysis and the determination of unburned carbon and fusing temperature.

CALORIFIC TESTS AND ANALYSES OF COAL

The quality of the fuel should be determined by calorific tests and analysis of the coal sample.

CALCULATION OF RESULTS

The methods to be followed in expressing and calculating those results which are not self-evident are explained as follows:

Efficiency. — The "efficiency of boiler, furnace and grate" is the relation between the heat absorbed per pound of coal fired and the calorific value of one pound of coal.

The "efficiency of boiler and furnace" is the relation between the heat absorbed per pound of combustible burned and the calorific value of one pound of combustible. This expression of efficiency furnishes a means for comparing one boiler and furnace with another, when the losses of unburned coal due to grates, cleanings, etc., are eliminated.

The "combustible burned" is determined by subtracting from the weight of coal supplied to the boiler the moisture in the coal, the weight of ash and unburned coal withdrawn from the furnace and ashpit, and the weight of dust, soot, and refuse, if any, withdrawn from the tubes, flues, and combustion chambers, including ash carried away in the gases, if any, determined from the analyses of coal and ash. The "combustible" used for determining the calorific value is the weight of coal less the moisture and ash found by analysis.

The "heat absorbed" per pound of coal or combustible is calculated by multiplying the equivalent evaporation from and at 212° per pound of coal or combustible by 970.4.

Correction for Live Steam, if any, used for Aiding Combustion. — If live steam is admitted into the furnace or ashpit for producing blast, injecting fuel or aiding combustion, it is to be deducted from the total evaporation, and the net evaporation used in the various calculations.

DATA AND RESULTS OF EVAPORATIVE TEST

SHORT FORM, CODE OF 1912

(1) Test of........................boiler located at........................
to determine....................conducted by......................
(2) Kind of furnace..
(3) Grate surface..sq. ft.
(4) Water-heating surface.......................................sq. ft.
(5) Superheating surface..sq. ft.
(6) Date...
(7) Duration...hr.
(8) Kind and size of coal...

AVERAGE PRESSURES, TEMPERATURES, ETC.

(9) Steam pressure by gauge...lb.
(10) Temperature of feed water entering boiler..........................deg.
(11) Temperature of escaping gases leaving boiler.........................deg.
(12) Force of draft between damper and boiler.............................in.
(13) Percentage of moisture in steam, or number of degrees of superheating..per cent or deg.

TOTAL QUANTITIES

(14) Weight of coal as fired...lb.
(15) Percentage of moisture in coal..................................per cent.
(16) Total weight of dry coal consumed....................................lb.
(17) Total ash and refuse...lb.
(18) Percentage of ash and refuse in dry coal.........................per cent.
(19) Total weight of water fed to the boiler..............................lb.
(20) Total water evaporated, corrected for moisture in steam...............lb.
(21) Total equivalent evaporation from and at 212°........................lb.

HOURLY QUANTITIES AND RATES

(22) Dry coal consumed per hour...lb.
(23) Dry coal per square foot of grate surface per hour....................lb.
(24) Water evaporated per hour corrected for quality of steam..............lb.
(25) Equivalent evaporation per hour from and at 212°......................lb.
(26) Equivalent evaporation per hour from and at 212° per square foot of water-heating surface..lb.

CAPACITY

(27) Evaporation per hour from and at 212° (same as Line 25)..............lb.
(28) Boiler horse-power developed (Item 27 ÷ 34½).....................bl.h.p.
(29) Rated capacity, in evaporation from and at 212° per hour..............lb.
(30) Rated boiler horse-power...bl.h.p.
(31) Percentage of rated capacity developed..........................per cent.

Economy Results

(32) Water fed per lb. of coal fired (Item 19 ÷ Item 14) .... lb.
(33) Water evaporated per lb. of dry coal (Item 20 ÷ Item 16) .... lb.
(34) Equivalent evaporation from and at 212° per lb. of dry coal (Item 21 ÷ Item 16) .... lb.
(35) Equivalent evaporation from and at 212° per lb. of combustible [Item 21 ÷ (Item 16 − Item 17)] .... lb.

Efficiency

(36) Calorific value of 1 lb. of dry coal .... B.t.u.
(37) Calorific value of 1 lb. of combustible .... B.t.u.
(38) Efficiency of boiler, furnace and grate $\left[100 \times \frac{\text{Item 34} \times 970.4}{\text{Item 36}}\right]$ per cent.
(39) Efficiency of boiler and furnace $\left[100 \times \frac{\text{Item 35} \times 970.4}{\text{Item 37}}\right]$ .... per cent.

Cost of Evaporation

(40) Cost of coal per ton of.... lb. delivered in boiler room .... dollars.
(41) Cost of coal required for evaporating 1000 lb. of water from and at 212° .... dollars.

RULES FOR CONDUCTING TESTS OF RECIPROCATING ENGINES

OBJECT AND PREPARATIONS

Determine the object, take the dimensions, note the physical conditions not only of the engine but of all parts of the plant that are concerned in the determinations, examine for leakages, install the testing appliances, and prepare for the test accordingly.

APPARATUS AND INSTRUMENTS

The apparatus and instruments required for a simple performance test of a steam engine, in which the steam consumption is determined by feed-water measurement, are:

(*a*) Tanks and platform scales for weighing water (or water meters calibrated in place).
(*b*) Graduated scales attached to the water glasses of the boilers.
(*c*) Pressure gauges, vacuum gauges, and thermometers.
(*d*) A steam calorimeter.
(*e*) A barometer.
(*f*) Steam-engine indicators.
(*g*) A planimeter.
(*h*) A tachometer or other speed-measuring apparatus.
(*i*) A friction brake or dynamometer.

The determination of the heat and steam consumption of an engine by feed-water test requires the measurement of the various supplies of water fed to the boiler, that of the water discharged by separators and drips not returned to the boiler, and that of water and steam which escapes by leakage of the boiler and piping; all of these last being deducted from the total feed water measured.

Where a surface condenser is provided and the steam consumption is determined from the water discharged by the air pump, no such measurement of drips and leakage is required, but assurance must be had that all the steam passing into the cylinders finds its way into the condenser. If the condenser leaks the defects causing it should be remedied, or suitable correction made for the leakage.

To ascertain the consumption of heat, the various feed temperatures are taken and heat calculations made accordingly. If the conditions imposed by the particular method adopted for carrying on the test depart from the usual practice, as for example where a colder supply of feed water is used than the ordinary supply, a preliminary or subsequent run should be made to ascertain the temperatures which obtain under the usual working conditions, and the heat measurements obtained under the test conditions appropriately corrected for such departures.

The steam consumed by steam-driven auxiliaries which are required for the operation of the engine should be included in the total steam from which the heat consumption is calculated and the quantity of steam thus used should be determined and reported.

OPERATING CONDITIONS

Determine what the operating conditions should be to conform to the object in view, and see that they prevail throughout the trial.

DURATION

A test for heat or steam consumption, with substantially constant load, should be continued for such time as may be necessary to obtain a number of successive hourly records, during which the results are reasonably uniform. For a test involving the measurement of feed water for this purpose, five hours is sufficient duration. Where a surface condenser is used and the measurement is that of the water discharged by the air pump, the duration may be somewhat shorter. In this case, successive half-hourly records may be compared and the time correspondingly reduced.

When the load varies widely at different times of the day, the duration should be such as to cover the entire period of variation.

The preliminary or subsequent trial for determining the working temperatures on a heat test, where the temperatures obtained under the test conditions depart from the usual temperatures, should be of such duration as may be required to secure working results.

STARTING AND STOPPING

The engine and appurtenances having been set to work and thoroughly heated under the prescribed conditions of test, except in cases where the object is to obtain the performance under working conditions, note the water levels in the boilers and feed reservoir, take the time and consider this the starting time. Then begin the measurements and observations and carry them forward until the end of the period determined on. When this time arrives, the water levels and steam pressure should be brought as near as practicable to the same points as at the start. This being done, again note the time and consider it the stopping time of the test. If there are differences in the water levels, proper corrections are to be applied.

Where a surface condenser is used, the collection of water discharged by the air pump begins at the starting time, and the water is thereafter measured or weighed until the end of the test; no observations of the boilers being required.

Care should be taken in cases where the activity of combustion in the boiler furnaces affects the height of water in the gauge glasses that the same conditions of fire and drafts are operating at the end as at the beginning. For this reason it is best to start and stop a test without interfering with the regularity of the operation of the feed pump, provided the latter can be regulated to run so as to supply the feed water at a uniform rate. In some cases where the supply of feed water is irregular, as, for example, where an injector is used of a larger capacity than is required, the supply of feed water should be temporarily shut off.

Suitable care should be observed in noting the average height of the water in the glasses, taking sufficient time to satisfactorily judge of the full extent of the fluctuation of the water line, and thereby its mean position.

RECORDS

The general data should be recorded as pointed out in the Code. Half-hourly readings of the instruments are sufficient, excepting where there are wide fluctuations. A set of indicator diagrams should be obtained at intervals of 20 minutes, and at more frequent intervals if the nature of the test makes it necessary. Mark on each card the cylinder and the end at which it was taken, also the time of day. Record on one card of each set the readings of the pressure gauges concerned, taken at the same time. These records should subsequently be entered on the general log, together with the areas, pressures, lengths, etc., measured from the diagrams when they are worked up.

DATA AND RESULTS OF HEAT AND FEED-WATER TESTS OF STEAM ENGINE

Short Form, Code of 1912

(1) Test of .... engine located at ....
to determine .... conducted by ....

(2) Type and class of engine and auxiliaries ....

(3) Dimensions of main engine:	1st Cyl.	2d Cyl.	3d Cyl.
(a) Diameter of cylinder .... in.	......	......	......
(b) Stroke of piston .... ft.	......	......	......
(c) Diameter of piston rod each end .... in.	......	......	......
(d) Average clearance .... per cent	......	......	......

	1st Cyl.	2d Cyl.	3d Cyl.
(e) Cylinder ratio........................	......	......	......
(f) Horse-power constant for 1 lb. m.e.p. and 1 r.p.m........................	......	......	......

(4) Dimensions and type of auxiliaries..

(5) Date..

(6) Duration...hr.

AVERAGE PRESSURES AND TEMPERATURES

(7) Pressure in steam pipe near throttle by gauge..........................lb.

(8) Barometric pressure of atmosphere in in. of mercury...................in.

(9) Pressure in receivers by gauge.......................................lb.

(10) Vacuum in condenser in in. of mercury...............................in.

(11) Pressure in jackets and reheaters by gauge..........................lb.

(12) Temperature of main supply of feed water..........................deg.

(13) Temperature of additional supplies of feed water....................deg.

TOTAL QUANTITIES

(14) Total water fed to boilers from main source of supply..................lb.

(15) Total water fed from additional supplies.............................lb.

(16) Total water fed to boilers from all sources...........................lb.

(17) Moisture in steam or superheating near throttle...........per cent or deg.

(18) Factor of correction for quality of steam..............................

(19) Total dry steam consumed for all purposes............................lb.

HOURLY QUANTITIES

(20) Water fed from main source of supply...............................lb.

(21) Water fed from additional supplies.................................lb.

(22) Total water fed to boilers per hour.................................lb.

(23) Total dry steam consumed per hour.................................lb.

(24) Loss of steam and water per hour due to drips from main steam pipes and to leakage of plant...lb.

(25) Net dry steam consumed per hour by engine and auxiliaries.............lb.

(26) Net dry steam consumed per hour:

- (a) By engine alone...lb.
- (b) By auxiliaries..lb.

HEAT DATA

(27) Heat units per lb. of dry steam, based on temperature of Line 12....B.t.u.

(28) Heat units per lb. of dry steam, based on temperature of Line 13.....B.t.u.

(29) Heat units consumed per hour, main supply of feed................B.t.u.

(30) Heat units consumed per hour, additional supplies of feed............B.t.u.

(31) Total heat units consumed per hour for all purposes.................B.t.u.

(32) Loss of heat per hour due to leakage of plant, drips, etc.............B.t.u.

(33) Net heat units consumed per hour:

- (a) By engine and auxiliaries...............................B.t.u.
- (b) By engine alone...B.t.u.
- (c) By auxiliaries..B.t.u.

Indicator Diagrams

	1st Cyl.	2d Cyl.	3d Cyl.
(34) Commercial cut-off in per cent of stroke........	......	......	......
(35) Initial pressure in lb. per sq. in. above atmosphere................................	......	......	......
(36) Back pressure at lowest point above or below atmosphere in lb. per sq. in..............	......	......	......
(37) Mean effective pressure in lb. per sq. in.........	......	......	......
(38) Steam accounted for by indicator in lb. per I.H.P. per hour:			
(*a*) Near cut-off........................	......	......	......
(*b*) Near release........................	......	......	......

Speed

(39) Revolutions per minute..rev.

(40) Piston speed in ft. per min..ft.

Power

(41) Indicated horse-power developed by main-engine cylinders:

1st cylinder..I.H.P.

2d cylinder..I.H.P.

3d cylinder..I.H.P.

Whole engine..I.H.P.

(42) Brake horse-power..B.H.P.

Economy Results

(43) Heat units consumed by engine and auxiliaries per hour:

(*a*) Per indicated horse-power...............................B.t.u.

(*b*) Per brake horse-power...................................B.t.u.

(44) Dry steam consumed per indicated horse-power per hour:

(*a*) By engine and auxiliaries..................................lb.

(*b*) By main engine alone......................................lb.

(*c*) By auxiliaries..lb.

(45) Dry steam consumed per brake horse-power per hour:

(*a*) By engine and auxiliaries..................................lb.

(*b*) By main engine alone......................................lb.

(*c*) By auxiliaries..lb.

(46) Percentage of steam used by main-engine cylinders accounted for by indicator diagrams:

(*a*) Near cut-off..per cent.

(*b*) Near release..per cent.

INDEX

SHORT-TITLE CATALOGUE

OF THE

PUBLICATIONS

OF

JOHN WILEY & SONS

NEW YORK

London: CHAPMAN & HALL, Limited

Montreal, Can.: RENOUF PUB. CO.

ARRANGED UNDER SUBJECTS

Descriptive circulars sent on application. Books marked with an asterisk (*) are sold at *net* prices only. All books are bound in cloth unless otherwise stated.

AGRICULTURE—HORTICULTURE—FORESTRY.

ARMSBY—Principles of Animal Nutrition..........8vo, $4 00
BOWMAN—Forest Physiography..........8vo, *5 00
BUDD and HANSEN—American Horticultural Manual:
  Part I. Propagation, Culture, and Improvement..........12mo, 1 50
  Part II. Systematic Pomology..........12mo, 1 50
ELLIOTT—Engineering for Land Drainage..........12mo, 2 00
  Practical Farm Drainage. (Second Edition, Rewritten)..........12mo, 1 50
FULLER—Domestic Water Supplies for the Farm..........8vo, *1 50
GRAVES—Forest Mensuration..........8vo, 4 00
  Principles of Handling Woodlands..........Small 8vo, *1 50
GREEN—Principles of American Forestry..........12mo, 1 50
GROTENFELT—Principles of Modern Dairy Practice. (WOLL.)..........12mo, 2 00
HAWLEY and HAWES—Forestry in New England..........8vo, *3 50
HERRICK—Denatured or Industrial Alcohol..........8vo, *4 00
KEMP and WAUGH—Landscape Gardening. (New Edition, Rewritten)..12mo, *1 50
LARSEN and WHITE—Dairy Technology..........Small 8vo, *1 50
McKAY and LARSEN—Principles and Practice of Butter-making..........8vo, *1 50
MAYNARD—Landscape Gardening as Applied to Home Decoration...12mo, 1 50
RECORD—Identification of the Economic Woods of the United States...8vo, *1 25
RECKNAGEL—Theory and Practice of Working Plans (Forest Organization)..........(*In Press.*)
SANDERSON—Insect Pests of Farm, Garden, and Orchard..........Small 8vo, *3 00
  Insects Injurious to Staple Crops..........12mo, 1 50
SCHWARZ—Longleaf Pine in Virgin Forest..........12mo, *1 25
SOLOTAROFF—Field Book for Street-tree Mapping..........12mo, *0 75
  In lots of one dozen.......... *8 00
  Shade Trees in Towns and Cities..........8vo, *3 00
STOCKBRIDGE—Rocks and Soils..........8vo, 2 50
WINTON—Microscopy of Vegetable Foods..........Large 8vo, 7 50
WOLL—Handbook for Farmers and Dairymen..........16mo, 1 50

ARCHITECTURE.

ATKINSON—Orientation of Buildings or Planning for Sunlight..........8vo, *2 00
BALDWIN—Steam Heating for Buildings..........12mo, 2 50
BERG—Buildings and Structures of American Railroads..........4to, 5 00

Birkmire—Architectural Iron and Steel....8vo, $3 50
Compound Riveted Girders as Applied in Buildings....8vo, 2 00
Planning and Construction of High Office Buildings....8vo, 3 50
Skeleton Construction in Buildings....8vo, 3 00
Briggs—Modern American School Buildings....8vo, 4 00
Byrne—Inspection of Materials and Workmanship Employed in Construction....16mo, 3 00
Carpenter—Heating and Ventilating of Buildings....8vo, 4 00
Corthell—Allowable Pressure on Deep Foundations....12mo, *1 25
Eckel—Building Stones and Clays....8vo, *3 00
Freitag—Architectural Engineering....8vo, 3 50
Fire Prevention and Fire Protection....16mo, mor., *4 00
Fireproofing of Steel Buildings....8vo, 2 50
Gerhard—Guide to Sanitary Inspections....12mo, 1 50
Modern Baths and Bath Houses....8vo, *3 00
Sanitation of Public Buildings....12mo, 1 50
Theatre Fires and Panics....12mo, 1 50
The Water Supply, Sewerage and Plumbing of Modern City Buildings, 8vo, *4 00
Greene—Elements of Heating and Ventilation....8vo, *2 50
Johnson—Statics by Algebraic and Graphic Methods....8vo, 2 00
Kellaway—How to Lay Out Suburban Home Grounds....8vo, 2 00
Kidder—Architects' and Builders' Pocket-book....16mo, mor., 5 00
Merrill—Stones for Building and Decoration....8vo, 5 00
Monckton—Stair-building....4to, 4 00
Patton—Practical Treatise on Foundations....8vo, 5 00
Peabody—Naval Architecture....8vo, 7 50
Rice—Concrete-block Manufacture....8vo, 2 00
Richey—Handbook for Superintendents of Construction....16mo, mor., 4 00
Building Foreman's Pocket Book and Ready Reference....16mo, mor., 5 00
Building Mechanics' Ready Reference Series: *
Carpenters' and Woodworkers' Edition....16mo, mor., *1 50
Cement Workers' and Plasterers' Edition....16mo, mor., *1 50
Plumbers', Steam-Fitters', and Tinners' Edition....16mo, mor., *1 50
Stone- and Brick-masons' Edition....16mo, mor., *1 50
Ries—Building Stones and Clay Products....8vo, *3 00
Sabin—House Painting (Glazing, Paper Hanging and Whitewashing).12mo, 1 00
Siebert and Biggin—Modern Stone-cutting and Masonry....8vo, 1 50
Snow—Principal Species of Wood....8vo, 3 50
Wait—Engineering and Architectural Jurisprudence....8vo, 6 00
Sheep, 6 50
Law of Contracts....8vo, 3 00
Law of Operations Preliminary to Construction in Engineering and Architecture....8vo, 5 00
Sheep, 5 50
Wilson—Air Conditioning....12mo, 1 50
Worcester and Atkinson—Small Hospitals, Establishment and Maintenance, Suggestions for Hospital Architecture, with Plans for a Small Hospital....12mo, 1 25

ASSAYING.

Betts—Lead Refining by Electrolysis....8vo, 4 00
Fletcher—Practical Instructions in Quantitative Assaying with the Blowpipe....16mo, mor., 1 50
Furman and Pardoe—Manual of Practical Assaying....8vo, 3 00
Lodge—Notes on Assaying and Metallurgical Laboratory Experiments. 8vo, 3 00
Low—Technical Methods of Ore Analysis....8vo, 3 00
Miller—Cyanide Process....12mo, 1 00
Manual of Assaying....12mo, 1 00
Minet—Production of Aluminum and its Industrial Use. (Waldo)..12mo, 2 50
Price and Meade—The Technical Analysis of Brass and the Non-Ferrous Alloys....12mo, *2 00
Ricketts and Miller—Notes on Assaying....8vo, 3 00
Robine and Lenglen—Cyanide Industry. (Le Clerc.)....8vo, 4 00
Seamon—Manual for Assayers and Chemists....Small 8vo, *2 50

ULKE—Modern Electrolytic Copper Refining........................8vo, $3 00
WILSON—Chlorination Process..................................12mo, 1 50
Cyanide Processes..12mo, 1 50

ASTRONOMY.

COMSTOCK—Field Astronomy for Engineers..........................8vo, 2 50
CRAIG—Azimuth 4to, 3 50
CRANDALL—Text-book on Geodesy and Least Squares................8vo, 3 00
DOOLITTLE—Treatise on Practical Astronomy.......................8vo, 4 00
HAYFORD—Text-book of Geodetic Astronomy.........................8vo, 3 00
HOSMER—Azimuth ....16mo, mor., 1 00
Text-book on Practical Astronomy.............................8vo, *2 00
MERRIMAN—Elements of Precise Surveying and Geodesy 8vo, 2 50
MICHIE and HARLOW—Practical Astronomy.........................8vo, *3 00
RUST—Ex-meridian Altitude, Azimuth and Star-Finding Tables.......8vo, 5 00
WHITE—Elements of Theoretical and Descriptive Astronomy........12mo, *2 00

BIOLOGY.

COHNHEIM—Enzymes..12mo, *1 50
DAVENPORT—Statistical Methods with Special Reference to Biological Variation....................................16mo, mor., 1 50
EFFRONT and PRESCOTT—Enzymes and Their Applications............8vo, 3 00
EULER and POPE—General Chemistry of the Enzymes...............8vo, *3 00
MAST—Light and Behavior of Organisms..........................8vo, *2 50
PRESCOTT and WINSLOW—Elements of Water Bacteriology, with Special Reference to Sanitary Water Analysis....................12mo, 1 50
WARD and WHIPPLE—Freshwater Biology ....(*In Press.*)
WHIPPLE—The Microscopy of Drinking Water......................8vo, 3 50
WINSLOW—The Systematic Relationship of the Coccaceæ......Small 8vo, 2 50

CHEMISTRY.

ABDERHALDEN—Physiological Chemistry in Thirty Lectures. (HALL and DEFREN)...8vo, *5 00
ABEGG—Theory of Electrolytic Dissociation. (VON ENDE.)..........12mo, *1 25
ALEXEYEFF—General Principles of Organic Syntheses. (MATTHEWS.)..8vo, 3 00
ALLEN—Exercises in General Chemistry, Loose Leaf Laboratory Manual. Oblong 4to, paper, *1 00
Tables for Iron Analysis.................................... 8vo, 3 00
ARMSBY—Principles of Animal Nutrition..........................8vo, 4 00
ARNOLD—Compendium of Chemistry. (MANDEL.).............Small 8vo, 3 50
ASSOCIATION OF STATE AND NATIONAL FOOD AND DAIRY DEPARTMENTS, Hartford Meeting, 1906 ....8vo, 3 00
Jamestown Meeting, 1907....................................8vo, 3 00
AUSTEN—Notes for Chemical Students............................12mo, 1 50
BERNADOU—Smokeless Powder.—Nitro-cellulose, and Theory of the Cellulose Molecule...12mo, 2 50
BILTZ—Introduction to Inorganic Chemistry. (HALL and PHELAN.)..12mo, *1 25
Laboratory Methods of Inorganic Chemistry. (HALL and BLANCHARD.) 8vo, 3 00
BINGHAM and WHITE—Laboratory Manual of Inorganic Chemistry....12mo, *1 00
BLANCHARD—Synthetic Inorganic Chemistry......................12mo, *1 00
BOTTLER—German and American Varnish Making. (SABIN.)... Small 8vo, *3 50
BROWNE—Handbook of Sugar Analysis............................8vo, *6 00
Sugar Tables for Laboratory Use..............................8vo, *1 25
BROWNING—Introduction to the Rarer Elements....................8vo, *1 50
BRUNSWIG—Explosives. (MUNROE and KIBLER).............Small 8vo, *3 00
CLAASSEN—Beet-sugar Manufacture. (HALL and ROLFE.)............8vo, *3 00
CLASSEN—Quantitative Analysis by Electrolysis. (HALL.).......(*In Press.*)

COHN—Indicators and Test-papers................................12mo, $2 00
Tests and Reagents..8vo, 3 00
COOPER—Constitutional Analysis by Physico-chemical Methods (Inorganic). (*In Press.*)
DANNEEL—Electrochemistry. (MERRIAM.)........................12mo, *1 25
DANNERTH—Methods of Textile Chemistry........................12mo, 2 00
DUHEM—Thermodynamics and Chemistry. (BURGESS.)..............8vo, 4 00
EISSLER—Modern High Explosives..................................8vo, 4 00
EKELEY—Laboratory Manual of Inorganic Chemistry..............12mo, *1 00
FLETCHER—Practical Instructions in Quantitative Assaying with the Blowpipe..16mo, mor., 1 50
FOWLER—Sewage Works Analyses...................................12mo, 2 00
FRESENIUS—Manual of Qualitative Chemical Analysis. (WELLS.).....8vo, 5 00
Manual of Qualitative Chemical Analysis. Part I. Descriptive. (WELLS.)..8vo, 3 00
Quantitative Chemical Analysis. (COHN.) 2 vols...............8vo, 12 50
When Sold Separately, Vol. I, $6. Vol. II, $8.
FUERTES—Water and Public Health...............................12mo, 1 50
FULLER—Qualitative Analysis of Medicinal Preparations...........12mo, *1 50
FURMAN and PARDOE—Manual of Practical Assaying..................8vo, 3 00
GETMAN—Exercises in Physical Chemistry.........................12mo, *2 00
GILL—Gas and Fuel Analysis for Engineers.......................12mo, 1 25
GOOCH—Methods in Chemical Analysis.............................8vo, *4 00
and BROWNING—Outlines of Qualitative Chemical Analysis.. Small 8vo, *1 25
GROTENFELT—Principles of Modern Dairy Practice. (WOLL.).......12mo, 2 00
GROTH—Introduction to Chemical Crystallography. (MARSHALL.)...12mo, 1 25
HAMMARSTEN—Text-book of Physiological Chemistry. (MANDEL.)....8vo, *4 00
HANAUSEK—Microscopy of Technical Products. (WINTON.)..........8vo, 5 00
HASKINS—Organic Chemistry..12mo, *2 00
HERRICK—Denatured or Industrial Alcohol..........................8vo, *4 00
HINDS—Inorganic Chemistry...8vo, 3 00
Laboratory Manual for Students.............................12mo, *1 00
HOLLEMAN—Laboratory Manual of Organic Chemistry for Beginners. (WALKER.)...12mo, *1 00
Text-book of Inorganic Chemistry. (COOPER.)..................8vo, 2 50
Text-book of Organic Chemistry. (WALKER AND MOTT.).........8vo, 2 50
(EKELEY) Laboratory Manual to Accompany Holleman's Text-book of Inorganic Chemistry...12mo, *1 00
HOLLEY—Analysis of Paint and Varnish Products.............Small, 8vo, *2 50
Lead and Zinc Pigments...................................Small 8vo, *3 00
HOPKINS—Oil-chemists' Handbook..................................8vo, 3 00
JACKSON—Directions for Laboratory Work in Physiological Chemistry..8vo, 1 25
JOHNSON—Rapid Methods for the Chemical Analysis of Special Steels, Steelmaking Alloys and Graphite...........................Small 8vo, 3 00
LANDAUER—Spectrum Analysis. (TINGLE.).........................8vo, 3 00
LASSAR-COHN—Application of Some General Reactions to Investigations in Organic Chemistry. (TINGLE.)..............................12mo, 1 00
LEACH—Food Inspection and Analysis..............................8vo, 7 50
LÖB—Electrochemistry of Organic Compounds. (LORENZ.)...........8vo, 3 00
LODGE—Notes on Assaying and Metallurgical Laboratory Experiments..8vo, 3 00
LOW—Technical Method of Ore Analysis.............................8vo, 3 00
LOWE—Paint for Steel Structures..................................12mo, 1 00
LUNGE—Techno-chemical Analysis. (COHN.).........................12mo, 1 00
MCKAY and LARSEN—Principles and Practice of Butter-making.......8vo, *1 50
MAIRE—Modern Pigments and their Vehicles........................12mo, 2 00
MANDEL—Handbook for Bio-chemical Laboratory....................12mo, 1 50
MARTIN—Laboratory Guide to Qualitative Analysis with the Blowpipe.12mo, *0 60
MASON—Examination of Water. (Chemical and Bacteriological.)....12mo, 1 25
Water-supply. (Considered Principally from a Sanitary Standpoint.) 8vo, 4 00
MATHEWSON—First Principles of Chemical Theory...................8vo, *1 00
MATTHEWS—Laboratory Manual of Dyeing and Textile Chemistry.....8vo, 3 50
Textile Fibres...8vo, 4 00
MEYER—Determination of Radicles in Carbon Compounds. (TINGLE.) Third Edition..12mo, *1 25
MILLER—Cyanide Process...12mo, 1 00
Manual of Assaying...12mo, 1 00

Minet—Production of Aluminum and its Industrial Use. (Waldo.)..12mo, $2 50
Mittelstädt—Technical Calculations for Sugar Works. (Bourbakis.) 12mo, *1 50
Mixter—Elementary Text-book of Chemistry......12mo, 1 50
Morgan—Elements of Physical Chemistry......12mo, 3 00
Physical Chemistry for Electrical Engineers......12mo, *1 50
Moore—Experiments in Organic Chemistry......12mo, *0 50
Outlines of Organic Chemistry......12mo, *1 50
Morse—Calculations used in Cane-sugar Factories......16mo, mor., 1 50
Muir—History of Chemical Theories and Laws......8vo, *4 00
Mulliken—General Method for the Identification of Pure Organic Compounds. Vol. I. Compounds of Carbon with Hydrogen and Oxygen......Large 8vo, 5 00
Vol. II. Nitrogenous Compounds......(*In Preparation.*)
Vol. III. The Commercial Dyestuffs......Large 8vo, 5 00
Nelson—Analysis of Drugs and Medicines......12mo, *3 00
Ostwald—Conversations on Chemistry. Part One. (Ramsey.)....12mo, 1 50
Part Two. (Turnbull.).12mo, 2 00
Introduction to Chemistry. (Hall and Williams.)......Small 8vo, *1 50
Owen and Standage—Dyeing and Cleaning of Textile Fabrics......12mo, 2 00
Palmer—Practical Test Book of Chemistry......12mo, *1 00
Pauli—Physical Chemistry in the Service of Medicine. (Fischer.)..12mo, *1 25
Pictet—Alkaloids and their Chemical Constitution. (Biddle.)......8vo, 5 00
Prescott and Winslow—Elements of Water Bacteriology, with Special Reference to Sanitary Water Analysis......12mo, 1 50
Reisig—Guide to Piece-Dyeing......8vo, *25 00
Richards and Woodman—Air, Water, and Food from a Sanitary Standpoint......8vo, 2 00
Ricketts and Miller—Notes on Assaying......8vo, 3 00
Rideal—Disinfection and the Preservation of Food......8vo, 4 00
Riggs—Elementary Manual for the Chemical Laboratory......8vo, 1 25
Robine and Lenglen—Cyanide Industry. (Le Clerc.)......8vo, 4 00
Ruddiman—Incompatibilities in Prescriptions......8vo, 2 00
Whys in Pharmacy......12mo, 1 00
Ruer—Elements of Metallography. (Mathewson.)......8vo, *3 00
Sabin—Industrial and Artistic Technology of Paint and Varnish......8vo, 3 00
Salkowski—Physiological and Pathological Chemistry. (Orndorff.)...8vo, 2 50
Schimpf—Essentials of Volumetric Analysis......Small 8vo, *1 50
Manual of Volumetric Analysis......8vo, 5 00
Qualitative Chemical Analysis......8vo, *1 25
Seamon—Manual for Assayers and Chemists......Small 8vo, *2 50
Smith—Lecture Notes on Chemistry for Dental Students......8vo, *2 50
Spencer—Handbook for Cane Sugar Manufacturers......16mo, mor., 3 00
Handbook for Chemists of Beet-sugar Houses......16mo, mor., 3 00
Stockbridge—Rocks and Soils......8vo, 2 50
Stone—Practical Testing of Gas and Gas Meters......8vo, 3 50
Tillman—Descriptive General Chemistry......8vo, *3 00
Elementary Lessons in Heat......8vo, *1 50
Treadwell—Qualitative Analysis. (Hall.)......8vo, 3 00
Quantitative Analysis. (Hall.)......8vo, 4 00
Turneaure and Russell—Public Water-supplies......8vo, 5 00
Van Deventer—Physical Chemistry for Beginners. (Boltwood.)..12mo, 1 50
Venable—Methods and Devices for Bacterial Treatment of Sewage...8vo, 3 00
Ward and Whipple—Freshwater Biology......(*In Press.*)
Ware—Beet-sugar Manufacture and Refining. Vol. I......8vo, 4 00
Vol. II......8vo, 5 00
Washington—Manual of the Chemical Analysis of Rocks......8vo, 2 00
Weaver—Military Explosives......8vo, *3 00
Wells—Laboratory Guide in Qualitative Chemical Analysis......8vo, 1 50
Short Course in Inorganic Qualitative Chemical Analysis for Engineering Students......12mo, 1 50
Text-book of Chemical Arithmetic......12mo, 1 25
Whipple—Microscopy of Drinking-water......8vo, 3 50
Wilson—Chlorination Process......12mo, 1 50
Cyanide Processes......12mo, 1 50
Winton—Microscopy of Vegetable Foods......8vo, 7 50
Zsigmondy—Colloids and the Ultramicroscope. (Alexander.). Large 12mo, 3 00

CIVIL ENGINEERING.

BRIDGES AND ROOFS. HYDRAULICS. MATERIALS OF ENGINEERING. RAILWAY ENGINEERING.

American Civil Engineers' Pocket Book. (Mansfield Merriman, Editor-in-chief.) .... 16mo, mor., *$5 00
Baker—Engineers' Surveying Instruments 12mo, 3 00
Bixby—Graphical Computing Table .... Paper 19½ × 24¼ inches, 0 25
Breed and Hosmer—Principles and Practice of Surveying.
Vol. I. Elementary Surveying 8vo, 3 00
Vol. II. Higher Surveying 8vo, 2 50
Burr—Ancient and Modern Engineering and the Isthmian Canal 8vo, *3 50
Comstock—Field Astronomy for Engineers 8vo, 2 50
Corthell—Allowable Pressure on Deep Foundations 12mo, *1 25
Crandall—Text-book on Geodesy and Least Squares 8vo, 3 00
Davis—Elevation and Stadia Tables 8vo, 1 00
Elliott—Engineering for Land Drainage 12mo, 2 00
Fiebeger—Treatise on Civil Engineering 8vo, *5 00
Flemer—Phototopographic Methods and Instruments 8vo, 5 00
Folwell—Sewerage. (Designing and Maintenance.) 8vo, 3 00
Freitag—Architectural Engineering 8vo, 3 50
Hauch and Rice—Tables of Quantities for Preliminary Estimates ... 12mo, *1 25
Hayford—Text-book of Geodetic Astronomy 8vo, 3 00
Hering—Ready Reference Tables (Conversion Factors.) .... 16mo, mor., 2 50
Hosmer—Azimuth .... 16mo, mor., 1 00
Text-book on Practical Astronomy 8vo, *2 00
Howe—Retaining Walls for Earth 12mo, 1 25
Ives—Adjustments of the Engineer's Transit and Level .... 16mo, bds., *0 25
Ives and Hilts—Problems in Surveying, Railroad Surveying and Geodesy .... 16mo, mor., 1 50
Johnson (J. B.) and Smith—Theory and Practice of Surveying. Small 8vo, *3 50
Johnson (L. J.)—Statics by Algebraic and Graphic Methods 8vo, 2 00
Kinnicutt, Winslow and Pratt—Sewage Disposal 8vo, *3 00
Mahan—Descriptive Geometry 8vo, *1 50
Merriman—Elements of Precise Surveying and Geodesy 8vo, 2 50
Merriman and Brooks—Handbook for Surveyors .... 16mo, mor., 2 00
Nugent—Plane Surveying 8vo, 3 50
Ogden—Sewer Construction 8vo, 3 00
Sewer Design 12mo, 2 00
Ogden and Cleveland—Practical Methods of Sewage Disposal for Residences, Hotels, and Institutions 8vo, *1 50
Parsons—Disposal of Municipal Refuse 8vo, 2 00
Patton—Treatise on Civil Engineering .... 8vo, half leather, 7 50
Reed—Topographical Drawing and Sketching 4to, 5 00
Riemer—Shaft-sinking under Difficult Conditions. (Corning and Peele.) 8vo, 3 00
Siebert and Biggin—Modern Stone-cutting and Masonry 8vo, 1 50
Smith—Manual of Topographical Drawing. (McMillan.) 8vo, 2 50
Soper—Air and Ventilation of Subways 12mo, 2 50
Tracy—Exercises in Surveying .... 12mo, mor., *1 00
Plane Surveying .... 16mo, mor., 3 00
Venable—Garbage Crematories in America 8vo, 2 00
Methods and Devices for Bacterial Treatment of Sewage 8vo, 3 00
Wait—Engineering and Architectural Jurisprudence 8vo, 6 00
Sheep, 6 50
Law of Contracts 8vo, 3 00
Law of Operations Preliminary to Construction in Engineering and Architecture 8vo, 5 00
Sheep, 5 50
Warren—Stereotomy—Problems in Stone-cutting 8vo, 2 50
Waterbury—Vest-Pocket Hand-book of Mathematics for Engineers.
2⅞ × 5⅜ inches, mor., *1 00
Enlarged Edition, Including Tables .... mor., *1 50
Webb—Problems in the Use and Adjustment of Engineering Instruments.
16mo, mor., 1 25
Wilson—Topographic, Trigonometric and Geodetic Surveying 8vo, 3 50

BRIDGES AND ROOFS.

BISHOP—Structural Details of Hip and Valley Rafters... Oblong large 8vo. *$1 75
BOLLER—Practical Treatise on the Construction of Iron Highway Bridges 8vo, 2 00
Thames River Bridge..........Oblong paper, *5 00
BURR and FALK—Design and Construction of Metallic Bridges........8vo, 5 00
Influence Lines for Bridge and Roof Computations..........8vo, 3 00
DU BOIS—Mechanics of Engineering. Vol. II..........Small 4to, 10 00
FOSTER—Treatise on Wooden Trestle Bridges..........4to, 5 00
FOWLER—Ordinary Foundations..........8vo, 3 50
GREENE—Arches in Wood, Iron, and Stone..........8vo, 2 50
Bridge Trusses..........8vo, 2 50
Roof Trusses..........8vo, 1 25
GRIMM—Secondary Stresses in Bridge Trusses..........8vo, 2 50
HELLER—Stresses in Structures and the Accompanying Deformations..8vo, 3 00
HOWE—Design of Simple Roof-trusses in Wood and Steel..........8vo, 2 00
Symmetrical Masonry Arches..........8vo, 2 50
Treatise on Arches..........8vo, 4 00
HUDSON—Deflections and Statically Indeterminate Stresses......Small 4to, *3 50
Plate Girder Design..........8vo, *1 50
JACOBY—Structural Details, or Elements of Design in Heavy Framing, 8vo, *2 25
JOHNSON, BRYAN and TURNEAURE—Theory and Practice in the Designing of Modern Framed Structures. New Edition.
Part I. Stresses in Simple Structures..........8vo, *3 00
Part II. Statically Indeterminate Structures and Secondary Stresses 8vo, *4 00
MERRIMAN and JACOBY—Text-book on Roofs and Bridges:
Part I. Stresses in Simple Trusses..........8vo, 2 50
Part II. Graphic Statics..........8vo, 2 50
Part III. Bridge Design..........8vo, 2 50
Part IV. Higher Structures..........8vo, 2 50
RICKER—Design and Construction of Roofs..........8vo, 5 00
SONDERICKER—Graphic Statics, with Applications to Trusses, Beams, and Arches..........8vo, *2 00
WADDELL—De Pontibus, Pocket-book for Bridge Engineers....16mo, mor., 2 00
Specifications for Steel Bridges..........12mo, *0 50

HYDRAULICS.

BARNES—Ice Formation..........8vo, 3 00
BAZIN—Experiments upon the Contraction of the Liquid Vein Issuing from an Orifice. (TRAUTWINE.)..........8vo, 2 00
BOVEY—Treatise on Hydraulics..........8vo, 5 00
CHURCH—Diagrams of Mean Velocity of Water in Open Channels. Oblong 4to, paper, 1 50
Hydraulic Motors..........8vo, 2 00
Mechanics of Fluids (Being Part IV of Mechanics of Engineering)..8vo, 3 00
COFFIN—Graphical Solution of Hydraulic Problems..........16mo, mor., 2 50
FLATHER—Dynamometers, and the Measurement of Power..........12mo, 3 00
FOLWELL—Water-supply Engineering..........8vo, 4 00
FRIZELL—Water-power..........8vo, 5 00
FUERTES—Water and Public Health..........12mo, 1 50
FULLER—Domestic Water Supplies for the Farm..........8vo, *1 50
GANGUILLET and KUTTER—General Formula for the Uniform Flow of Water in Rivers and Other Channels. (HERING and TRAUTWINE.)...8vo, 4 00
HAZEN—Clean Water and How to Get It..........Small 8vo, 1 50
Filtration of Public Water-supplies..........8vo, 3 00
HAZELHURST—Towers and Tanks for Water-works..........8vo, 2 50
HERSCHEL—115 Experiments on the Carrying Capacity of Large, Riveted, Metal Conduits..........8vo, 2 00
HOYT and GROVER—River Discharge..........8vo, 2 00
HUBBARD and KIERSTED—Water-works Management and Maintenance, 8vo, 4 00
LYNDON—Development and Electrical Distribution of Water Power..8vo, *3 00
MASON—Water-supply. (Considered Principally from a Sanitary Standpoint.)..........8vo, 4 00
MERRIMAN—Elements of Hydraulics..........12mo, *1 00
Treatise on Hydraulics. 9th Edition, Rewritten..........8vo, *4 00

MOLITOR—Hydraulics of Rivers, Weirs and Sluices..................8vo, *$2 00
MORRISON and BRODIE—High Masonry Dam Design..................8vo, *1 50
SCHUYLER—Reservoirs for Irrigation, Water-power, and Domestic Water supply. Second Edition, Revised and Enlarged.......Large 8vo, 6 00
THOMAS and WATT—Improvement of Rivers........................4to, *6 00
TURNEAURE and RUSSELL—Public Water-supplies....................8vo, 5 00
WEGMANN—Design and Construction of Dams. 6th Ed., enlarged.....4to, *6 00
Water Supply of the City of New York from 1658 to 1895.........4to, 10 00
WHIPPLE—Value of Pure Water..............................Small 8vo, 1 00
WILLIAMS and HAZEN—Hydraulic Tables..........................8vo, 1 50
WILSON—Irrigation Engineering................................8vo, 4 00
WOOD—Turbines..8vo, 2 50

MATERIALS OF ENGINEERING.

BAKER—Roads and Pavements...................................8vo, 5 00
Treatise on Masonry Construction.............................8vo, 5 00
BLACK—United States Public Works........................Oblong 4to, 5 00
BLANCHARD—Bituminous Surfaces and Bituminous Pavements. (*In Preparation.*)
and DROWNE—Highway Engineering, as Presented at the Second International Road Congress, Brussels, 1910..................8vo, *2 00
Text-book on Highway Engineering.......................(*In Press.*)
BOTTLER—German and American Varnish Making. (SABIN.)...Small 8vo, *3 50
BURR—Elasticity and Resistance of the Materials of Engineering......8vo, 7 50
BYRNE—Highway Construction..................................8vo, 5 00
Inspection of the Materials and Workmanship Employed in Construction. 16mo, 3 00
CHURCH—Mechanics of Engineering..............................8vo, 6 00
Mechanics of Solids (Being Parts I, II, III of Mechanics of Engineering)..8vo, 4 50
Mechanics of Fluids (Being Part IV of Mechanics of Engineering).8vo, 3 00
DU BOIS—Mechanics of Engineering:
Vol. I. Kinematics, Statics, Kinetics.....................Small 4to, 7 50
Vol. II. The Stresses in Framed Structures, Strength of Materials and Theory of Flexures.............................Small 4to, 10 00
ECKEL—Building Stones and Clays..............................8vo, *3 00
Cements, Limes, and Plasters..................................8vo, *6 00
FOWLER—Ordinary Foundations..................................8vo, 3 50
FULLER and JOHNSTON—Applied Mechanics:
Vol. I. Theory of Statics and Kinetics...................(*In Press.*)
Vol. II. Strength of Materials.................... (*In Preparation.*)
GREENE—Structural Mechanics...................................8vo, *2 50
HOLLEY—Analysis of Paint and Varnish Products.............Small 8vo, *2 50
Lead and Zinc Pigments.................................Small 8vo, *3 00
HUBBARD—Dust Preventives and Road Binders.....................8vo, *3 00
JOHNSON (J. B.)—Materials of Construction.................Large 8vo, 6 00
KEEP—Cast Iron..8vo, 2 50
KING—Elements of the Mechanics of Materials and of Power of Transmission..8vo, *2 50
LANZA—Applied Mechanics.....................................8vo, 7 50
LOWE—Paints for Steel Structures.............................12mo, 1 00
MAIRE—Modern Pigments and their Vehicles......................12mo, 2 00
MAURER—Technical Mechanics....................................8vo, 4 00
MERRILL—Stones for Building and Decoration.....................8vo, 5 00
MERRIMAN—Mechanics of Materials................................8vo, 5 00
Strength of Materials..12mo, *1 00
METCALF—Steel. A Manual for Steel-users.......................12mo, 2 00
MORRISON—Highway Engineering....................................8vo, 2 50
MURDOCK—Strength of Materials.................................12mo, *2 00
PATTON—Practical Treatise on Foundations........................8vo, 5 00
RICE—Concrete Block Manufacture..................................8vo, 2 00
RICHARDSON—Modern Asphalt Pavement..............................8vo, 3 00
RICHEY—Building Foreman's Pocket Book and Ready Reference. 16mo, mor., 5 00
Cement Workers' and Plasterers' Edition (Building Mechanics' Ready Reference Series)...............................16mo, mor., *1 50
Handbook for Superintendents of Construction..........16mo, mor., 4 00
Stone and Brick Masons' Edition (Building Mechanics' Ready Reference Series)...16mo, mor., *1 50

RIES—Building Stones and Clay Products..........................8vo,*$3 00
Clays: Their Occurrence, Properties, and Uses..................8vo, *5 00
and LEIGHTON—History of the Clay-working Industry of the United States..8vo, *2 50
SABIN—Industrial and Artistic Technology of Paint and Varnish.......8vo, 3 00
SMITH—Strength of Material....................................12mo, *1 25
SNOW—Principal Species of Wood..................................8vo, 3 50
SPALDING—Hydraulic Cement.......................................12mo, 2 00
Text-book on Road and Pavements..........................12mo, *2 00
TAYLOR and THOMPSON—Concrete Costs.......................Small 8vo, *5 00
Extracts on Reinforced Concrete Design........................8vo, *2 00
Treatise on Concrete, Plain and Reinforced.....................8vo, 5 00
THURSTON—Materials of Engineering. In Three Parts..............8vo, 8 00
Part I. Non-metallic Materials of Engineering and Metallurgy...8vo, 2 00
Part II. Iron and Steel..8vo, 3 50
Part III. A Treatise on Brasses, Bronzes, and Other Alloys and their Constituents..8vo, 2 50
TILLSON—Street Pavements and Paving Materials...................8vo, *4 00
TURNEAURE and MAURER—Principles of Reinforced Concrete Construction. 8vo, 3 50
WATERBURY—Cement Laboratory Manual........................12mo, 1 00
Laboratory Manual for Testing Materials of Construction.......12mo, *1 50
WOOD (DE V.) Treatise on the Resistance of Materials, and an Appendix on the Preservation of Timber.................................8vo, 2 00
(M. P.)—Rustless Coatings: Corrosion and Electrolysis of Iron and Steel..8vo, 4 00

RAILWAY ENGINEERING.

BERG—Buildings and Structures of American Railroads..............4to, 5 00
BROOKS—Handbook of Street Railroad Location..............16mo, mor., 1 50
BURT—Railway Station Service..................................12mo, *2 00
BUTTS—Civil Engineer's Field-book..........................16mo, mor., 2 50
CRANDALL—Railway and Other Earthwork Tables...................8vo, 1 50
and BARNES—Railroad Surveying..........................16mo, mor., 2 00
CROCKETT—Methods for Earthwork Computations...................8vo, *1 50
DREDGE—History of the Pennsylvania Railroad. (1879)...........Paper, 5 00
FISH—Earthwork Haul and Overhaul.........................(*In Press.*)
FISHER—Table of Cubic Yards..............................Cardboard, 0 25
GILBERT, WIGHTMAN and SAUNDERS—Subways and Tunnels of New York. 8vo, *4 00
GODWIN—Railroad Engineers' Field-book and Explorers' Guide..16mo, mor., 2 50
HUDSON—Tables for Calculating the Cubic Contents of Excavations and Embankments...8vo, 1 00
IVES and HILTS—Problems in Surveying, Railroad Surveying and Geodesy. 16mo, mor., 1 50
MOLITOR and BEARD—Manual for Resident Engineers..............16mo, 1 00
NAGLE—Field Manual for Railroad Engineers................16mo, mor., 3 00
ORROCK—Railroad Structures and Estimates........................8vo, *3 00
PHILBRICK—Field Manual for Engineers......................16mo, mor., 3 00
RAYMOND—Elements of Railroad Engineering.........................8vo, 3 50
Railroad Engineer's Field Book....................(*In Preparation.*)
Railroad Field Geometry..................................16mo, mor., 2 00
ROBERTS—Track Formulæ and Tables...........................16mo, mor., 3 00
SEARLES—Field Engineering..................................16mo, mor., 3 00
Railroad Spiral..16mo, mor., 1 50
TAYLOR—Prismoidal Formulæ and Earthwork..........................8vo, 1 50
WEBB—Economics of Railroad Construction....................Small 8vo, 2 50
Railroad Construction.....................................16mo, mor., 5 00
WELLINGTON—Economic Theory of the Location of Railways....Small 8vo, 5 00
WILSON—Elements of Railroad-Track and Construction............12mo, 2 00

DRAWING.

BARR and WOOD—Kinematics of Machinery	8vo,	$2
BARTLETT—Mechanical Drawing. Third Edition	8vo,	*3
" " Abridgment of the Second Edition	8vo,	*1
and JOHNSON—Engineering Descriptive Geometry	8vo,	*1
BISHOP—Structural Details of Hip and Valley Rafters	Oblong large 8vo,	*1
BLESSING and DARLING—Descriptive Geometry	8vo,	*1
Elements of Drawing	8vo,	*1
COOLIDGE—Manual of Drawing	8vo, paper,	1
and FREEMAN—Elements of General Drafting for Mechanical Engineers	Oblong 4to,	2
DURLEY—Kinematics of Machines	8vo,	4
EMCH—Introduction to Projective Geometry and its Application	8vo,	2
FRENCH and IVES—Stereotomy	8vo,	2
HILL—Text-book on Shades and Shadows, and Perspective	8vo,	2
JAMISON—Advanced Mechanical Drawing	8vo,	*2
Elements of Mechanical Drawing	8vo,	2
JONES—Machine Design:		
Part I. Kinematics of Machinery	8vo,	1
Part II. Form, Strength, and Proportions of Parts	8vo,	3
KIMBALL and BARR—Machine Design	8vo,	*3
MACCORD—Elements of Descriptive Geometry	8vo,	3
Kinematics; or, Practical Mechanism	8vo,	5
Mechanical Drawing	4to,	4
Velocity Diagrams	8vo,	1
MCLEOD—Descriptive Geometry	Small 8vo,	1
MAHAN—Descriptive Geometry and Stone-cutting	8vo,	*1
Industrial Drawing. (THOMPSON.)	8vo,	3
MOYER—Descriptive Geometry	8vo,	2
REED—Topographical Drawing and Sketching	4to,	5
REID—Mechanical Drawing. (Elementary and Advanced.)	8vo,	*2
Text-book of Mechanical Drawing and Elementary Machine Design	8vo,	3
ROBINSON—Principles of Mechanism	8vo,	3
SCHWAMB and MERRILL—Elements of Mechanism	8vo,	3
SMITH (A. W.) and MARX—Machine Design	8vo,	3
(R. S.)—Manual of Topographical Drawing. (MCMILLAN.)	8vo,	2
TITSWORTH—Elements of Mechanical Drawing	Oblong large 8vo,	*1
WARREN—Elements of Descriptive Geometry, Shadows, and Perspective	8vo,	3
Elements of Machine Construction and Drawing	8vo,	7
Elements of Plane and Solid Free-hand Geometrical Drawing	12mo,	1
General Problems of Shades and Shadows	8vo,	3
Manual of Elementary Problems in the Linear Perspective of Forms and Shadows	12mo,	1
Manual of Elementary Projection Drawing	12mo,	1
Plane Problems in Elementary Geometry	12mo,	1
WEISBACH—Kinematics and Power of Transmission. (HERRMANN and KLEIN.)	8vo,	5
WILSON (H. M.)—Topographic, Trigonometric and Geodetic Surveying	8vo,	3
(V. T.) Descriptive Geometry	8vo,	*1
Free-hand Lettering	8vo,	1
Free-hand Perspective	8vo,	2
WOOLF—Elementary Course in Descriptive Geometry	Large 8vo,	3

ELECTRICITY AND PHYSICS.

ABEGG—Theory of Electrolytic Dissociation. (VON ENDE.)	12mo,	*1
ANDREWS—Hand-book for Street Railway Engineers	3×5 inches, mor.,	1
ANTHONY and BALL—Lecture-notes on the Theory of Electrical Measurements	12mo,	1
and BRACKETT—Text-book of Physics. (MAGIE.)	Small 8vo,	3
BENJAMIN—History of Electricity	8vo,	3
BETTS—Lead Refining and Electrolysis	8vo,	4
BURGESS and LE CHATELIER—Measurement of High Temperatures. Third Edition	8vo,	*4
CLASSEN—Quantitative Analysis by Electrolysis. (HALL.)	*(In Press.)*	

COLLINS—Manual of Wireless Telegraphy and Telephony............12mo, *$1 50
CREHORE and SQUIER—Polarizing Photo-chronograph................8vo, 3 00
DANNEEL—Electrochemistry. (MERRIAM.)........................12mo, *1 25
DAWSON—"Engineering" and Electric Traction Pocket-book..16mo, mor., 5 00
DOLEZALEK—Theory of the Lead Accumulator (Storage Battery). (VON ENDE.)..........12mo, 2 50
DUHEM—Thermodynamics and Chemistry. (BURGESS.)..............8vo, 4 00
FLATHER—Dynamometers, and the Measurement of Power..........12mo, 3 00
GETMAN—Introduction to Physical Science.........................12mo, *1 50
GILBERT—De Magnete. (MOTTELAY.)...............................8vo, 2 50
HANCHETT—Alternating Currents.................................12mo, *1 00
HERING—Ready Reference Tables (Conversion Factors)........16mo, mor., 2 50
HOBART and ELLIS—High-speed Dynamo Electric Machinery..........8vo, *6 00
HOLMAN—Precision of Measurements...............................8vo, 2 00
Telescope-Mirror-scale Method, Adjustments, and Tests....Large 8vo, 0 75
HUTCHINSON—High-Efficiency Electrical Illuminants and Illumination. Small 8vo, *2 50
JONES—Electric Ignition for Combustion Motors....................8vo, *4 00
KARAPETOFF—Experimental Electrical Engineering:
Vol. I..8vo, *3 50
Vol. II...8vo, *2 50
KINZBRUNNER—Testing of Continuous-current Machines..............8vo, 2 00
KOCH—Mathematics of Applied Electricity..................Small 8vo, *3 00
LANDAUER—Spectrum Analysis. (TINGLE.)..........................8vo, 3 00
LAUFFER—Electrical Injuries....................................16mo, *0 50
LÖB—Electrochemistry of Organic Compounds. (LORENZ.)............8vo, 3 00
LYNDON—Development and Electrical Distribution of Water Power....8vo, *3 00
LYONS—Treatise on Electromagnetic Phenomena. Vols. I and II, 8vo, each, *6 00
MARTIN—Measurement of Induction Shocks.........................12mo, *1 25
MICHIE—Elements of Wave Motion Relating to Sound and Light......8vo, *4 00
MORGAN—Physical Chemistry for Electrical Engineers.............12mo, *1 50
NORRIS—Introduction to the Study of Electrical Engineering.........8vo, *2 50
PARSHALL and HOBART—Electric Machine Design..........4to, half mor., *12 50
REAGAN—Locomotives: Simple, Compound, and Electric........Small 8vo, 3 50
RODENHAUSER and SCHOENAWA—Electric Furnaces in the Iron and Steel Industry (VOM BAUR.)..........................(In Press.)
ROSENBERG—Electrical Engineering. (HALDANE GEE—KINZBRUNNER.).8vo, *2 00
RYAN—Design of Electrical Machinery:
Vol. I. Direct Current Dynamos..................................8vo, *1 50
Vol. II. Alternating Current Transformers........................8vo, *1 50
Vol. III. Alternators, Synchronous Motors, and Rotary Converters. 8vo, *1 50
SCHAPPER—Laboratory Guide for Students in Physical Chemistry....12mo, 1 00
TILLMAN—Elementary Lessons in Heat.............................8vo, *1 50
TIMBIE—Answers to Problems in Elements of Electricity.......12mo, Paper, *0 25
Elements of Electricity.................................Small 8vo, *2 00
Essentials of Electricity.......................................12mo, *1 25
TORY and PITCHER—Manual of Laboratory Physics...........Small 8vo, 2 00
ULKE—Modern Electrolytic Copper Refining........................8vo, 3 00
WATERS—Commercial Dynamo Design................................8vo, *2 00

LAW.

BRENNAN—Hand-book of Useful Legal Information for Business Men. 16mo, mor., *5 00
DAVIS—Elements of Law...8vo, *2 50
Treatise on the Military Law of United States....................8vo, *7 00
DUDLEY—Military Law and the Procedure of Courts-martial...Small 8vo, *2 50
MANUAL FOR COURTS MARTIAL.............................16mo, mor., 1 50
WAIT—Engineering and Architectural Jurisprudence..................8vo, 6 00
Sheep, 6 50
Law of Contracts..8vo, 3 00
Law of Operations Preliminary to Construction in Engineering and Architecture..8vo, 5 00
Sheep, 5 50

MATHEMATICS.

Baker—Elliptic Functions..8vo, $1 50
Briggs—Elements of Plane Analytic Geometry. (Bôcher.)........12mo, 1 00
Buchanan—Plane and Spherical Trigonometry.....................8vo, *1 00
Byerly—Harmonic Functions...8vo, 1 00
Chandler—Elements of the Infinitesimal Calculus................12mo, 2 00
Coffin—Vector Analysis...12mo, *2 50
Compton—Manual of Logarithmic Computations..................12mo, 1 50
Dickson—College Algebra.................................Small 8vo, *1 50
Introduction to the Theory of Algebraic Equations.........Small 8vo, *1 25
Emch—Introduction to Projective Geometry and its Application.......8vo, 2 50
Fiske—Functions of a Complex Variable.........................8vo, 1 00
Halsted—Elementary Synthetic Geometry........................8vo, 1 50
Elements of Geometry..8vo, 1 75
Rational Geometry...12mo, *1 50
Synthetic Projective Geometry.....................................8vo, 1 00
Hancock—Lectures on the Theory of Elliptic Functions..............8vo, *5 00
Hyde—Grassmann's Space Analysis.................................8vo, 1 00
Johnson (J. B.) Three-place Logarithmic Tables: Vest-pocket size, paper, *0 15
100 copies, *5 00
Mounted on heavy cardboard, 8 ×10 inches, *0 25
10 copies, *2 00
(W. W.) Abridged Editions of Differential and Integral Calculus.
Small 8vo, 1 vol., 2 50
Curve Tracing in Cartesian Co-ordinates.........................12mo, 1 00
Differential Equations...8vo, 1 00
Elementary Treatise on Differential Calculus..............Small 8vo, 1 50
Elementary Treatise on the Integral Calculus.............Small 8vo, 1 50
Theoretical Mechanics...12mo, *3 00
Theory of Errors and the Method of Least Squares............12mo, 1 50
Treatise on Differential Calculus..........................Small 8vo, 3 00
Treatise on the Integral Calculus..........................Small 8vo, 3 00
Treatise on Ordinary and Partial Differential Equations ...Small 8vo, 3 50
Karapetoff—Engineering Applications of Higher Mathematics:
Part I. Problems on Machine Design.....................Small 8vo, *0 75
Koch—Mathematics of Applied Electricity.................Small 8vo, *3 00
Laplace—Philosophical Essay on Probabilities. (Truscott and Emory.)
12mo, 2 00
Le Messurier—Key to Professor W. W. Johnson's Differential Equations.
Small 8vo, *1 75
Ludlow—Logarithmic and Trigonometric Tables....................8vo, *1 00
and Bass—Elements of Trigonometry and Logarithmic and Other
Tables..8vo, *3 00
Trigonometry and Tables published separately.................Each, *2 00
Macfarlane—Vector Analysis and Quaternions......................8vo, 1 00
McMahon—Hyperbolic Functions.....................................8vo, 1 00
Manning—Irrational Numbers and their Representation by Sequences and
Series..12mo, 1 25
Marsh—Industrial Mathematics..............................Small 8vo, *2 00
Mathematical Monographs. Edited by Mansfield Merriman and
Robert S. Woodward............................Octavo, each, 1 00

No. 1. History of Modern Mathematics, by David Eugene Smith.
No. 2. Synthetic Projective Geometry, by George Bruce Halsted.
No. 3. Determinants, by Laenas Gifford Weld.
No. 4. Hyperbolic Functions, by James McMahon.
No. 5. Harmonic Functions, by William E. Byerly.
No. 6. Grassmann's Space Analysis, by Edward W. Hyde.
No. 7. Probability and Theory of Errors, by Robert S. Woodward.
No. 8. Vector Analysis and Quaternions, by Alexander Macfarlane.
No. 9. Differential Equations, by William Woolsey Johnson.
No. 10. The Solution of Equations, by Mansfield Merriman.
No. 11. Functions of a Complex Variable, by Thomas S. Fiske.

Maurer—Technical Mechanics..8vo, 4 00
Merriman—Method of Least Squares................................8vo, 2 00
Solution of Equations..8vo, 1 00
Moritz—Elements of Plane Trigonometry...........................8vo, *2 00
High School Edition..................................Small 8vo, *1 00

RICE and JOHNSON—Differential and Integral Calculus. 2 vols. in one. Small 8vo, $1 50

Elementary Treatise on the Differential Calculus..........Small 8vo, 3 00

SMITH—History of Modern Mathematics..........................8vo, 1 00

VEBLEN and LENNES—Introduction to the Real Infinestimal Analysis of One Variable..8vo, *2 00

WATERBURY—Vest Pocket Hand-book of Mathematics for Engineers. 2¼×5⅜ inches, mor., *1 00

Enlarged Edition, Including Tables..........................mor., *1 50

WELD—Determinants..8vo, 1 00

WOOD—Elements of Co-ordinate Geometry.........................8vo, 2 00

WOODWARD—Probability and Theory of Errors.....................8vo, 1 00

MECHANICAL ENGINEERING.

MATERIALS OF ENGINEERING, STEAM-ENGINES AND BOILERS.

BACON—Forge Practice..12mo, 1 50

BALDWIN—Steam Heating for Buildings...........................12mo, 2 50

BARR and WOOD—Kinematics of Machinery.........................8vo, 2 50

BARTLETT—Mechanical Drawing. Third Edition....................8vo, 3 00

Abridgment of the Second Edition....8vo, *1 50

BURR—Ancient and Modern Engineering and the Isthmian Canal......8vo, *3 50

CARPENTER—Heating and Ventilating Buildings....................8vo, 4 00

and DIEDERICHS—Experimental Engineering 8vo, *6 00

CLERK—The Gas, Petrol and Oil Engine..........................8vo, *4 00

COMPTON—First Lessons in Metal Working........................12mo, 1 50

and DE GROODT—Speed Lathe.....................................12mo, 1 50

COOLIDGE—Manual of Drawing..............................8vo, paper, 1 00

and FREEMAN—Elements of General Drafting for Mechanical Engineers.......................................Oblong 4to, 2 50

CROMWELL—Treatise on Belts and Pulleys........................12mo, 1 50

Treatise on Toothed Gearing...................................12mo, 1 50

DINGEY—Machinery Pattern Making...............................12mo, 2 00

DURLEY—Kinematics of Machines.................................8vo, 4 00

FLANDERS—Gear-cutting Machinery.........................Small 8vo, 3 00

FLATHER—Dynamometers and the Measurement of Power..........12mo, 3 00

Rope Driving..12mo, 2 00

FULLER and JOHNSTON—Applied Mechanics:

Vol. I. Theory of Statics and Kinetics..................(*In Press.*)

Vol. II. Strength of Materials....................(*In Preparation.*)

GILL—Gas and Fuel Analysis for Engineers......................12mo, 1 25

GOSS—Locomotive Sparks..8vo, 2 00

GREENE—Elements of Heating and Ventilation....................8vo, *2 50

Pumping Machinery..8vo, *4 00

HERING—Ready Reference Tables (Conversion Factors).......16mo, mor., 2 50

HOBART and ELLIS—High Speed Dynamo Electric Machinery.........8vo, *6 00

HUTTON—Gas Engine...8vo, 5 00

JAMISON—Advanced Mechanical Drawing...........................8vo, 2 00

Elements of Mechanical Drawing................................8vo, 2 50

JONES—Gas Engine..8vo, 4 00

Machine Design:

Part I. Kinematics of Machinery................................8vo, 1 50

Part II. Form, Strength, and Proportions of Parts..............8vo, 3 00

KAUP—Machine Shop Practice..............................Small 8vo, *1 25

KENT—Mechanical Engineers' Pocket-Book.................16mo, mor., *5 00

KERR—Power and Power Transmission..............................8vo, 2 00

KIMBALL and BARR—Machine Design................................8vo, *3 00

KING—Elements of the Mechanics of Materials and of Power of Transmission...8vo, *2 50

LANZA—Dynamics of Machinery....................................8vo, *2 50

LEONARD—Machine Shop Tools and Methods.........................8vo, 4 00

LEVIN—Modern Gas Engine and the Gas Producer...................8vo, *4 00

LORENZ—Modern Refrigerating Machinery. (POPE, HAVEN, and DEAN.) 8vo, *4 00

MacCord—Kinematics; or, Practical Mechanism 8vo, $5 00
Mechanical Drawing 4to, 4 00
Velocity Diagrams 8vo, 1 50
MacFarland—Standard Reduction Factors for Gases 8vo, 1 50
Mahan—Industrial Drawing. (Thompson.) 8vo, 3 50
Mehrtens—Gas Engine Theory and Design .... Small 8vo, 2 50
Oberg—Handbook of Small Tools .... Small 8vo, 2 50
Parshall and Hobart—Electric Machine Design ... Small 4to, half leather, *12 50
Peele—Compressed Air Plant. Second Edition, Revised and Enlarged. 8vo, *3 50
Poole—Calorific Power of Fuels 8vo, 3 00
Porter—Engineering Reminiscences, 1855 to 1882 .... 8vo, *3 00
Reid—Mechanical Drawing. (Elementary and Advanced.) 8vo, *2 00
Text-book of Mechanical Drawing and Elementary Machine Design. 8vo, 3 00
Richards—Compressed Air 12mo, 1 50
Robinson—Principles of Mechanism 8vo, 3 00
Schwamb and Merrill—Elements of Mechanism 8vo, 3 00
Smith (O.)—Press-working of Metals 8vo, 3 00
(A. W.) and Marx—Machine Design 8vo, 3 00
Sorel—Carbureting and Combustion in Alcohol Engines. (Woodward and Preston.) .... Small 8vo, 3 00
Stone—Practical Testing of Gas and Gas Meters 8vo, 3 50
Thurston—Animal as a Machine and Prime Motor, and the Laws of Energetics 12mo, 1 00
Treatise on Friction and Lost Work in Machinery and Mill Work .. 8vo, 3 00
Tillson—Complete Automobile Instructor 16mo, *1 50
Titsworth—Elements of Mechanical Drawing .... Oblong 8vo, *1 25
Warren—Elements of Machine Construction and Drawing 8vo, 7 50
Waterbury—Vest Pocket Hand-book of Mathematics for Engineers. $2\frac{7}{8}\times5\frac{3}{8}$ inches, mor., *1 00
Enlarged Edition, Including Tables .... mor., *1 50
Weisbach—Kinematics and the Power of Transmission. (Herrmann—Klein.) 8vo, 5 00
Machinery of Transmission and Governors. (Herrmann—Klein.). 8vo, 5 00
Wood—Turbines 8vo, 2 50

MATERIALS OF ENGINEERING.

Bottler—German and American Varnish Making. (Sabin.) ... Small 8vo, *3 50
Burr—Elasticity and Resistance of the Materials of Engineering 8vo, 7 50
Church—Mechanics of Engineering 8vo, 6 00
Mechanics of Solids (Being Parts I, II, III of Mechanics of Engineering). 8vo, 4 50
Fuller and Johnston—Applied Mechanics:
Vol. I. Theory of Statics and Kinetics .... (*In Press.*)
Vol. II. Strength of Materials .... (*In Preparation.*)
Greene—Structural Mechanics 8vo, *2 50
Holley—Analysis of Paint and Varnish Products .... Small 8vo, *2 50
Lead and Zinc Pigments .... Small 8vo, *3 00
Johnson (C. M.)—Rapid Methods for the Chemical Analysis of Special Steels, Steel-making Alloys and Graphite .... Small 8vo, 3 00
(J. B.) Materials of Construction 8vo, 6 00
Keep—Cast Iron 8vo, 2 50
King—Elements of the Mechanics of Materials and of Power of Transmission 8vo, *2 50
Lanza—Applied Mechanics 8vo, 7 50
Lowe—Paints for Steel Structures 12mo, 1 00
Maire—Modern Pigments and their Vehicles 12mo, 2 00
Martin—Text-Book of Mechanics:
Vol. I. Statics 12mo, *1 25
Vol. II. Kinematics and Kinetics 12mo, *1 50
Vol. III. Mechanics of Materials 12mo, *1 50
Vol. IV. Applied Statics 12mo, *1 50
Maurer—Technical Mechanics 8vo, 4 00
Merriman—Mechanics of Materials 8vo, 5 00
Strength of Materials 12mo, *1 00
Metcalf—Steel. A Manual for Steel-users 12mo, 2 00
Murdock—Strength of Materials 12mo, *2 00

SABIN—Industrial and Artistic Technology of Paint and Varnish.......8vo, $3 00
SMITH (A. W.)—Materials of Machines............................12mo, 1 00
(H. E.)—Strength of Material..............................12mo, *1 25
THURSTON—Materials of Engineering........................3 vols., 8vo, 8 00
Part I. Non-metallic Materials of Engineering..................8vo, 2 00
Part II. Iron and Steel..8vo, 3 50
Part III. A Treatise on Brasses, Bronzes, and Other Alloys and their Constituents..8vo, 2 50
WATERBURY—Laboratory Manual for Testing Materials of Construction. 12mo, *1 50
WOOD (DE V.)—Elements of Analytical Mechanics...................8vo, 3 00
Treatise on the Resistance of Materials and an Appendix on the Preservation of Timber..8vo, 2 00
(M. P.) Rustless Coatings. Corrosion and Electrolysis of Iron and Steel...8vo, 4 00

STEAM-ENGINES AND BOILERS.

ABRAHAM—Steam Economy in the Sugar Factory. (BAYLE.)...(*In Press.*)
BERRY—Temperature-entropy Diagram. Third Edition Revised and Enlarged..12mo, 2 50
CARNOT—Reflections on the Motive Power of Heat. (THURSTON.)...12mo, 1 50
CHASE—Art of Pattern Making...................................12mo, 2 50
CREIGHTON—Steam-engine and other Heat Motors....................8vo, 5 00
DAWSON—"Engineering" and Electric Traction Pocket-book...16mo, mor., 5 00
GEBHARDT—Steam Power Plant Engineering..........................8vo, *6 00
GOSS—Locomotive Performance.....................................8vo, 5 00
HEMENWAY—Indicator Practice and Steam-engine Economy.........12mo, 2 00
HIRSHFELD and BARNARD—Heat Power Engineering.................8vo, *5 00
HUTTON—Heat and Heat-engines....................................8vo, 5 00
Mechanical Engineering of Power Plants........................8vo, 5 00
KENT—Steam Boiler Economy.......................................8vo, 4 00
KING—Steam Engineering.................................(*In Press.*)
KNEASS—Practice and Theory of the Injector......................8vo, 1 50
MACCORD—Slide-valves..8vo, 2 00
MEYER—Modern Locomotive Construction............................4to, 10 00
MILLER, BERRY, and RILEY—Problems in Thermodynamics....8vo, paper, 0 75
MOYER—Steam Turbines..8vo, 4 00
PEABODY—Manual of the Steam-engine Indicator...................12mo, 1 50
Tables of the Properties of Steam and Other Vapors and Temperature-Entropy Table..8vo, 1 00
Thermodynamics of the Steam-engine and Other Heat-engines.....8vo, 5 00
Thermodynamics of the Steam Turbine..........................8vo, *3 00
Valve-gears for Steam-engines.................................8vo, 2 50
and MILLER—Steam-boilers......................................8vo, 4 00
PERKINS—Introduction to General Thermodynamics...............12mo, *1 50
PUPIN—Thermodynamics of Reversible Cycles in Gases and Saturated Vapors. (OSTERBERG.)......................................12mo, 1 25
REAGAN—Locomotives: Simple, Compound, and Electric. New Edition. Small 8vo, 3 50
SINCLAIR—Locomotive Engine Running and Management..........12mo, 2 00
SMART—Handbook of Engineering Laboratory Practice............12mo, 2 50
SNOW—Steam-boiler Practice......................................8vo, 3 00
SPANGLER—Notes on Thermodynamics..............................12mo, 1 00
Valve-gears...8vo, 2 50
GREENE, and MARSHALL—Elements of Steam-engineering........8vo, 3 00
THOMAS—Steam-turbines...8vo, 4 00
THURSTON—Manual of Steam-boilers, their Designs, Construction, and Operation...8vo, 5 00
Manual of the Steam-engine...........................2 vols., 8vo, 10 00
Part I. History, Structure, and Theory......................8vo, 6 00
Part II. Design, Construction, and Operation................8vo, 6 00
WEHRENFENNIG—Analysis and Softening of Boiler Feed-water. (PATTERSON.)...8vo, 4 00
WEISBACH—Heat, Steam, and Steam-engines. (DU BOIS.)...........8vo, 5 00
WHITHAM—Steam-engine Design.....................................8vo, 5 00
WOOD—Thermodynamics, Heat Motors, and Refrigerating Machines...8vo, 4 00

MECHANICS PURE AND APPLIED.

Church—Mechanics of Engineering.... ....8vo, $6 00
Mechanics of Solids (Being Parts I, II, III of Mechanics of Engineering). 8vo, 4 50
Mechanics of Fluids (Being Part IV of Mechanics of Engineering).8vo, 3 00
Mechanics of Internal Work..................................8vo, *1 50
Notes and Examples in Mechanics......... ... . 8vo, 2 00
Dana—Text-book of Elementary Mechanics for Colleges and Schools..12mo, 1 50
Du Bois—Elementary Principles of Mechanics:
Vol. I. Kinematics..8vo, 3 50
Vol. II. Statics..8vo, 4 00
Mechanics of Engineering. Vol. I.......................Small 4to, 7 50
Vol. II......................Small 4to, 10 00
Fuller and Johnston—Applied Mechanics:
Vol. I. Theory of Statics and Kinetics...................(*In Press.*)
Vol. II. Strength of Materials.....................(*In Preparation.*)
Greene—Structural Mechanics.....................................8vo, *2 50
Hartmann—Elementary Mechanics for Engineering Students.......12mo, *1 25
James—Kinematics of a Point and the Rational Mechanics of a Particle. Small 8vo, 2 00
Johnson (W. W.) Theoretical Mechanics.........................12mo, *3 00
King—Elements of the Mechanics of Materials and of Power of Transmission..8vo, *2 50
Kottcamp—Exercises for the Applied Mechanics Laboratory, Loose Leaf Laboratory Manual........................Oblong 4to, paper, *1 00
Lanza—Applied Mechanics...8vo, 7 50
Martin—Text Book of Mechanics:
Vol. I. Statics..12mo, *1 25
Vol. II. Kinematics and Kinetics...........................12mo, *1 50
Vol. III. Mechanics of Materials...........................12mo, *1 50
Vol. IV. Applied Statics...................................12mo, *1 50
Maurer—Technical Mechanics.......................................8vo, 4 00
Merriman—Elements of Mechanics..............................12mo, *1 00
Mechanics of Materials..8vo, 5 00
Michie—Elements of Analytical Mechanics..........................8vo, *4 00
Robinson—Principles of Mechanism.................................8vo, 3 00
Sanborn—Mechanics Problems..............................Small 8vo, *1 50
Schwamb and Merrill—Elements of Mechanism.......................8vo, 3 00
Wood—Elements of Analytical Mechanics............................8vo, 3 00
Principles of Elementary Mechanics..........................12mo, 1 25

MEDICAL.

Abderhalden—Physiological Chemistry in Thirty Lectures. (Hall and Defren.)...8vo, *5 00
von Behring—Suppression of Tuberculosis. (Bolduan.)............12mo, 1 00
Bolduan—Immune Sera...12mo, *1 50
Bordet—Studies in Immunity. (Gay.)...............................8vo, 6 00
Chapin—The Sources and Modes of Infection.................Small 8vo, *3 00
Cohnheim—Enzymes..12mo, *1 50
Davenport—Statistical Methods with Special Reference to Biological Variations..16mo, mor., 1 50
Effront—Enzymes and Their Applications. (Prescott.)..............8vo, 3 00
Ehrlich—Studies on Immunity. (Bolduan.)..........................8vo, 6 00
Euler—General Chemistry of the Enzymes. (Pope.)..................8vo, *3 00
Fischer—Nephritis.......................................Small 8vo, *2 50
Oedema...8vo, *2 00
Physiology of Alimentation..............................Small 8vo, *2 00
de Fursac—Manual of Psychiatry. (Rosanoff and Collins.).Small 8vo, *2 50
Fuller—Qualitative Analysis of Medicinal Preparations...........12mo, *1 50
Hammarsten—Text-book on Physiological Chemistry. (Mandel.)....8vo, *4 00
Jackson—Directions for Laboratory Work in Physiological Chemistry.8vo, 1 25
Lassar-Cohn—Praxis of Urinary Analysis. (Lorenz.)..............12mo, 1 00
Lauffer—Electrical Injuries.....................................16mo, *0 50
Mandel—Hand-book for the Bio-Chemical Laboratory...............12mo, 1 50

MARTIN—Measurement of Induction Shocks.......................12mo, *1 25
NELSON—Analysis of Drugs and Medicines.......................12mo, *3 00
PAULI—Physical Chemistry in the Service of Medicine. (FISCHER.)...12mo, *1 25
POZZI-ESCOT—Toxins and Venoms and their Antibodies. (COHN.)...12mo, *1 00
ROSTOSKI—Serum Diagnosis. (BOLDUAN.).......................12mo, 1 00
RUDDIMAN—Incompatibilities in Prescriptions.......................8vo, 2 00
Whys in Pharmacy.......................12mo, 1 00
SALKOWSKI—Physiological and Pathological Chemistry. (ORNDORFF.). 8vo, 2 50
SATTERLEE—Outlines of Human Embryology.......................12mo, *1 25
SMITH—Lecture Notes on Chemistry for Dental Students.............8vo, *2 50
WHIPPLE—Typhoid Fever.......................Small 8vo, *3 00
WOODHULL—Military Hygiene for Officers of the Line..........Small 8vo, *1 50
Personal Hygiene.......................12mo, *1 00
WORCESTER and ATKINSON—Small Hospitals Establishment and Maintenance, and Suggestions for Hospital Architecture, with Plans for a Small Hospital.......................12mo, 1 25

METALLURGY.

BETTS—Lead Refining by Electrolysis.......................8vo, 4 00
BOLLAND—Encyclopedia of Founding and Dictionary of Foundry Terms used in the Practice of Moulding.......................12mo, 3 00
Iron Founder.......................12mo, 2 50
BORCHERS—Metallurgy. (HALL and HAYWARD.).......................8vo, *3 00
BURGESS and LE CHATELIER—Measurement of High Temperatures. Third Edition.......................8vo, *4 00
DOUGLAS—Untechnical Addresses on Technical Subjects............12mo, 1 00
GOESEL—Minerals and Metals: A Reference Book...........16mo, mor., 3 00
ILES—Lead-smelting.......................12mo, *2 50
JOHNSON—Rapid Methods for the Chemical Analysis of Special Steels, Steel-making Alloys and Graphite..................Large, 12mo, 3 00
KEEP—Cast Iron.......................8vo, 2 50
METCALF—Steel. A Manual for Steel-users.......................12mo, 2 00
MINET—Production of Aluminum and its Industrial Use. (WALDO.)..12mo, 2 50
PALMER—Foundry Practice.......................Small 8vo, *2 00
PRICE and MEADE—Technical Analysis of Brass...................12mo, *2 00
RODENHAUSER and SCHOENAWA—Electric Furnaces in the Iron and Steel Industry. (VOM BAUR.).......................(*In Press*)
RUER—Elements of Metallography. (MATHEWSON.)................8vo, *3 00
SMITH—Materials of Machines.......................12mo, 1 00
TATE and STONE—Foundry Practice.......................12mo, 2 00
THURSTON—Materials of Engineering. In Three Parts...............8vo, 8 00
Part I. Non-metallic Materials of Engineering, see Civil Engineering, page 9.
Part II. Iron and Steel.......................8vo, 3 50
Part III. A Treatise on Brasses, Bronzes, and Other Alloys and Their Constituents.......................8vo, 2 50
ULKE—Modern Electrolytic Copper Refining.......................8vo, 3 00
WEST—American Foundry Practice.......................12mo, 2 50
Moulders' Text Book.......................12mo, 2 50

MILITARY AND MARINE ENGINEERING.

ARMY AND NAVY.

BERNADOU—Smokeless Powder, Nitro-cellulose, and the Theory of the Cellulose Molecule.......................12mo, 2 50
CHASE—Art of Pattern Making.......................12mo, 2 50
Screw Propellers and Marine Propulsion.......................8vo, 3 00
CLOKE—Enlisted Specialists' Examiner.......................8vo, *2 00
Gunner's Examiner.......................8vo, *1 50
CRAIG—Azimuth.......................4to, 3 50
CREHORE and SQUIER—Polarizing Photo-chronograph................8vo, 3 00
DAVIS—Elements of Law.......................8vo, *2 50
Treatise on the Military Law of United States....................8vo, *7 00

DUDLEY—Military Law and the Procedure of Courts-martial....Small 8vo, *2 50
DURAND—Resistance and Propulsion of Ships........................8vo, 5 00
DYER—Handbook of Light Artillery...............................12mo, *3 00
EISSLER—Modern High Explosives..................................8vo, 4 00
FIEBEGER—Text-book on Field Fortification.................Small 8vo, *2 00
HAMILTON and BOND—The Gunner's Catechism...................18mo, 1 00
HOFF—Elementary Naval Tactics.....................................8vo, *1 50
INGALLS—Handbook of Problems in Direct Fire.....................8vo, 4 00
Interior Ballistics...8vo, *3 00
LISSAK—Ordnance and Gunnery 8vo, *6 00
LUDLOW—Logarithmic and Trigonometric Tables.....................8vo, *1 00
LYONS—Treatise on Electromagnetic Phenomena. Vols. I. and II., 8vo, each, *6 00
MAHAN—Permanent Fortifications. (MERCUR)...........8vo, half mor., *7 50
MANUAL FOR COURTS-MARTIAL...........................16mo, mor., 1 50
MERCUR—Attack of Fortified Places............................12mo, *2 00
Elements of the Art of War..8vo, *4 00
NIXON—Adjutants' Manual...24mo, 1 00
PEABODY—Naval Architecture......................................8vo, 7 50
Propellers..8vo, 1 25
PHELPS—Practical Marine Surveying...............................8vo, *2 50
PUTNAM—Nautical Charts...8vo, 2 00
RUST—Ex-meridian Altitude, Azimuth and Star-Finding Tables.......8vo, 5 00
SELKIRK—Catechism of Manual of Guard Duty.....................24mo, *0 50
SHARPE—Art of Subsisting Armies in War ....18mo, mor., 1 50
TAYLOR—Speed and Power of Ships. 2 vols. Text 8vo, plates oblong 4to, *7 50
TUPES and POOLE—Manual of Bayonet Exercise and Musketry Fencing.
24mo, leather, *0 50
WEAVER—Military Explosives.......................................8vo, *3 00
WOODHULL—Military Hygiene for Officers of the Line........ . Small 8vo, *1 50

MINERALOGY.

BROWNING—Introduction to Rarer Elements.........................8vo, *1 50
BRUSH—Manual of Determinative Mineralogy. (PENFIELD.).........8vo, 4 00
BUTLER—Pocket Hand-book of Blowpipe Analysis.................16mo, *0 75
Pocket Hand-book of Minerals..........................16mo, mor., 3 00
CHESTER—Catalogue of Minerals............................8vo, paper, 1 00
Cloth, 1 25
CRANE—Gold and Silver...8vo, *5 00
DANA—First Appendix to Dana's New " System of Mineralogy.". Large 8vo, 1 00
Second Appendix to Dana's New " System of Mineralogy." Large 8vo, 1 50
Manual of Mineralogy. (FORD.).................................12mo, *2 00
Minerals, and How to Study Them..............................12mo, 1 50
System of Mineralogy.......................Large 8vo, half leather, 12 50
Text-book of Mineralogy..8vo, 4 00
DOUGLAS—Untechnical Addresses on Technical Subjects............12mo, 1 00
EAKLE—Mineral Tables...8vo, 1 25
ECKEL—Building Stones and Clays....................................8vo, *3 00
GOESEL—Minerals and Metals: A Reference Book............16mo, mor., 3 00
GROTH—The Optical Properties of Crystals. (JACKSON.).............8vo, *3 50
Introduction to Chemical Crystallography. (MARSHALL.)...... 12mo, 1 25
HAYES—Handbook for Field Geologists......................16mo, mor., *1 50
IDDINGS—Igneous Rocks...8vo, 5 00
Rock Minerals...8vo, 5 00
JOHANNSEN—Determination of Rock-forming Minerals in Thin Sections, 8vo,
With Thumb Index, 5 00
LEWIS—Determinative Mineralogy..........................Small 8vo, *1 50
MARTIN—Laboratory Guide to Qualitative Analysis with the Blowpipe.12mo, *0 60
MERRILL—Non-metallic Minerals: Their Occurrence and Uses........8vo, 4 00
Stones for Building and Decoration...............................8vo, 5 00
PENFIELD—Notes on Determinative Mineralogy and Record of Mineral
Tests..8vo, paper, *0 50
Tables of Minerals, Including the Use of Minerals and Statistics of
Domestic Production...8vo, 1 00
PIRSSON—Rocks and Rock Minerals...................................12mo, *2 50
RICHARDS—Synopsis of Mineral Characters..................12mo, mor., *1 25

RIES—Building Stones and Clay Products..........................8vo,*$3 00
Clays: Their Occurrence, Properties and Uses..................8vo, *5 00
and LEIGHTON—History of the Clay-working Industry of the United States..8vo, *2 50
ROWE—Practical Mineralogy Simplified..........................12mo, *1 25
TILLMAN—Text-book of Important Minerals and Rocks.............8vo, *2 00
WASHINGTON—Manual of the Chemical Analysis of Rocks...........8vo, 2 00

MINING.

BEARD—Mine Gases and Explosions........................Small 8vo, *3 00
BRUNSWIG—Explosives. (MUNROE and KIBLER.)......Ready Fall, 1912
CRANE—Gold and Silver...8vo, *5 00
Index of Mining Engineering Literature, Vol. I..................8vo, *4 00
8vo, mor., *5 00
Vol. II..................8vo, *3 00
8vo, mor., *4 00
Ore Mining Methods..8vo, *3 00
DANA and SAUNDERS—Rock Drilling..............................8vo, *4 00
DOUGLAS—Untechnical Addresses on Technical Subjects............12mo, 1 00
EISSLER—Modern High Explosives...............................8vo, 4 00
GILBERT, WIGHTMAN and SAUNDERS—Subways and Tunnels of New York. 8vo, *4 00
GOESEL—Minerals and Metals: A Reference Book.............16mo, mor., 3 00
IHLSENG—Manual of Mining......................................8vo, 5 00
ILES—Lead Smelting...12mo, *2 50
PEELE—Compressed Air Plant...................................8vo, *3 50
RIEMER—Shaft Sinking under Difficult Conditions. (CORNING and PEELE.) 8vo, 3 00
WEAVER—Military Explosives....................................8vo, *3 00
WILSON—Hydraulic and Placer Mining...........................12mo, 2 50
Treatise on Practical and Theoretical Mine Ventilation.........12mo, 1 25

SANITARY SCIENCE.

ASSOCIATION OF STATE AND NATIONAL FOOD AND DAIRY DEPARTMENTS, Hartford Meeting, 1906.................................8vo, 3 00
Jamestown Meeting, 1907.......................................8vo, 3 00
BASHORE—Outlines of Practical Sanitation......................12mo, *1 25
Sanitation of a Country House.................................12mo, 1 00
Sanitation of Recreation Camps and Parks.....................12mo, 1 00
FOLWELL—Sewerage. (Designing, Construction, and Maintenance.)...8vo, 3 00
Water-supply Engineering.......................................8vo, 4 00
FOWLER—Sewage Works Analyses..................................12mo, 2 00
FUERTES—Water-filtration Works.................................12mo, 2 50
GERHARD—Guide to Sanitary Inspections.........................12mo, 1 50
Modern Baths and Bath Houses...................................8vo, *3 00
Sanitation of Public Buildings.................................12mo, 1 50
The Water Supply, Sewerage, and Plumbing of Modern City Buildings. 8vo, *4 00
HAZEN—Clean Water and How to Get It.....................Small 8vo, 1 50
Filtration of Public Water-supplies............................8vo, 3.00
KINNICUTT, WINSLOW and PRATT—Sewage Disposal.................8vo, *3 00
LEACH—Inspection and Analysis of Food with Special Reference to State Control...8vo, 7 50
MASON—Examination of Water. (Chemical and Bacteriological.)....12mo, 1 25
Water-supply. (Considered principally from a Sanitary Standpoint.) 8vo, 4 00
MERRIMAN—Elements of Sanitary Engineering.....................8vo, *2 00
OGDEN—Sewer Construction......................................8vo, 3 00
Sewer Design..12mo, 2 00
OGDEN and CLEVELAND—Practical Methods of Sewage Disposal for Residences, Hotels and Institutions..........................8vo, *1 50
PARSONS—Disposal of Municipal Refuse..........................8vo, 2 00

PRESCOTT and WINSLOW—Elements of Water Bacteriology, with Special Reference to Sanitary Water Analysis....................12mo, $1 50
PRICE—Handbook on Sanitation................................12mo, *1 50
RICHARDS—Conservation by Sanitation............................8vo, 2 50
Cost of Cleanness..12mo, 1 00
Cost of Food. A Study in Dietaries.........................12mo, 1 00
Cost of Living as Modified by Sanitary Science.............12mo, 1 00
Cost of Shelter..12mo, 1 00
Laboratory Notes on Industrial Water Analysis...............8vo, *0 50
RICHARDS and WOODMAN—Air, Water, and Food from a Sanitary Standpoint..8vo, 2 00
RICHEY—Plumbers', Steam-fitters', and Tinners' Edition (Building Mechanics' Ready Reference Series) ....16mo, mor., *1 50
RIDEAL—Disinfection and the Preservation of Food.................8vo, 4 00
SOPER—Air and Ventilation of Subways..........................12mo, 2 50
TURNEAURE and RUSSELL—Public Water-supplies....................8vo, 5 00
VENABLE—Garbage Crematories in America.........................8vo, 2 00
Method and Devices for Bacterial Treatment of Sewage..........8vo, 3 00
WARD and WHIPPLE—Freshwater Biology....................(*In Press.*)
WHIPPLE—Microscopy of Drinking-water...........................8vo, 3 50
Typhoid Fever.......................................Small 8vo, *3 00
Value of Pure Water.................................Small 8vo, 1 00

MISCELLANEOUS.

BURT—Railway Station Service..................................12mo, *2 00
CHAPIN—How to Enamel..12mo, *1 00
EMMONS—Geological Guide-book of the Rocky Mountain Excursion of the International Congress of Geologists.................Large 8vo, 1 50
FERREL—Popular Treatise on the Winds...........................8vo, 4 00
FITZGERALD—Boston Machinist 18mo, 1 00
FRITZ—Autobiography of John....................................8vo, *2 00
GANNETT—Statistical Abstract of the World.....................24mo, 0 75
GREEN—Elementary Hebrew Grammar...............................12mo, 1 25
HAINES—American Railway Management...........................12mo, 2 50
HANAUSEK—The Microscopy of Technical Products. (WINTON.)......8vo, 5 00
JACOBS—Betterment Briefs. A Collection of Published Papers on Organized Industrial Efficiency.....................................8vo, 3 50
METCALFE—Cost of Manufactures, and the Administration of Workshops. 8vo, 5 00
PARKHURST—Applied Methods of Scientific Management 8vo, *2 00
PUTNAM—Nautical Charts...8vo, 2 00
RICKETTS—History of Rensselaer Polytechnic Institute, 1824–1894. Small 8vo, 3 00
ROTCH and PALMER—Charts of the Atmosphere for Aeronauts and Aviators. Oblong 4to, *2 00
ROTHERHAM—Emphasised New Testament...................Large 8vo, 2 00
RUST—Ex-Meridian Altitude, Azimuth and Star-finding Tables.......8vo, 5 00
STANDAGE—Decoration of Wood, Glass, Metal, etc.................12mo, 2 00
WESTERMAIER—Compendium of General Botany. (SCHNEIDER.)......8vo, 2 00
WINSLOW—Elements of Applied Microscopy.......................12mo, 1 50

CPSIA information can be obtained at www.ICGtesting.com
Printed in the USA
LVOW02s0923290815

452044LV00019B/611/P